Smart Nanomaterials Technology

Nanotechnology is a rapidly growing scientific field and has attracted a great interest over the last few years because of its abundant applications in different fields like biology, physics and chemistry. This science deals with the production of minute particles called nanomaterials having dimensions between 1 and 100 nm which may serve as building blocks for various physical and biological systems. On the other hand, there is the class of smart materials where the material that can stimuli by external factors and results a new kind of functional properties. The combination of these two classes forms a new class of smart nanomaterials, which produces unique functional material properties and a great opportunity to larger span of application. Smart nanomaterials have been employed by researchers to use it effectively in agricultural production, soil improvement, disease management, energy and environment, medical science, pharmaceuticals, engineering, food, animal husbandry and forestry sectors.

This book series in Smart Nanomaterials Technology aims to comprehensively cover topics in the fabrication, synthesis and application of these materials for applications in the following fields:

- Energy Systems—Renewable energy, energy storage (supercapacitors and electrochemical cells), hydrogen storage, photocatalytic water splitting for hydrogen production
- Biomedical—controlled release of drugs, treatment of various diseases, biosensors,
- Agricultural—agricultural production, soil improvement, disease management, animal feed, egg, milk and meat production/processing,
- Forestry—wood preservation, protection, disease management
- Environment—wastewater treatment, separation of hazardous contaminants from wastewater, indoor air filters.

Azamal Husen
Editor

Plant Response to Silver Nanoparticles

Plant Growth, Development, Production, and Protection

Springer

Editor
Azamal Husen
Wolaita Sodo University
Wolaita, Ethiopia

ISSN 3004-8273 ISSN 3004-8281 (electronic)
Smart Nanomaterials Technology
ISBN 978-981-97-7351-0 ISBN 978-981-97-7352-7 (eBook)
https://doi.org/10.1007/978-981-97-7352-7

This Springer imprint is published by the registered company Springer Nature Singapore Pte Ltd.
The registered company address is: 152 Beach Road, #21-01/04 Gateway East, Singapore 189721, Singapore

Preface

Silver nanoparticles (Ag-NPs) have attracted more attention than any other NP. Various methods namely chemical, physical, and biological are used to produce desirable Ag-NPs. The shape and size of NPs depend on the concentration of the precursor and the reducing agents regardless of the in vitro or in vivo process of synthesis. These particles are further utilized in almost all disciplines of science and technology. In this book, overall plant response to Ag-NPs has been examined. These particles often accelerate the growth and yield of plants. Although they have shown both beneficial and harmful effects on the plant system. These responses of Ag-NPs are primarily dependent on the concentration, plant species or cultivars, exposure time, shape, and size of NPs. In general, lower concentrations of Ag-NPs increase seed germination, rate of photosynthesis, and overall growth, but at higher concentrations, all these responses are declined in many plant species. Ag-NPs at higher concentrations induce stress and or phytotoxicity and produce reactive oxygen species which leads to the disruption of cellular metabolism. Plants are exposed to Ag-NPs, and their morphology, physiology, and biochemistry are significantly affected. The molecular and physiological observation revealed that Ag-NPs exposure resulted in oxidative stress and induced antioxidative and defense responses. These responses may differ from species to species depending on the concentration and exposure time of NPs. However, in a few investigations the energy production-related metabolic pathways in the tricarboxylic acid cycle and in sugar metabolism were stimulated stronger by Ag-NPs exposure, thus releasing more energy, and accelerating the physiological metabolic responses against stress in the Ag-NPs treatment while subsequently reducing the plant growth and yield at the maturation stage. Importantly, shikimate-phenylpropanoid biosynthesis and tryptophan and galactose metabolisms were regulated only by the Ag-NPs treatment, which was a specific effect of NPs. Further, Ag-NPs exposure increased the number of chromosomal aberrations, and micronuclei, and decreased the mitotic index in plant root tip cells. Proteomic study has shown that the exposure Ag-NPs resulted in an accumulation of protein precursors, indicative of the dissipation of a proton motive force. The identified proteins are involved in oxidative stress tolerance, Ca^{2+} regulation and signaling, transcription and protein degradation, cell wall and DNA/RNA/protein direct damage, cell division, and apoptosis.

Ag-NPs also influence the transcription of flowering key genes and thus delay flowering time. It has been also noticed that the level of seed germination and subsequent growth of those seedlings that germinated both decreased with increasing sizes and concentrations of Ag-NPs exposure. Based on the analysis of Ag-NPs accumulation in plant tissues, it implied that a higher uptake was found when the seeds were treated with the smaller Ag-NPs, but they are trapped in the roots rather than transported to the leaves. Perhaps, these resulted in less negative effects on seedling growth in some of the plant species, when compared to the seed soaking with the larger particle size. Further reports have shown that mycorrhizal colonization decreases silver accumulation in NP-exposed plants and moderates changes in the expression level of membrane transport proteins. Moreover, a beneficial role of arbuscular mycorrhizal fungi in influencing the effects of Ag-NPs on plant-microbe systems in a soil matrix has been also established. Besides the terrestrial plants, these particles have also influenced the growth of some wetlands and aquatic plants. The use of Ag-NPs and silver ions in tissue culture medium and plant growth performance has been also examined. Environmental safety, and ethical issues related to the use of plant and soil systems have been also explored.

Taken together, the book in hand provides an overview of the most important and selected topics related to plant response to Ag-NPs. This book provides numerous valuable information to scientists, researchers, and students, working specially on plant biology, plant nanobiotechnology, plant biochemistry, plant microbiology, agricultural, and other allied subjects, and or science. I am grateful to all contributors for readily accepting my invitation, sharing their knowledge in specialized areas of research, and readily adjusting the suggestions for improving the shape of their contributions. Finally, my special thanks go to Shagufta, Zaara, Mehwish, and Huzaifa for providing their time and overall extended support to put everything together.

Wolaita, Ethiopia Azamal Husen

Contents

About the Editor

Azamal Husen is served as Full Professor and Head of the Department of Biology, University of Gondar, Ethiopia; and currently working as a Foreign Delegate at Wolaita Sodo University, Wolaita, Ethiopia. He is also working as an Adjunct Professor at Graphic Era (Deemed to be University), Dehradun, Uttarakhand, India. He also worked as Full Professor at Sankalchand Patel University, Visnagar, India; and Visiting Faculty of the Forest Research Institute and the Doon College of Agriculture and Forest at Dehra Dun, India. His research and teaching experience of 25 years encompasses Nanobiotechnology, Plant Physiology, Environmental Botany, Medicinal Plants; and Clonal Propagation. He has conducted research sponsored by the World Bank, the NATP Project, the ICAR, the ICFRE, the JBIC, etc. He received various fellowships from India; and recognition awards from Ethiopia, for excellent teaching, research, and community service. Husen has given various orientation courses for academic staff. He is a Fellow of the Plantae group of the American Society of Plant Biologists, and a Member of the International Society of Root Research, Asian Council of Science Editors, and so on. He has published extensively including research papers, review articles and book chapters; edited books of international repute, presented papers in several conferences, and produced over a dozen manuals and monographs. Husen has been on the Editorial board, Guest Editor of special issues, and the panel of reviewers of several high-impact reputed journals. He is a founding Book Series Editor of *Exploring Medicinal Plants* (Taylor & Francis Group, USA); *Plant*

Biology, Sustainability, and *Climate Change* (Elsevier, USA); *Smart Nanomaterials Technology* (Springer Nature, Singapore); and *Nanotechnology and Its Applications* (Springer Nature, Singapore).

Plant Response to Silver Nanoparticles in Terms of Growth, Development, Production, and Protection: An Overview

Mayur Mukut Murlidhar Sharma, Divya Kapoor, Atul Loyal, Rahul Kumar, Pankaj Sharma, and Azamal Husen

Abstract Silver nanoparticles (AgNPs) have demonstrated their remarkable ability to significantly influence crucial processes in plants. These processes include but are not limited to seed germination, the photosynthetic apparatus, antioxidant system activity, and the expression of genes vital for plant function, operating under both optimal and adverse conditions. The variability in the effects of AgNPs on plant growth and development across different plant species highlights the importance of gaining a nuanced understanding of these interactions. Current research underscores that various factors, including dosage, size, shape, surface charge, and capping agent, exert a substantial impact on the uptake, translocation, distribution, and growth enhancement effects of AgNPs in plants. This chapter thoroughly explores the intricate interactions between silver nanoparticles and plants, shedding light on the underlying mechanisms that contribute to the observed diverse outcomes. Furthermore, the discussion extends to the profound influence of AgNPs on plant growth and development, underscoring their potential to enhance nutrient uptake and facilitate targeted

M. M. M. Sharma
Department of Agriculture and Industry, Kangwon National University, Chuncheon, Korea

D. Kapoor · P. Sharma (✉)
Department of Microbiology, CCS Haryana Agricultural University, Hisar, Haryana 125004, India
e-mail: sharmap@uic.edu

A. Loyal
Department of Genetics and Plant Breeding, CCS Haryana Agricultural University, Hisar, Haryana 125004, India

R. Kumar
Department of Zoology, College of Agriculture, RRS Bawal, Haryana 123501, India

P. Sharma
Department of Ophthalmology and Visual Sciences, University of Illinois Chicago, Chicago 60612, USA

A. Husen
Wolaita Sodo University, PO Box 138, Wolaita, Ethiopia

Department of Biotechnology, Graphic Era (Deemed to be University), Dehradun, Uttarakhand 248002, India

A. Husen (ed.), *Plant Response to Silver Nanoparticles*, Smart Nanomaterials Technology, https://doi.org/10.1007/978-981-97-7352-7_1

delivery. This contribution ultimately plays a pivotal role in increasing plant productivity and yield. Additionally, the comprehensive investigation into the impact of silver nanoparticles on plant defense mechanisms against pathogens and pests reveals their promising antimicrobial and antifungal properties. This presents AgNPs as a potential eco-friendly alternative to traditional pesticides. The multifaceted nature of these interactions emphasizes the transformative potential of silver nanoparticles in revolutionizing agricultural practices. Their ability to contribute to sustainable and environmentally friendly crop management further solidifies their position as agents of positive change in the realm of agriculture.

Keywords Plant growth · Plant protection · Seed germination · Seed yield · Nanopriming

1 Introduction

The advancement of nano-technological interventions presents a significant opportunity to reshape diagnostic and therapeutic modalities within the healthcare and agricultural sectors, driving the transition of theoretical propositions into demonstrably efficacious outcomes [7, 25–27, 63, 64, 74–78, 81]. The exponential growth and diversification of nanotechnological tools in recent decades have demonstrably revolutionized contemporary agricultural practices. Nanotechnology's unique ability to manipulate matter at the nanoscale allows for the customization of material properties, such as pore size, functional group integration, and high surface-to-volume ratio, resulting in a paradigm shift towards highly targeted and efficient agricultural interventions [36, 99–103]. The utilization of nano-enabled products in the agricultural domain is steadily growing to address prevailing challenges. Nano-technological interventions have emerged as a significant asset for the agricultural sector, evident in their efficacy in targeted delivery, controlled release mechanisms, enhanced solubility, and prolonged shelf-life, as assessed in various studies [45, 46, 68, 79–83, 60, 73]. Nanoparticles manifest in various formulations, such as nanoemulsions or nanocapsules, specifically harnessed for targeted delivery and enhanced productivity. With the rapid advancement in nanotechnology, metal nanoparticles have become focal points of research due to their diverse functionalities and ready accessibility. Nano fertilizers and nano pesticides stand out as primary sources for obtaining metal nanoparticles. These fertilizers play multifaceted roles, with metal nanoparticles like copper, iron, molybdenum, manganese, and zinc exhibiting the capacity to stimulate plant growth and enhance overall yield even at minimal doses [55]. The overarching objective of incorporating nanomaterials (NMs) in agriculture is to mitigate nutrient losses, enhance crop yields, decrease the quantities of plant protection products, and optimize production costs for maximal output [91]. Nano-silver stands out as the most extensively researched and applied nanoparticle within bio-systems. It is acknowledged for its robust inhibitory and bactericidal effects, encompassing a broad spectrum of antimicrobial activities. Notably, silver nanoparticles, characterized by their elevated surface area and a heightened proportion of surface atoms, exhibit enhanced antimicrobial efficacy compared to bulk silver. Furthermore, silver nanoparticles (Ag NPs) are recognized for their well-established antioxidant, antibacterial, antifungal, antiviral, and anti-inflammatory properties. While the utilization of Ag NPs in agriculture is predominantly theoretical at present, it is anticipated that researchers will uncover diverse applications for silver nanoparticles soon [33], Fig. 1.

Globally, the production of silver nanoparticles has reached 880 tons per annum, primarily driven by their prominent role in biomedical sciences, where they serve as drug carriers or immunomarkers. Nevertheless, there has been a recent upsurge in their utilization within the agricultural domain [14]. Silver nanoparticles (NPs) engage with the proteins, enzymes, metabolites, and other biomolecules within plants, contributing to an augmentation in biomass and the growth of roots and

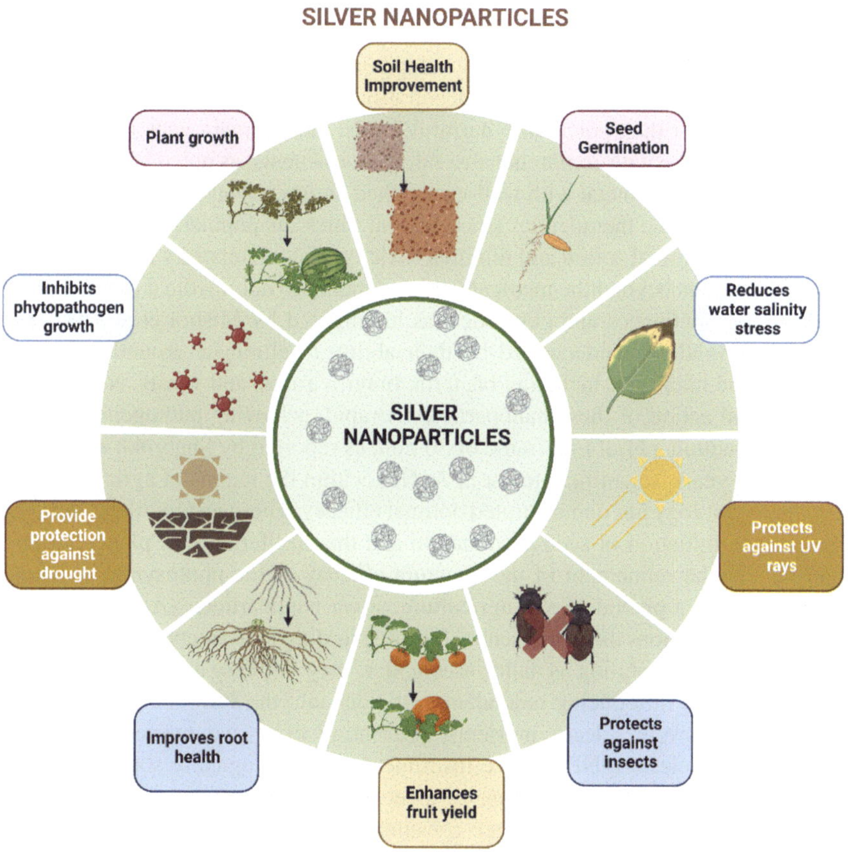

Fig. 1 A portrayal depicting the different beneficial attributes of silver nanoparticles to plant systems

shoots. Furthermore, silver NPs demonstrate distinctive actions, such as the reduction of ethylene content in onions, leading to an enhancement in antioxidant properties [84]. Despite reports highlighting the potential toxicity of Silver NPs, their prudent application has proven beneficial for both plants and growers. In a study by Mehta et al. [58], the application of silver NPs on various legume and cereal crops demonstrated an increase in biomass at specific concentrations, influencing bacterial diversity in the soil. Moreover, the incorporation of silver nanoparticles, alongside silver nitrate, in growth media for tissue culture has shown success. These nanoparticles have been effective in eliminating pathogenic contamination from explants while concurrently promoting positive outcomes in callus induction, organogenesis, somatic embryogenesis, somaclonal variation, genetic transformation, and the production of secondary metabolites [48].

These modern materials possess the capability to modify plant physiology upon entry into the intricate plant-soil system, providing a readily exploitable avenue for understanding their subsequent effects. Additionally, to facilitate the controlled delivery of active compounds, a profound understanding of the interactions between these nanomaterials (NMs) and plants whether positive or negative is imperative. This knowledge base holds promise for the development of innovative nanomaterial-based products. Despite the belief that the concentration of NMs in the natural environment is generally lower than potentially harmful levels, there remain gaps in our understanding that necessitate comprehensive safety assessments, as noted by Batley et al. [9]. Notably, silver, a metal with well-established antimicrobial properties, has been utilized in Ayurvedic therapeutics since ancient times. Its popularity stems from its efficient antimicrobial action and minimal toxicity. In contemporary contexts, silver finds applications beyond the medical field, extending to non-medical domains such as electronic appliances and cosmetics, as highlighted by Mishra et al. [54]. The efficacy of AgNPs is underscored by their ability to eliminate around 650 types of pathogenic microbes, including bacteria, fungi, viruses, and yeasts. Notably, the antimicrobial action of these nanoparticles is rapid, achieving pathogenic bacteria eradication within a brief timeframe of 30 min, as reported by Shahrokh et al. [71].

In recent years, scientific interest in AgNPs within the realms of agriculture and plant biotechnology has demonstrated optimal efficacy. These nanoparticles not only augment the initiation of seed germination and the proliferation of plants but also contribute to the refinement of the quantum efficacy of the photosynthetic mechanism. Exerting a pivotal role in agriculture, silver nanoparticles (AgNPs) provide numerous applications that are essential for safeguarding food safety and augmenting crop production. Operating as nano-insecticides, they dispense a pinpointed dosage to plants without introducing redundant pesticides into the surroundings. Nanofertilizers progressively release nutrients, preventing excessive nutrient depletion. Silver nanoparticles (AgNPs) prove instrumental for ecologically sound and non-hazardous pest management, acting as a secure instrument for controlling pests. They can be assimilated into either consumable or non-biodegradable polymers for dynamic food packaging. Furthermore, AgNPs exhibit a plethora of biological attributes, encompassing antiviral, antibacterial, and antifungal activities, thereby providing formidable protection to plants against deleterious microorganisms [41].

This chapter's goal is to conduct a thorough survey and summarization of recent literature, encompassing both the positive and negative effects of AgNPs on plant growth, along with an exploration of their diverse agricultural applications.

2 Effect on Seed Germination

In contradistinction to the acknowledged toxicity and antimicrobial characteristics of silver nanoparticles (AgNPs), contemporary investigations have delved into their function in enhancing seed germination and ameliorating crop yields. Notwithstanding their noteworthy importance in agriculture and beneficial impacts on seed germination, the utilization of silver nanoparticles (AgNPs) has also been linked to detrimental effects. The initiation of seed germination represents a crucial stage in instituting plants in agriculture, playing a pivotal role in determining crop quality [41]. Silver nanoparticles (AgNPs) exhibit the capacity to assimilate with seed pericarps, facilitating hydric absorption within seeds and catalyzing starch metabolism, thereby endorsing and bolstering the process of seed germination. Additionally, they are applied as nano-priming agents to enhance the viability of aged rice seeds and promote starch metabolism. One of the key objectives of AgNP priming is to upregulate the expression of aquaporin genes, thereby facilitating the diffusion of H_2O_2 and water [47]. The predominant method for initiating seed germination using nanoparticles is through nano-priming. Nano-priming is correlated with enhancing α-amylase activity, consequently leading to higher soluble sugar content, which in turn supports the growth of seedlings [28]. In a study conducted by Mehmood et al. [52], the seed germination of three plant species, namely corn (*Zea mays* L.), watermelon (*Citrullus lanatus*), and zucchini (*Cucurbita pepo L.),* was investigated to understand the effects of varying AgNPs dosages on germination parameters. Different concentrations, ranging from 0.05 to 2.5 mg/ml, were employed in the experiment. The findings suggested that the exposure to silver nanoparticles elevated the germination rates of all three species. Eminently, a noteworthy amelioration in germination percentage metrics was discerned, particularly in zucchini and watermelon flora, relative to untreated seeds. In an alternative investigation, extracellular silver nanoparticles (AgNPs) were generated through the utilization of Bacillus subtilis spizizenni. Bajra (*Pennisetum glaucum*) seeds treated with 1 mM AgNPs demonstrated commendable outcomes, manifesting improved seed germination within a mere 3-day span. Statistical scrutiny via one-way analysis of variance unveiled a noteworthy enhancement in both plumule and radicle length when juxtaposed with control seeds of Bajra [65]. In a study, AgNPs were utilized to assess their impact on mitigating salinity stress in lentil seeds. The application of AgNPs resulted in enhanced seedling growth and improved seed germination in lentils. Specifically, the germination percentage increased at concentrations of 10 μg/mL AgNPs, leading to a reduction in mean germination time for lentil plants. The application of AgNPs not only improved various germination parameters but also conferred improved forbearance in drought circumstances. Notably, AgNPs at a concentration of 10 μg/mL were found to significantly enhance seed germination of lentil even in the presence of drought stress [23]. In a study by Pandey et al. [59], AgNPs and $AgNO_3$ were applied at concentrations of 100, 500, and 1000 mg/L, and 100 and 500 mg/L, respectively, in *Brassica juncea* var. The Varuna mustard plant was used to assess the toxicity of these particles. The findings suggested that the introduction of silver nanoparticles (AgNPs) yielded a modest amelioration in the length of both the root and shoot. Furthermore, it heightened the chlorophyll and protein content at all concentrations in contrast to the control. In another study, the application of AgNPs and salt at concentrations of 0, 40, 60, and 80 ppm, and 0, 30, 60, 90, and 120 $mMol^{-1}$, respectively, demonstrated that the 80 ppm AgNPs treatment significantly improved ($P \leq 0.05$) the percentage germination of seeds. Silver nanoparticles (AgNPs) were discerned as imperative for augmenting salinity tolerance in *S. hortensis* seedlings, triggering diverse plant protection contrivances against the deleterious effects of salt toxicity [57]. While silver nanoparticles (AgNPs) have demonstrated myriad favorable effects on plant growth, certain botanical specimens have evinced toxicity in reaction to their administration, suggesting conceivable deleterious repercussions. As an exemplification, the administration of silver nanoparticles (AgNPs) at varying concentrations, namely 1000 μg/mL, 1600 μg/mL, and 1200 μg/mL on *Brassica campestris, Oryza sativa, and Vigna radiata*, respectively, culminated in root suppression at concentrations of 4500 μg/mL and 6000 μg/mL.. This resulted in a noteworthy diminution in root length to 1% and 0.5% in Oryza sativa and Brassica campestris, respectively, and 0.5% in *Vigna radiata* [51].

3 Effect on Plant Growth

Silver nanoparticles (AgNPs) have exhibited noteworthy effectiveness in promoting seed germination, and a multitude of studies have investigated their applications in various facets of plant growth (Table 1). Optimal concentrations of silver nanoparticles (AgNPs) have been demonstrated to notably enhance the growth and yield of crops, as multiple studies corroborate favorable effects with minimal injurious effects on plant growth. However, contrasting effects have also been reported for several plant species. The physiological ramifications of silver nanoparticles (AgNPs) on plants are intricately intertwined with variables such as particle dimensions, morphology, concentration, specific surface area, electrical charge, and surface attributes, along with the specific botanical species and environmental circumstances. These circumstances encompass the growth medium, pH levels, ion concentration, root circumference, soil microbial community, and other pertinent parameters. Furthermore, the duration of exposure and the manner of exposure constitute pivotal elements influencing the phytotoxicity of silver nanoparticles (AgNPs). For instance, a study carried out by Syu et al. [88] unveiled that decahedral and spherical silver nanoparticle (AgNPs), owing to their diminutive size and expansive specific surface area, exerted beneficial effects on Arabidopsis growth. Moreover, a contemporary investigation suggested that silver nanoparticles (AgNPs) of smaller dimensions (30 nm) exhibited greater toxicity to Lemna minor in comparison to other sizes of AgNPs. This observation implies that the optimal size of AgNPs is more easily assimilated by plants, eliciting a more robust response and potential for toxicity [85]. The outcomes imply that the uptake of nanoparticles is contingent upon silver nanoparticles (AgNPs) endowed with apt dimensions, charge, and surface characteristics, each of which assumes a critical role in the bioeffects of AgNPs. In a transient exposure study conducted with Nelumbo nucifera, the activities of antioxidant enzymes experienced an escalation after the initial day of silver nanoparticle (AgNP) exposure but exhibited a diminishing trend with prolonged exposure. Likewise, the elevation in the expression of detoxification-associated genes involved in glutathione (GSH) biosynthesis and glutathione S-transferase (GST) transpired, underscoring the time-dependent bioeffect of silver nanoparticles (AgNPs) [43]. AgNPs with surface coatings, especially those coated with polyvinylpyrrolidone (PVP), demonstrate comparatively lower toxicity, enhanced stability, and diminished aggregation of nanoparticles. AgNPs enveloped in disparate materials manifest divergent biological impacts. For instance, the utilization of AgNP-CTAB (cetyltrimethylammonium bromide) elevated the stimulation of catalase and ascorbate peroxidase, while treatments involving AgNP-PVP induced a substantial augmentation in proline and GSH content, associated with nonenzymatic antioxidants in plants [10]. The scientific literature is replete with numerous studies examining the impact of AgNPs on plant progress. For example, the exogenous application of AgNPs has been found to have diverse effects on plant growth. In the case of Oryza sativa, treatment with 30 μg/mL of AgNPs resulted in boosted growth of root, while treatment with 60 μg/mL halted root growth [53]. In a controlled experimental study, the ramifications of silver nanoparticles (AgNPs) on parameters such as mean germination time, germination rate, germination percentage, root length, and the dry/fresh weight of seedlings were systematically examined across three spice plant species. Silver nanoparticles (AgNPs) were administered at varying concentrations, spanning from 0.05 to 2.5 mg/L, throughout the seed germination phase. The three spice plant species exhibited heterogeneous responses to different doses of AgNPs in relation to the quantified growth parameters and germination percentage. Significantly, the values of germination rate for all three plants witnessed amelioration in response to the introduction of AgNPs [4]. In a scientific investigation, silver nanoparticles (AgNPs) were synthesized utilizing trisodium citrate salt (at a concentration of 9×10^{-4} M) and subsequently disseminated through aerosol application at concentrations of 0, 50, and 75 ppm on crops of distinct species, namely *Vigna sinensis, Triticum aestivum, and Brassica juncea.* The application of silver nanoparticles (AgNPs) at a concentration of 50 mg/L elicited heightened dry weight and length of wheat (cowpea) shoots. Moreover, an augmentation in shoot parameters was observed at 75 ppm in Brassica. In a distinct study, AgNPs were synthesized and introduced into the soil to assess their impact on a variety of plants, including *Lactuca sativa* (lettuce), *Avena sativa* (oat), and *Raphanus sativus* (radish). This application resulted in an elevated percentage of dry mass and enhanced root growth in the scrutinized botanical specimens [89]. The foliar application of silver nanoparticles (AgNPs) at concentrations of 20, 40, and 60 mg/L has manifested elevated growth parameters in fenugreek plants, including shoot dry weight, the abundance of plants/leaves, and shoot length. Furthermore, additional biochemical aspects have demonstrated enhancement, such as an augmentation in the concentration of indole acetic acid (IAA) and photosynthetic pigments like chlorophyll a, chlorophyll b, and carotenoids. This treatment has additionally precipitated improvements in yield magnitude (number of seeds/pod, seed index weight of seeds/plant, and number of pods/plant), antioxidant efficacy, and seed quality parameters (carbohydrate%, protein%, tannins, flavonoids, and phenolics contents) [66]. The utilization of silver nanoparticles (AgNPs) at various concentrations, ranging from 0 to 400 ppm, showcased a beneficial influence on the antioxidant status and growth of 7-day-old Brassica juncea seedlings. The treatment with AgNPs impacted crucial growth parameters, including the vitality index, shoot elongation, root extension, and fresh mass of seedlings. Notably, the elongation of root and the dynamism index experienced an escalation to 326% and 133% of the treated seedlings, indicating a substantial augmentation in growth. Additionally, there was clear evidence of amplified chlorophyll concentration and photosynthetic quantum efficiency in the foliage of the treated seedlings, further accentuating the constructive influence of silver nanoparticles (AgNPs) on the physiological facets of plant growth [72]. In a scientific inquiry, the application of silver nanoparticles (AgNPs) at concentrations ranging from 0 to 30 mM exhibited beneficial influences on plant growth and upheld the ionic equilibrium of cells (Na^+, K^+, and Na^+/K^+ ratio) by reducing the assimilation of Na and Cl, consequently ameliorating oxidative stress in plants subjected to salinity. The intervention also prompted catalase activities, superoxide dismutase, and diminished peroxidase activity. Furthermore, a decline in malondialdehyde (MDA) and hydrogen peroxide (H_2O_2) contents was noted in plants experiencing salt stress. The dose-additive administration of silver nanoparticles (AgNPs) heightened the yield, height, and photosynthesis pathway of plants subjected to salt stress [39]. In an alternate exploration, silver nanoparticles (AgNPs) were fabricated utilizing the extract derived from waste pistachio seed coats. The ensuing solution of AgNPs at a concentration of 75 mg/L was administered through aerial spraying on eggplants, culminating in elevated carotenoid/chlorophyll contents and enhanced vegetal development. Moreover, the inclusion of fly ash further spurred plant growth. The application of AgNPs at a concentration of 75 ppm through soil spraying amended with 20% fly ash conspicuously augmented plant growth [40]. The toxicity of AgNPs to plants in terrestrial ecosystems, especially crops, is contingent on concentration and size. It has been observed to reduce germination and stunt the growth of seeds, impacting the length and mass of roots and shoots. The deposition of silver nanoparticles (AgNPs) in leaves and roots initiates defensive mechanisms at the cellular and tissue strata, culminating in modifications to proteomic expression, antioxidant activities, and metabolic processes.. Botanical transformations, including the generation of reactive oxygen species, modifications in total chlorophyll, heightened superoxide dismutase activities, increased levels of H_2O_2, glutathione, carotenoids, ascorbate, proline contents, and similar alterations, have been meticulously documented post-exposure to silver nanoparticles (AgNPs). These procedures add to the repression of photosynthesis and transition, anomalous morphological modifications, and other overt symptoms in plants [11]. In a study where AgNPs were synthesized using Aloe vera extract, the application of AgNPs and $AgNO_3$ at concentrations of 40, 60, 80, 100, and 120 mM resulted in a reduction in Brassica seedling growth. This reduction was attributed to the higher accumulation of AgNPs and $AgNO_3$, leading to a severe inhibition of photosynthesis. The engagement of silver nanoparticles (AgNPs) with the metabolic processes and growth dynamics of mustard seedlings induced a state of subtle stress upon the plants [92]. In an alternative research endeavor, silver nanoparticles (AgNPs) were synthesized employing the citrate reduction methodology and subsequently rendered stable through the incorporation of chitosan. *Lactuca sativa* L. seeds were subjected to treatment with varying concentrations of silver nanoparticles (AgNPs) at 12.5, 25, 50, and 100 ppm. This experimental procedure aimed to assess the potential deleterious effects of AgNPs, with the percentage of germinated seeds and observed alterations in root morphology serving as discerning parameters for toxicity evaluation. At a concentration of 100 ppm, a discernible decrease in root elongation was observed in contrast to the positive control, represented by distilled water [13]. Within *Arabidopsis thaliana*, the exogenous administration of silver nanoparticles (AgNPs) at a concentration of 0.2 μg/L resulted in the inhibition of root hair development. The silver nanoparticles (AgNPs) employed in this instance were synthesized and stabilized through the utilization of polyvinylpyrrolidone (PVP). In the context of soil application, silver nanoparticles (AgNPs) were administered at concentrations spanning from 100 to 900 μg/kg. Significantly, at the 800 μg/kg concentration in the soil, a discernible reduction in germination rate was witnessed, thereby impacting the growth of *Vicia faba*, as well as impeding the growth of *Rhizobium leguminosarum bv. viciae and Glomus aggregatum* [1].

4 Molecular Mechanism Underlying Plant Response to AgNPs

The intricate mechanisms underpinning both the deleterious and advantageous effects of silver nanoparticles (AgNPs) on plants have not yet been fully elucidated. Emerging research indicates that silver nanoparticles (AgNPs) are presumed to modulate stress responses in plants through diverse pathways. One of these pathways involves the generation of reactive oxygen species (ROS), which can inflict damage upon cellular structures, nucleic acids, and other pivotal biological macromolecules. Silver nanoparticles (AgNPs) may exert influence on mitosis and orchestrate plant hormone signal transduction, thereby indirectly modulating the expression of downstream target genes. Two other prospective pathways involve the regulation of cellular energy metabolism and the transformation of energy [98]. Within the aquatic botanical organism *Spirodela polyrhiza*, silver nanoparticles (AgNPs) were

Table 1 Review of literature highlighting the effect of silver nanoparticles on plant growth

Nanoparticles	Activity	Species	Key references
Silver NPs ranged from 10 to 50 nm	Seed germination and Plant growth	*Mimusops laurifolia*	Alshehddi and Bokhari [5]
10–30 nm	Improved plant growth parameters	*phaseolus vulgaris L. and (Zea mays L.)*	Salama [67]
8.1 ± 1.6 nm	Increased seed germination and growth parameters	*Pisum sativum*	Rahman et al. [62]
29 nm	Increased plant growth	*Brassica juncea*	Sharma et al. [72]
25 nm	Reduced seed germination and plant growth	*Hordeum Vulgarae*	Fayez et al. [18]
7–25 nm	Inhibited several Phytopathogenic fungi	*Alternaria alternate, Alternaria brassicicola, Alternaria solani, Botrytis cinerea, Corynespora cassiicola, Didymella bryoniae, F. oxysporum, Pythium aphanidermatum*	Kim et al. [42]
15–30 nm	Abiotic stress ameiloration	*Triticum aestivum*	Abou-Zeid et al. [3]
10–50 nm	Improved root and shoot growth	*Vitis vinifera*	Kara et al. [37]
10–35 nm	Stimulates seedling growth of wheat seedlings	*Triticum aestivum*	Kannaujia et al. [35]
8–28 nm	Enhanced production	*Citrus reticulata*	Hussain et al. [24]

ascertained to impede the functionality of ribulose-1,5-bisphosphate carboxylase-oxygenase (Rubisco), culminating in a decelerated pace of carbon dioxide assimilation. This resulted in increased free electrons and ROS in the chloroplast, ultimately reducing photosynthesis [31]. Silver nanoparticles (AgNPs), characterized by a pronounced attraction to sulfhydryl groups on cysteines, have the potential to impact the electron transport chain within the mitochondrial membrane. This impact is manifested through the perturbation of membrane proteins, modification of mitochondrial membrane permeability, and the initiation of reactive oxygen species (ROS) liberation from mitochondria. Subsequently, this sequence culminates in cellular demise [20]. In the leaves of broad bean (Vicia faba), silver nanoparticles (AgNPs) were noted to diminish stomatal conductance (gs) and CO_2 assimilation, concurrently eliciting an augmented production of reactive oxygen species (ROS) originating from chloroplasts [17]. The liberation of reactive oxygen species (ROS) within chloroplasts, mitochondria, and peroxisomes undergoes alterations in the presence of silver nanoparticles (AgNPs). Diminished concentrations of AgNPs exhibit the capacity to foster plant growth through the provocation of diminished ROS accumulation, whereas escalated concentrations engender an augmentation in ROS accumulation, thereby impeding plant growth [31]. Silver nanoparticles (AgNPs) wield substantial influence on mitotic processes within the root tips of plants, instigating chromosomal anomalies, disruptions, and the occurrence of multi-nucleation. This cascade ultimately leads to DNA impairment during the interphase. In the cellular milieu of wheat root tip cells, silver nanoparticles (AgNPs intricately impact the kinetics of mitotic spindles during the metaphase, instigating a decline in the mitotic index alongside a concomitant escalation in the heteronuclear index [2, 70]. Notwithstanding, the deleterious impact of silver nanoparticles (AgNPs) on the root tips of plants may, paradoxically, instigate the generation of primordial structures for lateral roots, which possess the capacity to mature into fully operational lateral root systems. Within the Arabidopsis root tip, there is an augmented cellular proliferation discernible under diminished concentrations of silver nanoparticles (AgNPs), whereas heightened concentrations impede the process of cell division, consequently limiting the expansiveness of root growth. Subtle concentrations of silver nanoparticles (AgNPs) elicit a rise in cellular numbers and the heightened expression of genes linked to the cell cycle, consequently propelling the expansion of root structures. In contradistinction, heightened concentrations of silver nanoparticles (AgNPs) manifest a contrary impact by diminishing the numerical abundance of cells and attenuating the expression of genes intricately linked to the cell cycle within the root apical meristem [94]. Silver nanoparticles (AgNPs) can exert an impact on the growth and developmental processes of plants by intricately overseeing the biosynthesis or signaling cascades of hormones, including but not limited to auxin, ethylene, and salicylic acid. In a scholarly investigation undertaken by Syu et al. [88], an intricate analysis of the transcriptional levels of genes pertaining to ethylene synthesis exposed a discernible down-regulation in the expression of genes encoding two crucial rate-limiting enzymes, specifically ACC (the precursor for ethylene, 1-aminocyclopropanecarboxylic acid) synthase 7 (ACS7) and ACC oxidase 2 (ACO2). This posits that silver nanoparticles (AgNPs) possess the capability to enhance the growth and maturation of plants through the suppression of ethylene synthesis. The conjectural postulation revolves around the notion that the biological repercussions of AgNPs are, to some extent, ascribed to Ag + , emanating from the oxidative transformation of AgNPs [93]. Nitric oxide (NO), a gaseous entity intricately implicated in orchestrating intricate signaling cascades amid abiotic stress within plants, assumes a pivotal role in the growth and biochemical modifications induced by silver nanoparticles (AgNPs). In the investigation by Tripathi et al. [90], it was discerned that silver nanoparticles (AgNPs) instigated a reduction in the inherent accumulation of nitric oxide (NO), orchestrated by ethylene. This led to a decrement in NO levels, exhibiting a correlation with heightened plant resilience against the stress induced by AgNPs. Moreover, AgNPs demonstrate not only the capacity to impede the synthesis and responsiveness to ethylene but also the potential involvement in the processes related to the response and transportation of auxin. Within the root tips of Arabidopsis, silver nanoparticles (AgNPs) indirectly hinder the accrual of auxin, as discerned through the employment of an auxin reporter gene. This hindrance is accomplished by suppressing the signaling pathway of ethylene [87]. During silver nanoparticles (AgNPs) treatment, transcriptional scrutiny unveiled a diminution in the expression levels of constituents within the auxin receptor family, concomitant with an escalation in the expression levels of carrier proteins facilitating auxin transport. The transcriptional activity of the auxin-responsive gene OsIAA1 experienced an augmentation within the rice callus upon exposure to AgNPs of biosynthetic origin [49]. Consequently, the mechanisms governing auxin transport and signaling play a nuanced role in the biological repercussions orchestrated by silver nanoparticles (AgNPs). When indole-3-acetic acid (IAA) is present, AgNPs were discerned to ameliorate the deleterious impacts of cadmium (Cd) stress on diverse attributes associated with plant growth in the context of carrot plants (Daucus carota L.). The mechanism involves a reduction in the concentration of malondialdehyde (MDA) and augmentation in the enzymatic activity associated with phenol synthesis, oxidation, and peroxidase. These concerted actions play a pivotal role in the sequestration of reactive oxygen species (ROS) and the enhancement of antioxidant capacity in plants, thereby contributing to the amelioration of the deleterious effects of cadmium (Cd) toxicity [16]. The salicylic acid (SA) signaling cascade is widely acknowledged for its regulatory function in orchestrating the systemic acquired resistance response within plants. Additionally, in Arabidopsis, the supplementation of silver nanoparticles (AgNPs) with salicylic acid has been observed to effectively stimulate the advancement of root hair structures. This intimates a conceivable interrelation between the SA signaling pathway and the influences of AgNPs on the developmental processes of root hairs in Arabidopsis flora [22]. The fortification of plant resistance against diseases achieved by SiO2 NPs (silica nanoparticles) has been identified to transpire through the initiation of the plant's immune response, hinging on salicylic acid (SA), as documented by. El-Shetehy et al. [15]. In a preceding scholarly examination entailing the exposure of Arabidopsis root systems to AgNPs (silver nanoparticles), evident outcomes encompassed the heightened expression of genes intricately linked to the biosynthesis of salicylic acid (SA) and the phenomenon of homologous recombination (HR). Simultaneously, there was an amelioration in the status of transcriptional gene silencing (TGS). DNA damage, induced by a myriad of stressors, spanning both abiotic and biotic domains, can be rectified through the intricate process of homologous recombination (HR). In plants, HR is elicited, predominantly orchestrated by salicylic acid (SA), to initiate the systemic defense response in response to stressors. The application of salicylic acid (SA) was established to heighten tolerance to silver nanoparticles (AgNPs) by mitigating the buildup of reactive oxygen species (H_2O_2) and malondialdehyde (MDA). This impact was additionally evidenced in the modulation of enzymes intricately linked to carbon metabolism and the synthesis of β-carotene [8]. The intricate relationship between the initiation of the salicylic acid (SA)-dependent defense response and the plant's aptitude to withstand silver nanoparticles (AgNPs) implies a sophisticated interplay between exposure to nanomaterials and the plant's stress response mechanisms. AgNPs have been ascertained to influence plant growth by modulating the expression of proteins intricately associated with glycolysis and the tricarboxylic acid cycle, both being indispensable pathways governing cellular energy metabolism. In the instance of soybean seedlings, the administration of AgNPs facilitated the oxidative conversion of pyruvate to acetyl-CoA, thereby energizing the tricarboxylic acid cycle and augmenting seedling growth by amplifying the energy supply to cellular entities [56]. Similarly, within the Arabidopsis context, the stimulation by silver nanoparticles (AgNPs) induced the initiation of the tricarboxylic acid cycle and pathways associated with sugar metabolism, resulting in the liberation of augmented energy reservoirs to bolster the progression of plant growth [38]. In addition, AgNPs chemically amalgamated were noted for their capacity to enhance shoot and root development in wheat plants. This enhancement was attributed to the reduction in the levels of glyceraldehyde-3-phosphate dehydrogenase and phosphoenolpyruvate carboxylase, accompanied by the suppression of glycolytic processes [29]. The highlighted discoveries underscore the multifaceted influence of silver nanoparticles (AgNPs) on plant metabolic processes and development, underscoring the imperative necessity to comprehend the intricate molecular mechanisms dictating the interactions between nanoparticles and plants. In the roots of Nicotiana tabacum exposed to 100 μM AgNPs, the analysis of Gene Ontology (GO) showcased a reduction in the levels of proteins linked to energy metabolism, suggesting an association with the phytotoxic consequences of AgNPs. AgNPs exhibit the capability to interact with a myriad of proteins participating in amino acid, carbohydrate, and energy metabolism. The interaction between polyvinylpyrrolidone (PVP)-coated AgNPs and the pivotal enzyme of the Calvin cycle, ribulose 1,5-diphosphate carboxylase, induced oxidative stress damage in the aquatic plant Spirodela polyrhiza, consequently resulting in the accumulation of reactive oxygen species (ROS) within chloroplasts [86]. Furthermore, the interaction between silver nanoparticles (AgNPs) and glyceraldehyde-3-phosphate dehydrogenase (GAPDH) and malate dehydrogenase (MDH) induced a substantial reduction in the levels of unbound sulfhydryl moieties, resulting in a concomitant attenuation of enzymatic activity [32]. Similarly, the intricate interplay between silver nanoparticles (AgNPs) and catalase (CAT) instigated a discernible transformation in the structural conformation of catalase, thereby exerting an influence on its biological efficacy [44]. The aptitude of silver nanoparticles (AgNPs) to influence the interaction between signal molecules and membrane proteins or hamper the operation of membrane transporters has the potential to impede signal transmission, thereby disrupting physiological and biochemical processes within plants. This intricate interplay underscores the paramount importance of attaining a comprehensive understanding of the molecular events elicited by AgNPs in the context of plant biology.

5 Effect on Production and Yield

The application of AgNPs as a nano-fertilizer represents an efficient strategy for the precise management of plant nutrient supply. Conventional fertilizers often result in excessive nutrient loss in the field, leading to adverse environmental impacts. Nano-fertilizers, in contrast, can bind to fertilizers and release an optimal amount of nutrients to plants, reducing the overall need for excess fertilizer application in the field. This approach offers a more targeted and sustainable solution for nutrient management in agriculture. The significance of AgNPs extends to various facets of agriculture. Nano-fertilizers synthesized from onion extracts have proven effective for tomato and brinjal plants. AgNPs, acting as nano-fertilizers, exhibit a slow and controlled release of nutrients, preventing nutrient loss and enhancing the absorption of nutrients by plants from the soil [97]. The integration of nanotechnology in agriculture holds the potential to revolutionize nutrient management practices, offering more efficient and environmentally friendly solutions. A notable example involves combining AgNPs with a magnetic field, resulting in the highest fruit yield (16.420 tons/ha) in muskmelon (*Cucumis melo L.).* The application of AgNPs alone demonstrated a 150% increase in fruit yield compared to the control group, highlighting the positive impact of nanotechnology in enhancing agricultural productivity [19]. In another study, various concentrations of AgNPs (0, 25, 50, 75, 100, 125, and 150 ppm) were employed for different seedling varieties of wheat. The use of AgNPs contributed to improvements in growth, yield, NPK uptake quality, and nutrient efficiency in wheat. Notably, the wheat plants treated with 25 ppm of AgNPs exhibited remarkable leaf growth and yield, whereas the application of 75 ppm showed a comparatively lower yield [30].

In a different research endeavor, the aqueous extract obtained from the root bark of *Berberis lycium* Royle was utilized in the synthesis of silver nanoparticles (AgNPs). The impacts of AgNPs at concentrations of 0, 30, 60, and 90 ppm were employed to scrutinize the carbohydrate and protein contents of *Pisum sativum L.* seeds. The seeds garnered from this investigation demonstrated the utmost documented levels of carbohydrates and proteins [52]. In a different study, eco-friendly nano-fertilizers were developed by synthesizing silver nanoparticles (AgNPs) through green methods, and these were applied to onion crops in field conditions at a concentration of 2000 ppm. The results revealed that foliar application with a lower concentration of 20 ppm of AgNPs led to the highest mean values for various morphological aspects, yield, yield components, and quality characteristics. In comparison to higher concentrations and commercial products utilizing doses ranging from 2000 to 2500 ppm, the 20 ppm concentration demonstrated superior performance. This approach not only proved effective in enhancing crop parameters but also emphasized the importance of eco-friendly nano-fertilizers to avoid potential genotoxicity and mutations that could impact human health in the future [21]. In a research study aimed at evaluating the effectiveness of silver nanoparticles (AgNPs) in enhancing plant growth and mitigating salt-induced damage in pearl millet, the application of AgNPs led to improved superoxide dismutase and catalase activities, while reducing peroxidase activity. Additionally, it decreased the levels of hydrogen peroxide (H_2O_2) and malondialdehyde (MDA) in plants subjected to salt stress. Overall, the use of AgNPs resulted in increased plant height, yield, and photosynthesis in salt-stressed plants, with the effects being dose-dependent [39, 40]. While the positive effects of silver nanoparticles (AgNPs) on plant traits and yield have been demonstrated, their imprudent use can lead to significant adverse consequences. For instance, a study involving the growth of wheat (Triticum aestivum L.) in soils amended with different concentrations of AgNPs (20, 200, and 2000 mg kg^{-1}) for four months revealed severe phytotoxicity with increasing exposure doses. This included lower biomass, reduced plant height, and decreased grain weight. Ag accumulation in roots was significantly higher than in shoots and grains. Furthermore, there was a decrease in the content of micronutrients (Fe, Cu, and Zn) in Ag-NP-treated grains, suggesting a negative impact on crop quality [95].

6 Role in Plant Protection

The utilization of nanoparticles, including silver nanoparticles (AgNPs), as effective agents for pest management has shown promise due to their non-toxic, safe, and improved pest control properties. Nano-based pesticides offer precise and high doses to target plants, making them a potential alternative for pest control [61]. Nano-pesticidal formulations originating from inorganic substrates have been scrutinized for the management of pests and insects infesting stored grains. Significantly, nanoparticles derived from plant sources are acknowledged for their environmental friendliness and are deemed relatively straightforward to fabricate, presenting a notable departure from the chemical-driven synthesis approaches utilized in the creation of nano-pesticides [12]. In examinations evaluating lethality, repellent efficacy, antifeedant characteristics, and studies on deterring oviposition, AgNPs formulated with malathion demonstrated noteworthy pesticidal potency against pests, particularly the Tribolium castaneum species. Tribolium castaneum is a distinguished beetle acknowledged for its propensity to infest stored grains and has displayed resilience against traditional synthetic insecticides. The amalgamation of AgNPs and malathion at varied concentrations, encompassing 50 ppm, 75 ppm, and 100 ppm, displayed effectiveness in pest management, resulting in a 75% mortality rate for AgNPs and 95% for malathion at 75 ppm. Consequently, AgNPs have exhibited efficacy in the control of pests and beetles [6]. The synthesis of silver nanoparticles (AgNPs) using marine bacterium *Shewanella* algae *bangaramma* has demonstrated bactericidal and larvicidal properties. The study involved determining the LC50 and LC90 values with a 95% confidence limit for III-instar larvae Burmeister (*Lepidiota mansueta*). The results showed maximum LC50 and LC90 values, and the mortality of the larvae was significantly enhanced in the exposed groups at all concentrations. This suggests that AgNPs synthesized through marine bacteria-mediated methods in a culture medium exhibit efficient antifouling and larvicidal activities [96].

The study involved the utilization of *Bacillus thuringiensis* kurstaki (Btk), an entomopathogenic bacterium, to produce Btk-AgNPs. The insecticidal activity of these Btk-synthesized AgNPs was then tested against larvae of *Agrotis ipsilon* and *Trichoplusia n*i. The application of Btk-synthesized AgNPs, whether using Bt pellet or Bt supernatant, showed high virulence toward larvae of Trichoplusia ni compared to *Agrotis ipsilon*. This indicates that the synthesized AgNPs using Btk exhibit potent insecticidal activity, particularly against *Trichoplusia ni* larvae [69]. In this study, silver nanoparticles (AgNPs) with sizes ranging from 3 to 25 nm were synthesized using the entomopathogenic fungus Beauveria bassiana from two isolates. The isolates were categorized into two groups: 50% and 51–100% based on the mortality observed in 20 isolates. The results demonstrated a maximum mortality of 64% in the 90 isolates. The potency of these AgNPs against the mustard aphid *(Lipaphis erysimi)* indicated their potential for insect management in agriculture. The study suggests that AgNPs synthesized from *Beauveria bassiana* have insecticidal properties that can be harnessed for pest control in agricultural settings [34]. In this research, silver nanoparticles (AgNPs) were synthesized using *Leonotis nepetifolia*. The AgNPs exhibited potential antifeedant activity against the larvae of *Spodoptera litura* (78.77%) and *Helicoverpa armigera* (82.16%). Larval mortality rates were high, with maximum pupal mortality rates recorded at 78.49% and 72.70% for *S. litura* and *H. armigera*, respectively. Furthermore, when tested on mosquito larvae, the AgNPs showed LC50 values of 35.48 ppm for *Culex quinquefasciatus* and 47.44 ppm for *Aedes aegypti*. The study suggests that biosynthesized AgNPs using *L. nepetifolia* have potential insecticidal properties, making them environmentally favorable for pest management. Histological examinations confirmed severe tissue damage in the goblet and epithelial cells of the larval midgut region in *H. armigera*, *S. litura*, *C. quinquefasciatus*, and *A. aegypti* with increasing concentrations of the nanomaterial. This reinforces the potential of AgNPs for controlling insect pests across different species [50].

7 Conclusion

Our exploration of the interaction between silver nanoparticles and plants reveals a complex interplay, one brimming with promising avenues and potential pitfalls. These microscopic entities can act as potent tools for agriculture, stimulating root growth, enhancing chlorophyll content, and facilitating nutrient uptake. Their antimicrobial properties provide a glimpse into the future of eco-friendly pest control. However, at higher concentrations, silver nanoparticles can cast a dark shadow. They can disrupt vital plant processes, impairing growth and yield. Their potential migration through the food chain raises concerns about ecological and human health implications. Therefore, navigating this nanotech-agricultural terrain necessitates a careful balancing act. We must delve deeper into the underlying mechanisms governing plant responses to silver nanoparticles, paving the way for targeted and responsible applications. Rigorous risk assessments and robust regulatory frameworks are crucial to ensure the safe integration of these technologies into agricultural practices. In essence, the story of plant-nanoparticle interactions is a dynamic one, woven with both opportunity and caution. It underscores the delicate interplay between innovation and responsibility. By embracing a multidisciplinary approach that bridges biology, chemistry, and environmental science, we can unlock the full potential of this technology while safeguarding our ecosystems. In doing so, we can cultivate a future where silver nanoparticles dance in harmony with plants, fostering sustainability and bountiful harvests.

References

1. Abbasi Khalaki M, Ghorbani A, Moameri M (2016) Effects of silica and silver nanoparticles on seed germination traits of Thymus kotschyanus in laboratory conditions. J Rangeland Sci 6(3):221–231
2. Abdelsalam NR, Abdel-Megeed A, Ali HM, Salem MZ, Al-Hayali MF, Elshikh MS (2018) Genotoxicity effects of silver nanoparticles on wheat (Triticum aestivum L.) root tip cells. Ecotoxicol Environ Saf 155:76–85
3. Abou-Zeid H, Ismail G (2018) The role of priming with biosynthesized silver nanoparticles in the response of Triticum aestivum L to salt stress. Egypt J Bot 58(1):73–85
4. Almutairi ZM, Alharbi A (2015) Effect of silver nanoparticles on seed germination of crop plants. Int J Nuclear Quantum Eng 9(6):689–693
5. Alshehddi LA, Bokhari N (2020) Influence of gold and silver nanoparticles on the germination and growth of Mimusops laurifolia seeds in the South-Western regions in Saudi Arabia. Saudi J Biol Sci 27(1):574–580
6. Aa AS (2019) Comparative bioassay of silver nanoparticles and malathion on infestation of red flour beetle, Tribolium castaneum. J Basic Appl Zool 80(1):1
7. Bachheti RK, Bachheti A, Husen A (2024) Metal and metal-oxide based nanomaterials (Synthesis, Agricultural, Biomedical and Environmental Interventions). Springer Nature Singapore Pte Ltd., 152 Beach Road, #21-01/04 Gateway East, Singapore 189721, Singapore
8. Bahador E, Einali A, Azizian-Shermeh O, Sangtarash MH (2019) Metabolic responses of the green microalga Dunaliella salina to silver nanoparticles-induced oxidative stress in the presence of salicylic acid treatment. Aquat Toxicol 217:105356
9. Batley GE, Kirby JK, McLaughlin MJ (2013) Fate and risks of nanomaterials in aquatic and terrestrial environments. Acc Chem Res 46(3):854–862
10. Biba R, Cvjetko P, Tkalec M, Košpić K, Štefanić PP, Šikić S, Domijan AM, Balen B (2022) Effects of silver nanoparticles on physiological and proteomic responses of tobacco (Nicotiana tabacum) seedlings are coating-dependent. Int J Mol Sci 23(24):15923
11. Budhani S, Egboluche NP, Arslan Z, Yu H, Deng H (2019) Phytotoxic effect of silver nanoparticles on seed germination and growth of terrestrial plants. J Environ Sci Health C 37(4):330–355
12. Chen X, Chen Y, Zou L, Zhang X, Dong Y, Tang J, McClements DJ, Liu W (2019) Plant-based nanoparticles prepared from proteins and phospholipids consisting of a core–multilayer-shell structure: fabrication, stability, and foamability. J Agric Food Chem 67(23):6574–6584
13. de Paiva Pinheiro SK, de Medeiros CM, Miguel TB, de Freitas Barros FC, Farias CP, Ferreira OP, de Castro ME (2020) Toxic effects of silver nanoparticles on the germination and root development of lettuce (Lactuca sativa). Aust J Bot 68(2):127–136
14. Dykman L, Shchyogolev S (2018) The effect of gold and silver nanoparticles on plant growth and development. In: Metal nanoparticles. Nova, New York, pp 63–300
15. El-Shetehy M, Moradi A, Maceroni M, Reinhardt D, Petri-Fink A, Rothen-Rutishauser B, Mauch F, Schwab F (2021) Silica nanoparticles enhance disease resistance in Arabidopsis plants. Nat Nanotechnol 16(3):344–353
16. Faiz S, Shah AA, Naveed NH, Nijabat A, Yasin NA, Batool AI, Ali HM, Javed T, Simon PW, Ali A (2022) Synergistic application of silver nanoparticles and indole acetic acid alleviate cadmium induced stress and improve growth of Daucus carota L. Chemosphere 290:133200
17. Falco WF, Scherer MD, Oliveira SL, Wender H, Colbeck I, Lawson T, Caires AR (2020) Phytotoxicity of silver nanoparticles on Vicia faba: evaluation of particle size effects on photosynthetic performance and leaf gas exchange. Sci Total Environ 701:134816

18. Fayez KA, El-Deeb BA, Mostafa NY (2017) Toxicity of biosynthetic silver nanoparticles on the growth, cell ultrastructure and physiological activities of barley plant. Acta Physiol Plant 39:1–3
19. Feizi H, Pour SJ, Rad KH (2013) Biological response of muskmelon (Cucumis melo L.) to magnetic field and silver nanoparticles. Ann Res Rev Biol: 794–804
20. Flores-López LZ, Espinoza-Gómez H, Somanathan R (2019) Silver nanoparticles: electron transfer, reactive oxygen species, oxidative stress, beneficial and toxicological effects. Mini review. J Appl Toxicol 39(1):16–26
21. Fouda MM, Abdelsalam NR, El-Naggar ME, Zaitoun AF, Salim BM, Bin-Jumah M, Allam AA, Abo-Marzoka SA, Kandil EE (2020) Impact of high throughput green synthesized silver nanoparticles on agronomic traits of onion. Int J Biol Macromol 149:1304–1317
22. García-Sánchez S, Bernales I, Cristobal S (2015) Early response to nanoparticles in the Arabidopsis transcriptome compromises plant defence and root-hair development through salicylic acid signalling. BMC Genomics 16(1):1–7
23. Hojjat SS, Ganjali A (2016) The effect of silver nanoparticle on lentil seed germination under drought stress. Int J Farm Allied Sci 5(3):208–212
24. Hussain M, Iqbal Raja N, Mashwani ZU, Iqbal M, Ejaz M, Aslam S (2019) Green synthesis and evaluation of silver nanoparticles for antimicrobial and biochemical profiling in Kinnow (Citrus reticulata L.) to enhance fruit quality and productivity under biotic stress. IET Nanobiotechnol 13(3):250–256
25. Husen A (2022) Engineered nanomaterials for sustainable agricultural production, soil improvement and stress management. Elsevier Inc., 50 Hampshire St., 5th Floor, Cambridge, MA 02139, USA. https://doi.org/10.1016/C2021-0-00054-7
26. Husen A (2023) Nanomaterials and nanocomposites exposures to plants (Response, Interaction, Phytotoxicity and Defense Mechanisms). Springer Nature Singapore Pte Ltd., 152 Beach Road, #21-01/04 Gateway East, Singapore 189721, Singapore. https://doi.org/10.1007/978-981-99-2419-6
27. Husen A, Siddiqi KS (2023) Advances in smart nanomaterials and their applications. Elsevier Inc., 50 Hampshire St., 5th Floor, Cambridge, MA 02139, USA. https://doi.org/10.1016/C2021-0-02202-1
28. Imtiaz H, Shiraz M, Mir AR, Siddiqui H, Hayat S (2023) Nano-priming techniques for plant physio-biochemistry and stress tolerance. J Plant Growth Regul:1–21
29. Jhanzab HM, Razzaq A, Bibi Y, Yasmeen F, Yamaguchi H, Hitachi K, Tsuchida K, Komatsu S (2019) Proteomic analysis of the effect of inorganic and organic chemicals on silver nanoparticles in wheat. Int J Mol Sci 20(4):825
30. Jhanzab HM, Razzaq A, Jilani G, Rehman A, Hafeez A, Yasmeen F (2015) Silver nanoparticles enhance the growth, yield and nutrient use efficiency of wheat. Int J Agron Agri Res 7(1):15–22
31. Jiang HS, Yin LY, Ren NN, Zhao ST, Li Z, Zhi Y, Shao H, Li W, Gontero B (2017) Silver nanoparticles induced reactive oxygen species via photosynthetic energy transport imbalance in an aquatic plant. Nanotoxicology 11(2):157–167
32. Jiang HS, Zhang Y, Lu ZW, Lebrun R, Gontero B, Li W (2019) Interaction between silver nanoparticles and two dehydrogenases: Role of thiol groups. Small 15(27):1900860
33. Kale SK, Parishwad GV, Patil AS (2021) Emerging agriculture applications of silver nanoparticles. ES Food Agrofor 3:17–22
34. Kamil D, Prameeladevi T, Ganesh S, Prabhakaran N, Nareshkumar R, Thomas SP (2017) Green synthesis of silver nanoparticles by entomopathogenic fungus Beauveria bassiana and their bioefficacy against mustard aphid (Lipaphis erysimi Kalt.)
35. Kannaujia R, Srivastava CM, Prasad V, Singh BN, Pandey V (2019) Phyllanthus emblica fruit extract stabilized biogenic silver nanoparticles as a growth promoter of wheat varieties by reducing ROS toxicity. Plant Physiol Biochem 142:460–471
36. Kapoor D, Yadav S, Sharma MM, Sharma P (2023) Interaction between metal nanoparticles and PGPR on the plant growth and development. In: Nanomaterials and nanocomposites exposures to plants: response, interaction, phytotoxicity and defense mechanisms. Springer Nature Singapore, Singapore, pp 327–351

37. Kara Z, Sabır A, Koç F, Sabır FK, Avcı A, Koplay M, Doğan O (2021) Silver nanoparticles synthesis by grape seeds (Vitis vinifera L.) extract and rooting effect on grape cuttings. Erwerbs-Obstbau 63(Suppl 1):1–8
38. Ke M, Qu Q, Peijnenburg WJ, Li X, Zhang M, Zhang Z, Lu T, Pan X, Qian H (2018) Phytotoxic effects of silver nanoparticles and silver ions to Arabidopsis thaliana as revealed by analysis of molecular responses and of metabolic pathways. Sci Total Environ 644:1070–1079
39. Khan I, Awan SA, Raza MA, Rizwan M, Tariq R, Ali S, Huang L (2021a) Silver nanoparticles improved the plant growth and reduced the sodium and chlorine accumulation in pearl millet: a life cycle study. Environ Sci Pollut Res 28:13712–13724
40. Khan M, Khan AU, Moon IS, Felimban R, Alserihi R, Alsanie WF, Alam M (2021b) Synthesis of biogenic silver nanoparticles from the seed coat waste of pistachio (Pistacia vera) and their effect on the growth of eggplant. Nanotechnol Rev 10(1):1789–1800
41. Khan S, Zahoor M, Khan RS, Ikram M, Islam NU (2023) The impact of silver nanoparticles on the growth of plants: the agriculture applications. Heliyon
42. Kim SW, Jung JH, Lamsal K, Kim YS, Min JS, Lee YS (2012) Antifungal effects of silver nanoparticles (AgNPs) against various plant pathogenic fungi. Mycobiology 40(1):53–58
43. Li S, Chen S, Zhang Z, Huang Y, Li G, Li Y, Deng X, Li J (2022) Short-term exposure to silver nano-particles alters the physiology and induces stress-related gene expression in Nelumbo nucifera. Plant Physiol Biochem 177:38–45
44. Liu W, Worms I, Slaveykova VI (2020) Interaction of silver nanoparticles with antioxidant enzymes. Environ Sci Nano 7(5):1507–1517
45. Loyal A, Pahuja SK, Rani N, Pooja, Srivastava RK, Sharma P (2023a) Titanium oxide nanoparticles: plant response, interaction, phytotoxicity, and defence mechanisms. In: Nanomaterials and nanocomposites exposures to plants: response, interaction, phytotoxicity and defense mechanisms 2023. Springer Nature Singapore, Singapore, pp 263–284
46. Loyal A, Pahuja SK, Sharma P, Malik A, Srivastava RK, Mehta S (2023b) Potential environmental and human health implications of nanomaterials used in sustainable agriculture and soil improvement. In: Engineered nanomaterials for sustainable agricultural production, soil improvement and stress management. Academic Press, pp 387–412
47. Mahakham W, Sarmah AK, Maensiri S, Theerakulpisut P (2017) Nanopriming technology for enhancing germination and starch metabolism of aged rice seeds using phytosynthesized silver nanoparticles. Sci Rep 7(1):8263
48. Mahendran D, Geetha N, Venkatachalam P (2019) Role of silver nitrate and silver nanoparticles on tissue culture medium and enhanced the plant growth and development. In: In vitro plant breeding towards novel agronomic traits: biotic and abiotic stress tolerance, pp 59–74
49. Manickavasagam M, Pavan G, Vasudevan V (2019) A comprehensive study of the hormetic influence of biosynthesized AgNPs on regenerating rice calli of indica cv. IR64. Sci Reports 9(1):8821
50. Manimegalai T, Raguvaran K, Kalpana M, Maheswaran R (2020) Green synthesis of silver nanoparticle using Leonotis nepetifolia and their toxicity against vector mosquitoes of Aedes aegypti and Culex quinquefasciatus and agricultural pests of Spodoptera litura and Helicoverpa armigera. Environ Sci Pollut Res 27:43103–43116
51. Mazumdar H, Ahmed GU (2011) Synthesis of silver nanoparticles and its adverse effect on seed germinations in Oryza sativa, Vigna radiata and Brassica campestris. Int J Adv Biotechnol Res 2(4):404–413
52. Mehmood A, Murtaza G (2017) Impact of biosynthesized silver nanoparticles on protein and carbohydrate contents in seeds of Pisum sativum L. Crop Breed Appl Biotechnol 17:334–340
53. Mirzajani F, Askari H, Hamzelou S, Farzaneh M, Ghassempour A (2013) Effect of silver nanoparticles on Oryza sativa L. and its rhizosphere bacteria. Ecotoxicol Environ Saf 88:48–54
54. Mishra S, Singh HB (2015) Biosynthesized silver nanoparticles as a nanoweapon against phytopathogens: exploring their scope and potential in agriculture. Appl Microbiol Biotechnol 99:1097–1107
55. Monreal CM, DeRosa M, Mallubhotla SC, Bindraban PS, Dimkpa C (2016) Nanotechnologies for increasing the crop use efficiency of fertilizer-micronutrients. Biol Fertil Soils 52:423–437

56. Mustafa G, Sakata K, Hossain Z, Komatsu S (2015) Proteomic study on the effects of silver nanoparticles on soybean under flooding stress. J Proteomics 122:100–118
57. Nejatzadeh F (2021) Effect of silver nanoparticles on salt tolerance of Satureja hortensis l. during in vitro and in vivo germination tests. Heliyon 7(2)
58. Pallavi Mehta CM, Srivastava R, Arora S, Sharma AK (2016) Impact assessment of silver nanoparticles on plant growth and soil bacterial diversity. 3 Biotech 6:1
59. Pandey C, Khan E, Mishra A, Sardar M, Gupta M (2014) Silver nanoparticles and its effect on seed germination and physiology in Brassica juncea L.(Indian mustard) plant. Adv Sci Lett 20(7–8):1673–1676
60. Peharec Štefanić P, Jarnević M, Cvjetko P, Biba R, Šikić S, Tkalec M, Cindrić M, Letofsky-Papst I, Balen B (2019) Comparative proteomic study of phytotoxic effects of silver nanoparticles and silver ions on tobacco plants. Environ Sci Pollut Res 26:22529–22550
61. Ragaei M, Sabry AK (2014) Nanotechnology for insect pest control. Int J Sci, Environ Technol 3(2):528–545
62. Rahman MS, Chakraborty A, Kibria A, Hossain MJ (2023) Effects of silver nanoparticles on seed germination and growth performance of pea (Pisum sativum). Plant Nano Biol 5:100042
63. Rani K, Sharma P, Kumar S, Wati L, Kumar R, Gurjar DS, Kumar D, Kumar R (2019) Legumes for sustainable soil and crop management. In: Sustainable management of soil and environment, pp 193–215
64. Rani K, Rani A, Sharma P, Dahiya A, Punia H, Kumar S, Sheoran S, Banerjee A (2022) Legumes for agroecosystem services and sustainability. In: Advances in legumes for sustainable intensification. Academic Press, pp 363–380
65. Sable SV, Ranade S, Joshi S (2018) Role of AgNPs in the enhancement of seed germination and its effect on plumule and radicle length of Pennisetum glaucum. IET Nanobiotechnol 12(7):922–926
66. Sadak MS (2019) Impact of silver nanoparticles on plant growth, some biochemical aspects, and yield of fenugreek plant (Trigonella foenum-graecum). Bull Natl Res Centre 43(1):1–6
67. Salama HM (2012) Effects of silver nanoparticles in some crop plants, common bean (Phaseolus vulgaris L.) and corn (Zea mays L.). Int Res J Biotechnol 3(10):190–197
68. Sangwan S, Sharma P, Wati L, Mehta S (2023) Effect of chitosan nanoparticles on growth and physiology of crop plants. In: Engineered nanomaterials for sustainable agricultural production, soil improvement and stress management. Academic Press, pp 99–123
69. Sayed AM, Kim S, Behle RW (2017) Characterisation of silver nanoparticles synthesised by Bacillus thuringiensis as a nanobiopesticide for insect pest control. Biocontrol Sci Tech 27(11):1308–1326
70. Scherer MD, Sposito JC, Falco WF, Grisolia AB, Andrade LH, Lima SM, Machado G, Nascimento VA, Gonçalves DA, Wender H, Oliveira SL (2019) Cytotoxic and genotoxic effects of silver nanoparticles on meristematic cells of Allium cepa roots: a close analysis of particle size dependence. Sci Total Environ 660:459–467
71. Shahrokh S, Emtiazi G (2009) Toxicity and unusual biological behavior of nanosilver on gram positive and negative bacteria assayed by microtiter-plate. Eur J Biol Sci 1(3):28–31
72. Sharma P, Bhatt D, Zaidi MG, Saradhi PP, Khanna PK, Arora S (2012) Silver nanoparticle-mediated enhancement in growth and antioxidant status of Brassica juncea. Appl Biochem Biotechnol 167:2225–2233
73. Sharma P, Sharma MM, Anamika, Kapoor D, Rani K, Singh D, Barkodia M (2020) Role of microbes for attaining enhanced food crop production. In: Microbial biotechnology: basic research and applications, pp 55–78
74. Sharma P, Pandey V, Sharma MM, Patra A, Singh B, Mehta S, Husen A (2021a) A review on biosensors and nanosensors application in agroecosystems. Nanoscale Res Lett 16:1–24
75. Sharma MM, Sharma P, Kapoor D, Beniwal P, Mehta S (2021b) Phytomicrobiome community: an agrarian perspective towards resilient agriculture. In: Plant performance under environmental stress: hormones, biostimulants and sustainable plant growth management, pp 493–534

76. Sharma P, Sharma MM, Malik A, Vashisth M, Singh D, Kumar R, Singh B, Patra A, Mehta S, Pandey V (2021c) Rhizosphere, rhizosphere biology, and rhizospheric engineering. In: Plant growth-promoting microbes for sustainable biotic and abiotic stress management. Springer International Publishing, Cham, pp 577–624
77. Sharma P, Meyyazhagan A, Easwaran M, Sharma MM, Mehta S, Pandey V, Liu WC, Kamyab H, Balasubramanian B, Baskaran R, Klemeš JJ (2022a) Hydrogen Sulfide: A new warrior in assisting seed germination during adverse environmental conditions. Plant Growth Regul 98(3):401–420
78. Sharma P, Sangwan S, Kumari A, Singh S, Kaur H (2022b) Impact of climate change on soil microorganisms regulating nutrient transformation. In: Plant stress mitigators: action and application. Springer Nature Singapore, Singapore, pp 145–172
79. Sharma MM, Kapoor D, Rohilla R, Sharma P (2023a) Nanomaterials and their toxicity to beneficial soil microbiota and fungi associated plants rhizosphere. In: Nanomaterials and nanocomposites exposures to plants: response, interaction, phytotoxicity and defense mechanisms. Springer Nature Singapore, Singapore, pp 353–380
80. Sharma P, Patra A, Singh B, Mehta S (2023b) Microbial rejuvenation of soils for sustainable agriculture. In: Sustainable agriculture reviews 60: microbial processes in agriculture. Springer Nature Switzerland, Cham, pp 293–323
81. Sharma P, Sangwan S, Kaur H, Patra A, Anamika, Mehta S (2023c) Diversity and evolution of nitrogen fixing bacteria. In: Sustainable agriculture reviews 60: microbial processes in agriculture, pp 95–120
82. Sharma P, Sangwan S, Kaur H, Patra A, Mehta S (2023d) Microbial remediation of agricultural residues. In: Sustainable agriculture reviews 60: microbial processes in agriculture. Springer Nature Switzerland, Cham, pp 325–358
83. Sharma P, Sangwan S, Mehta S (2023e) Emerging role of phosphate nanoparticles in agriculture practices. In: Engineered nanomaterials for sustainable agricultural production, soil improvement and stress management. Academic Press, pp 71–97
84. Siddiqi KS, Husen A (2022) Plant response to silver nanoparticles: a critical review. Crit Rev Biotechnol 42(7):773–990. https://doi.org/10.1080/07388551.2021.1975091
85. Souza LR, Corrêa TZ, Bruni AT, da Veiga MA (2021) The effects of solubility of silver nanoparticles, accumulation, and toxicity to the aquatic plant Lemna minor. Environ Sci Pollut Res 28:16720–16733
86. Štefanić PP, Cvjetko P, Biba R, Domijan AM, Letofsky-Papst I, Tkalec M, Šikić S, Cindrić M, Balen B (2018) Physiological, ultrastructural and proteomic responses of tobacco seedlings exposed to silver nanoparticles and silver nitrate. Chemosphere 209:640–653
87. Sun J, Wang L, Li S, Yin L, Huang J, Chen C (2017) Toxicity of silver nanoparticles to Arabidopsis: inhibition of root gravitropism by interfering with auxin pathway. Environ Toxicol Chem 36(10):2773–2780
88. Syu YY, Hung JH, Chen JC, Chuang HW (2014) Impacts of size and shape of silver nanoparticles on Arabidopsis plant growth and gene expression. Plant Physiol Biochem 83:57–64
89. Tomacheski D, Pittol M, Simões DN, Ribeiro VF, Santana RM (2017) Impact of silver ions and silver nanoparticles on the plant growth and soil microorganisms. Global J Environ Sci Manag 3(4):341–350
90. Tripathi DK, Kandhol N, Rai P, Mishra V, Pandey S, Deshmukh R, Sahi S, Sharma S, Singh VP (2022) Ethylene renders silver nanoparticles stress tolerance in rice seedlings by regulating endogenous nitric oxide accumulation. Plant Cell Physiol 63(12):1954–1967
91. Usman M, Farooq M, Wakeel A, Nawaz A, Cheema SA, ur Rehman H, Ashraf I, Sanaullah M (2020) Nanotechnology in agriculture: current status, challenges and future opportunities. Sci Total Environ 721:137778
92. Vishwakarma K, Shweta, Upadhyay N, Singh J, Liu S, Singh VP, Prasad SM, Chauhan DK, Tripathi DK, Sharma S (2017) Differential phytotoxic impact of plant mediated silver nanoparticles (AgNPs) and silver nitrate (AgNO3) on Brassica sp. Front Plant Sci 8:1501

93. Wang J, Koo Y, Alexander A, Yang Y, Westerhof S, Zhang Q, Schnoor JL, Colvin VL, Braam J, Alvarez PJ (2013) Phytostimulation of poplars and Arabidopsis exposed to silver nanoparticles and Ag+ at sublethal concentrations. Environ Sci Technol 47(10):5442–5449
94. Wang L, Sun J, Lin L, Fu Y, Alenius H, Lindsey K, Chen C (2020) Silver nanoparticles regulate Arabidopsis root growth by concentration-dependent modification of reactive oxygen species accumulation and cell division. Ecotoxicol Environ Saf 190:110072
95. Yang J, Jiang F, Ma C, Rui Y, Rui M, Adeel M, Cao W, Xing B (2018) Alteration of crop yield and quality of wheat upon exposure to silver nanoparticles in a life cycle study. J Agric Food Chem 66(11):2589–2597
96. Yokesh Babu M, Janaki Devi V, Ramakritinan CM, Umarani R, Taredahalli N, Kumaraguru AK (2014) Application of biosynthesized silver nanoparticles in agricultural and marine pest control. Curr Nanosci 10(3):374–381
97. Yuvaraj M, Subramanian KS (2020) Novel slow release nanocomposite fertilizers. In: Nanotechnology and the environment. IntechOpen
98. Zhang N, Sun J, Yin L, Liu J, Chen C (2023) Silver nanoparticles: from in vitro green synthesis to in vivo biological effects in plants. Adv Agrochem
99. Singh A, Sharma P, Kumari A, Kumar R, Pathak DV (2019) Management of root-knot nematode in different crops using microorganisms. Plant Biotic Interactions: State of the Art 85–99
100. Singh S, Sangwan S, Sharma P, Devi P, Moond M (2021 Jun 1) Nanotechnology for sustainable agriculture: an emerging perspective. J Nanosci Nanotechnol 21(6):3453–3465
101. Kapoor D, Sharma P, Sharma MM, Yadav S, Husen A (2024 Sep 11) Exploring soil microbiota and their role in plant growth, stress tolerance, disease control and nutrient immobilizer. Biocatal Agric Biotechnol 103358
102. Sharma MM, Kapoor D, Loyal A, Kumar R, Sharma P, Husen A (2024 May 1) Environmental toxicity of engineered carbon nanoparticles. In: Carbon-Based nanomaterials: Synthesis, agricultural, biomedical, and environmental interventions. Springer Nature Singapore, Singapore, pp 337–353
103. Sharma MM, Kapoor D, Sharma P, Husen A (2024 May 1) Role of carbon nanomaterials in the prevention of plant disease. In: Carbon-based nanomaterials: synthesis, agricultural, biomedical, and environmental interventions. Springer Nature Singapore, Singapore, pp 105–121

Beneficial and Adverse Effects of Silver Nanoparticles on Plant

Noureddine Chaachouay, Lahcen Zidane, and Azamal Husen

Abstract Silver nanoparticles are exceptionally prominent with unique features, making them suitable for various applications. The beneficial effects of silver nanoparticles have been revealed in many areas of the economy, including agriculture. Silver nanoparticles have garnered significant scientific interest in agriculture and plant biotechnology due to their high effectiveness in recent years. It not only boosts the process of seed germination and plant development but even promotes the quantum efficiency of the photosynthetic function. Silver nanoparticles are essential in agriculture because they guarantee food security and enhance agricultural output. Furthermore, they function as nano-pesticides, delivering an adequate dosage to the desired plants while minimizing the release of excessive pesticides into the surrounding environment. Nano-fertilizers gradually provide nutrients to the plants, therefore controlling extreme nutrient depletion. The potential uses of silver nanoparticles in farming and horticulture production are substantial. There is a possibility of achieving higher output levels while using fewer resources. However, there is presently an absence of knowledge on the possible hazards to the health of plants and soil fauna. Current investigations have revealed that employing nanoparticles affects plant development and soil fauna, as seen in their interaction. Before the widespread use or commercialization of nanoparticle-based agrochemicals, it is crucial to carefully evaluate the potential negative impacts on plant development and soil ecology. This chapter thoroughly examines and summarises contemporary research on the

N. Chaachouay (✉)
Agri-Food and Health Laboratory (AFHL), Faculty of Sciences and Techniques of Settat, Hassan First University, P.O. Box 382, 26000 Settat, Morocco
e-mail: nour.chay@gmail.com

L. Zidane
Plant, Animal Productions and Agro-Industry Laboratory, Department of Biology, Faculty of Sciences, Ibn Tofail University, B.P. 133 14000, Kenitra, Morocco

A. Husen
Wolaita Sodo University, PO Box 138, Wolaita, Ethiopia

Department of Biotechnology, Graphic Era (Deemed to be University), Dehradun 248002, Uttarakhand, India

A. Husen (ed.), *Plant Response to Silver Nanoparticles*, Smart Nanomaterials Technology, https://doi.org/10.1007/978-981-97-7352-7_2

beneficial and adverse effects of Ag-NPs on plant development and their uses in agriculture.

Keywords Agriculture · Nanotechnology · Phytotoxicity · Plants · Silver nanoparticles

1 Introduction

Nanotechnology is a burgeoning scientific discipline that primarily uses nanoparticles to surpass size constraints and revolutionize the scientific paradigm [18, 40, 72]. It has transformed the agricultural industry and performed crucial roles in enhancing crop productivity, agriculture biotechnology, and food security [4, 38, 39, 41]. The unique physicochemical characteristics of nanoparticles are very effective in stimulating plant metabolism. Nevertheless, the intricacies of their interactions with plants remain unclear and need more comprehension. Conflicting data on nanoparticles have been documented regarding their absorption, accumulation, biotransformation, translocation, and toxicity in several plant taxa [77]. Due to their well-recognized antibacterial and antifungal characteristics, they may be used in many applications, such as home items, food packaging, textiles, medical equipment, antiseptics in healthcare settings, and personal healthcare. Silver nanoparticles (Ag-NPs) have been synthesized and have shown influence in the agricultural sector to enhance plant growth, act as fungicides to combat fungal diseases, and serve as agents to enhance fruit ripening [45, 84]. Incorporating Ag-NPs into products and their management, recycling, or disposal invariably increases the likelihood that these nanoparticles will be released into the environment [11]. Ag-NPs were used in agriculture as stimulants for plant growth, fungicides to combat fungal infections, and agents to improve fruit ripening [52]. The likelihood of Ag-NPs being discharged into the environment is inevitably increased with the expanding use of these particles, which occurs during their production, incorporation into products, and the handling, recycling, or disposal of such items [27]. Ag-NPs are anticipated to enter the environment via surface waters such as lakes, streams, and rivers. The primary route of entry is via the presence of biosolids derived from wastewater treatment. Ag-NPs are often found in water and soil and collected in significant amounts [22]. Furthermore, there is a substantial increase in the levels of Ag-NPs found in surface water and sewage treatment. In agriculture, Ag-NPs-contaminated water may seep into fields via fertilization and irrigation. Ag-NPs, once released, can penetrate many substances and ultimately reach the rhizosphere of plants [1]. Hence, the Ag-NPs are unavoidably absorbed by crops and readily incorporated into the food chain, affecting food production and quality and presenting a potential hazard to human health [32].

Ag-NPs can release silver ions (Ag^+), which are durable, capable of accumulating in organisms, and very poisonous. Hence, introducing Ag-NPs into ecosystems gives rise to significant apprehensions about their safety and environmental toxicity [36, 83]. Given that plants play a crucial role in ecosystems as the main trophic level and form the foundation of the food chain, it is essential to have a comprehensive knowledge of the effects of Ag-NPs on plants to evaluate their toxicity. Therefore, this chapter provides an account of the absorption and movement of Ag-NPs, as well as a comprehensive overview of the effects of Ag-NPs on plants. In addition, the chapter discusses the methods by which Ag-NPs cause phytotoxicity in plants and how plants reduce the adverse impact of Ag-NPs in accelerated tolerance under various situations.

2 Uptake and Transport

2.1 Mechanisms of Ag-NPs Uptake by Plant Roots

Ag-NPs in plants are transferred via intercellular gaps for short distances and through vascular tissue for considerable distances (Fig. 1). Upon exposure to plants, nanoparticles infiltrate the cell walls and plasma membranes of the outermost layers of the roots [20]. This is followed by a sequence of processes that allow the Ag-NPs to enter the plant's vascular tissues, namely the xylem, and ultimately reach the stele [91]. At first, Ag-NPs bind to the surface of the root hair cells. Subsequently, Ag-NPs of a size less than 10 nm traverse the cell wall and plasma membrane of the cell. Plasmodesmata, mediating the transport of Ag-NPs of diminutive size, employ channels characterized by pore dimensions spanning from 20 to 50 nm. Ag-NPs can bind to carrier proteins and penetrate cells via plasma membrane ion channels or aquaporins [35]. Ag-NPs of more excellent dimensions may lead to the creation of supplementary "pores" in the cellular membrane. In contrast, Ag-NPs covered in rhizosphere exudates or novel compounds with membrane transporters can be taken in internally [82].

Plants can selectively take up Ag-NPs from the soil. Individual and tiny Ag-NPs are more readily absorbed due to their ability to approach the cell wall closely and be directly taken up into the cells via the cell wall [62]. While not all Ag-NPs present in soil are separated, other factors may impact the uptake of Ag-NPs. When Ag-NPs and Sb were present, plants absorbed sulfide-Ag-NPs more readily. The plant cells readily absorb the coated Ag-NPs. Ag-NPs may be coated with rhizosphere exudates, naturally produced by plant roots, to create complexes that plant cells can subsequently absorb [71]. Likewise, plant cells readily absorbed artificial-coated Ag-NPs coated with polyvinyl pyrrolidone (PVP) and cetyltrimethylammonium bromide (CTAB) to alter their surface charge [93]. The collective impact of nanoparticles' dimensions, electrical order, and surface characteristics often influence cells' efficient uptake of Ag-NPs [64]. Furthermore, Ag-NPs may readily be absorbed and used by plants when they are evenly distributed in the surrounding environment.

Fig. 1 Mechanisms of Ag-NPs uptake and transport by plant

2.2 *Migration of Ag-NPs in Plant Tissues*

Ag-NP is disseminated within the plant structure utilizing the routes provided by the primary and secondary root networks. The migration of Ag-NPs through plant tissues can occur via two principal mechanisms: symplastic transport, which moves through the cytoplasm of adjacent plant cells connected by plasmodesmata, and apoplastic transport, which involves movement through the extracellular spaces and cell walls. Research has uncovered that Ag-NPs traverse the intercellular reaches between cells in *Arabidopsis* roots. Transport pathways for molecules within plants differ based on size; smaller molecules typically travel directly through the symplast and the connecting plasmodesmata channels, whereas larger, macromolecular substances are translocated via the apoplastic spaces between cells [8, 16]. The Casparian strip, located on the surface of the endodermis, acts as a natural barrier that impedes the movement of large molecules [37]. Ag-NPs encounter substantial obstacles when attempting to penetrate the Casparian strip, a barrier within the plant root structure. Only a minimal proportion of Ag-NPs can navigate through the successive root layers, initiating at the epidermis, advancing through the cortex, moving toward the endodermis, and ultimately reaching the vascular bundle. This intricate journey involves both apoplastic routes, moving through the cell walls and intercellular spaces, and symplastic paths, traveling within the cytoplasm of cells interconnected by plasmodesmata. The complexity of this transport process reflects the selective permeability and regulatory functions of the root's cellular architecture [49].

Research indicates that specific nanoparticles can infiltrate the vascular cylinder through gaps manifest in the Casparian strip during the development of secondary roots. This observation suggests that these openings serve as conduits for nanoparticle movement into the central vascular system of the plant. Additionally, evidence has been found that Ag-NPs gradually progress toward the developing lateral root primordia, corresponding with the elongation of the primary roots in *Arabidopsis* plants. This implies a targeted movement or attraction of Ag-NPs to areas of new root growth, which could be crucial for understanding how nanoparticles interact with plant development and physiology [55]. Nanoparticles can migrate from the plant root system to the shoot through the xylem channels within the secondary roots' vascular bundles. This translocation process can be influenced by the surface electrical properties of the nanoparticles, which may interact with the plant's biological environment. These interactions can be theoretically analyzed using the classical Derjaguin-Landau-Verwey-Overbeek theory, which provides insights into the forces between charged particles in a fluid medium. Upon entering the xylem and phloem of the plant's vascular system, Ag-NPs can be transported in significant amounts to various parts of the plant, effectively distributing them throughout the organism [76].

2.3 Accumulation of Ag-NPs in Plant Tissues

Ag-NPs accumulate at diverse locations within the plant's cellular structure following their absorption and transport within the plant system. These nanoparticles are found extensively distributed across various cellular components such as the cell walls, plasmodesmata, and the spaces between cells. Moreover, Ag-NPs can be located within the cytoplasm and are also associated with several organelles, including mitochondria, vacuoles, and chloroplasts, indicating their widespread penetration into the plant's cellular machinery [37]. Ag-NPs in plants may occur in nanoparticles, Ag ions, or Ag^0 (zero valent Ag). The connection between nanoparticles and plants, like the relationship between donors and receptors, mainly dictates the extent of this accumulation. Firstly, the solubility of Ag-NPs may vary in the physiological conditions of various plant species, resulting in variations in the collection of silver [58]. Furthermore, plants can break down and remove the buildup of toxic compounds as a means of self-protection. Research has shown that manufactured nanoparticles, as well as other organic and inorganic nanoparticles, may undergo degradation inside the body of an organism [24]. Nevertheless, the direct elimination of Ag-NPs in plants is a challenge, contributing to the buildup of Ag-NPs [33]. In the context of Ag-NPs acting as donors within biological systems, both the size and concentration of these nanoparticles play crucial roles in their accumulation within plants. A surface coating on these particles may not necessarily affect their aggregation levels. When Ag-NPs are assimilated by edible plants, which are then consumed, they have the potential to ascend through the food chain, thereby exerting a significant impact on human health. This underscores the importance of understanding the dynamics of nanoparticle interaction with plant systems, particularly those within the food supply [37]. Hence, it is crucial to contemplate strategies for diminishing or eradicating the buildup of Ag-NPs in plants.

2.4 Biotransformation of Ag-NPs in Plant Cells

Ag-NPs may undergo chemical transformations in plants, forming various chemical forms such as cysteine complex, Ag^0, Ag_2S, Ag^2O Ag, etc. Ag-NPs primarily undergo biotransformation via redox reactions or dissociation to convert into other forms under certain circumstances. Ag-NPs undergo conversion into silver ions, silver oxide, and silver sulfide in plant cells when exposed to oxidative conditions [20]. Simultaneously, Ag-NPs may undergo partial dissociation and release Ag+ ions. Nevertheless, when exposed to environmental conditions promoting reduction, unbound Ag+ ions convert into Ag atoms or nanometallic particles. Environmental factors, including the ionic strength and pH of the recipient cells, significantly influence the process [76]. Furthermore, Ag-NPs may convert into Ag cysteine complexes using surface functional groups (-SH). Alternatively, they can be loaded onto AgCl silver ion microtubules by mediating root exudates under a light irradiation system [5]. Hence, the biotransformation process of Ag-NPs in plants may be seen as a dynamic equilibrium, contributing to a better understanding of the toxicological mechanisms of Ag-NPs in plants [61]. Changes in environmental circumstances induce morphological transformations of Ag-NPs inside plant cells, leading to alterations in plants' biological impacts.

3 Beneficial Effects of AG-NPS on Plants

Ag-NPs have garnered significant interest in recent times owing to their wide-ranging uses in different industries, including agriculture. This talk will explore the advantageous impacts of Ag-NPs on plants, emphasizing four crucial aspects [45, 60, 65] (Fig. 2):

- Augmented growth and productivity
- Heightened absorption and integration of nutrients
- Enhanced resilience to stress and defense mechanisms
- Their possible use as nano-fertilizers.

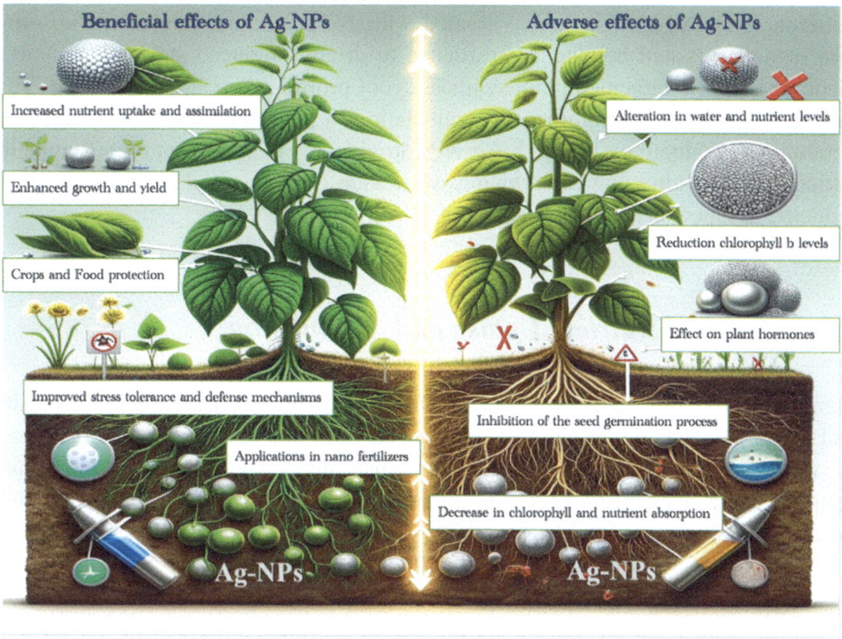

Fig. 2 Beneficial and adverse effects of Ag-NPs on plant

3.1 Enhanced Growth and Yield

Ag-NPs have shown a higher effectiveness in promoting seed germination. Prior research has demonstrated the use of these applications in plant growth parameters. The optimal concentrations of Ag-NPs have significantly enhanced the growth and productivity of crops [48]. Multiple studies have shown that applying Ag-NPs benefited crop growth while causing little or no harm to plant development due to phototoxicity [91]. Ag-NPs lead to improved growth and productivity thanks to their distinctive features. The compact dimensions and expansive surface area of Ag-NPs enable them to infiltrate plant cells and exert their beneficial impacts easily [23]. Ag-NPs can interact with different parts of plant cells, such as proteins and enzymes, resulting in chemical reactions [83]. These reactions often include the activation of genes linked to growth and development, leading to higher rates of photosynthesis—enhanced photosynthesis results in increased energy generation and biomass buildup in plants.

Moreover, it has been shown that Ag-NPs can improve root growth, specifically in increasing root length and branching. The enlarged root system enables plants to explore a greater area of soil effectively, enhancing their ability to get water and nutrients [31]. As a result, plants treated with Ag-NPs exhibit enhanced absorption of nutrients and heightened resilience to drought-induced stress. The synergistic impact of augmented photosynthesis, root proliferation, and nutrient absorption finally results in a significant augmentation in plant growth, blossoming, and agricultural productivity [14, 24]. Ag-NPs provide a viable solution for sustainable agriculture by addressing the growing worldwide need for food production while reducing reliance on conventional chemical fertilizers.

3.2 Increased Nutrient Uptake and Assimilation

Plants can adjust or endure adverse circumstances such as heat stress, cold, salt, and drought. The cellular and molecular aspects of plant response to abiotic stress are well investigated. Recent developments in nanomaterial engineering suggest that nano fertilizers have the potential to enhance crop output even under challenging environmental conditions [42, 46, 78]. Salt stress significantly restricts crop productivity in around 23% of the world's farmed areas. The capacity of Ag-NPs to augment the absorption and integration of nutrients in plants is a crucial component of their usefulness in agriculture. Ag-NPs are efficient transporters of vital nutrients, namely critical metals and micronutrients [83]. By binding these nutrients to the surface of Ag-NPs, they are protected from soil interactions that might reduce their accessibility to plant roots. The Ag-NPs serve to preserve the nutrients, ensuring that they maintain their bioavailability throughout transportation to the root zone [24, 37].

Consequently, plants can acquire these nutrients more efficiently, enhancing nutrient absorption. Furthermore, Ag-NPs can alter the soil environment, providing favorable nutrient availability and absorption conditions [47]. They can modify the soil's physicochemical characteristics, such as pH and ion exchange capacity, creating a better environment for plant roots to absorb nutrients. This alteration improves the entire process of acquiring nutrients, enabling plants to absorb the necessary nutrients more efficiently and use them for growth and development [14, 48]. In general, the enhanced absorption and integration of nutrients provided by Ag-NPs play a crucial role in promoting plant well-being and vitality, which are fundamental to attaining maximum crop productivity in contemporary farming practices.

3.3 Improved Stress Tolerance and Defense Mechanisms

Ag-NPs have been discovered to substantially improve plant stress tolerance and defense systems, providing many advantages in agriculture. Ag-NPs enhance stress tolerance by functioning as powerful antioxidants [10, 30]. Plants commonly produce an abundance of reactive oxygen species (ROS) when they encounter environmental stresses, including drought, excessive salinity, or exposure to heavy metals. These ROS may lead to cellular damage [15]. Multiple investigations have shown convincing evidence of the advantageous impacts of Ag-NPs on the ability of plants to withstand stress and enhance their defensive systems. A study investigated the effects of Ag-NPs on tomato plants exposed to salt stress. It was discovered that using Ag-NPs effectively decreased the oxidative stress caused by excessive salinity [15]. Ag-NPs have shown the ability to remove ROS and increase the effectiveness of antioxidant enzymes such as superoxide dismutase and catalase [53]. Consequently, applying Ag-NP therapy led to enhanced photosynthesis, higher chlorophyll content, and improved plant development compared to salt-stressed plants without such treatment. The results emphasize the potential of Ag-NPs in reducing the detrimental impacts of salt stress, thereby enhancing crop resilience in saline soils [79].

Moreover, studies have investigated using Ag-NPs to augment plant resistance against diseases. Scientists examined the impact of Ag-NPs on *Arabidopsis thaliana* plants that were infected with the bacterial disease *Pseudomonas syringae* [3, 68, 81]. It was noted that using Ag-NPs substantially reduced illness symptoms and inhibited bacterial development in the afflicted plants. Ag-NPs activated pathogenesis-related (PR) genes and the buildup of PR proteins, crucial constituents of the plant's defensive mechanism. This work showcases the capacity of Ag-NPs to not only safeguard plants from pathogenic assaults but also to activate their inherent defensive systems [46]. The current discoveries highlight the critical function of Ag-NPs in enhancing plant resilience to stress and strengthening defense systems. This contributes to promoting sustainable agriculture in response to evolving environmental circumstances and disease challenges.

3.4 Potential Applications in Nano-fertilizers

Significant quantities of commercial fertilizers, such as urea and nitrate, are often used. However, it is essential to note that these fertilizers may harm both plants and the beneficial microflora [86]. They also contribute to environmental contamination. Nanomaterials may be advantageous in this field because they gradually release agrochemicals, thanks to their larger particle surface area than traditional materials [9]. Ag-NPs possess several distinctive characteristics, and their use as a plant nutrient may enhance the efficiency of nutrient absorption by plants [7]. Thus, by using this method, the overutilization of fertilizer is reduced in the field. Ag-NPs have shown significant significance in several facets of agriculture [54]. Using onion extract in producing nano fertilizers is very productive for tomato and brinjal plants. Ag-NPs function as nanoscale fertilizers, gradually and efficiently releasing nutrients to mitigate nutrient loss. It aids the plant in augmenting the uptake of nutrients from the soil. Excessive fertilizers were lost in the field, resulting in adverse environmental impacts. Nanoparticles adhere to the fertilizers and then dispense appropriate nutrients to the plants [12, 45]. Therefore, using this method reduces the excessive fertilizer use in the field. Nano-silver exhibits enhanced nutrient absorption from the soil compared to its bulk counterpart [13]. Thus, it may be used in both crop protection and crop protection. Nanofertilizers may be used. Ag-NPs can improve seeds' germination ability in several plant species [44, 89]. In addition, nanofertilizers facilitate the gradual and efficient release of nutrients, mitigating nutrient losses. Furthermore, nano fertilizers may optimize plants' assimilation of nutrients from the soil.

3.5 *Ag-NPs in Crops and Food Protection*

The vitality of plants is crucial for the well-being of animals and humans, although it is often overlooked and underrepresented in health literature. Plants serve as the primary and vital source of nourishment for cattle, supplying almost 80% of the food humans consume [73]. The presence of pests and plant diseases poses a significant risk to the availability of plants and the safety of their consumption by humans and animals [28]. On a global scale, the loss of essential staple crops accounts for 30% of total output, resulting in billions of dollars in financial losses [80]. Nanotechnology is a highly efficient and secure method that improves food quality and enhances the accuracy of agricultural practices [63]. They may eradicate target germs from hydroponics, soil, and plants by safeguarding plants via numerous means, such as functioning as nano-antiviral, antibacterial, and nano-antifungal agents [56].

4 Adverse Effects of AG-NPS on Plants

The effects of Ag-NPs on plants in land-based ecosystems, namely crops, depend on their concentration and size [69]. It impeded seed germination and inhibited seed growth. Furthermore, it influenced the roots' and shoots' dimensions and weight. The accumulation of this chemical in leaves and roots initiates the cellular and tissue defence mechanism, resulting in alterations in proteome expression, antioxidant activities, and metabolism [6]. Botanical modifications, including the production of reactive oxygen species, alterations in total chlorophyll levels, changes in superoxide dismutase activities, higher levels of H_2O_2, glutathione, carotenoid, ascorbate, proline contents, etc., were observed after exposure to Ag-NPs [66]. These changes either increase or decrease depending on the specific parameter. These mechanisms inhibited photosynthesis/transition, aberrant morphological alterations, and other symptoms (Fig. 3.1).

Many investigations indicate that Ag-NPs negatively affects seed germination and seedling development. However, other studies cite no impact or even some favorable characteristics. The extent of this impact mainly relies on the concentration of Ag-NPs and the specific plant type [48, 61, 83]. The phytotoxicity of Ag-NPs to plants at the physiological level may be anticipated by observing a decrease in chlorophyll and nutrient absorption, a fall in transpiration rate, and changes in hormone levels [33, 53, 91]. Ag-NPs can interfere with leaf chlorophyll production, impacting plants' photosynthetic process. Recent research has shown that exposure to Ag-NPs induced alterations in the thylakoid structure of *Physcomitrella* patens.

Additionally, Ag-NPs reduced chlorophyll b levels and disrupted the equilibrium of several vital components in the leafy gametophytes [91]. *Lupinus termis* seedlings saw significant decreases in shoot and root growth, fresh weights, chlorophyll levels, and total protein contents after exposure to Ag-NPs for ten days. Exposure to Ag-NPs resulted in a significant reduction in the transpiration rate of *Cucurbita pepo* [74]. Furthermore, Ag-NPs can modify the membrane's consistency and porosity, hence influencing the process of water and nutrient absorption. An inquiry uncovered that the presence of Ag-NPs correlated with the dosage, reducing the water content of radish sprouts [70]. The concentration of essential nutrients, such as copper (Cu), magnesium (Mg), calcium (Ca), zinc (Zn), boron (B), and manganese (Mn), was drastically reduced. These findings indicate that Ag-NPs might influence plant development by modifying water and nutrient concentrations.

Ag-NPs have been shown to affect plant hormones, as described. Sun et al. [87] discovered that the gravitational response of *Arabidopsis* seedling roots was hindered when exposed to Ag-NPs, with the inhibition depending on the dosage. Further analysis found that the presence of Ag-NPs resulted in a reduction in the accumulation of auxin. Additionally, the investigation on gene expression showed that genes associated with auxin receptors were downregulated after exposure to Ag-NPs. Silver nanoparticles have the potential to cause harm to plants at the cellular and molecular levels [75, 91]. A recent investigation has shown that the root tip cells of wheat may easily absorb the Ag-NPs [57]. After the Ag-NPs were absorbed by the cells, several chromosomal abnormalities were seen in the root tip. These abnormalities included misalignment of chromosomes during metaphase, chromosome breakage, malfunctioning of the spindle mechanism, uneven separation of chromosomes, chromosome fragmentation, and dispersed and lagging chromosomes [2]. These abnormalities significantly disrupted the normal functioning of the cells. Besides, multiple investigations have shown that changes in both cell structure and cell division accompany the suppression of plant growth after exposure to Ag-NPs. The primary mechanism responsible for the phytotoxic effects of Ag-NPs is the overproduction of ROS triggered by Ag-NPs, leading to oxidative stress in plant cells [92, 94]. Several studies have shown that ROS formation in plants increases dramatically after exposure to Ag-NPs.

Upon introducing Ag-NPs into soil, they may interact with soil microorganisms, resulting in direct and indirect effects on soil microbial populations. Ag-NPs possess antibacterial characteristics that directly hinder the development of certain soil bacteria and fungi [19, 34]. This may disturb the equilibrium of microbial communities, reducing the variety and quantity of advantageous soil microorganisms, such as those engaged in nutrient cycling and support of plant development. Furthermore, Ag-NPs can gradually build up in the soil, which might pose a persistent danger to soil microorganisms [43]. The presence of Ag-NPs in the soil might indirectly modify the physicochemical characteristics of the soil matrix, including pH and nutrient accessibility [25, 59]. These modifications may also impact the function and structure of soil microbial communities. Ag-NPs can potentially affect the solubility and movement of vital elements such as phosphorus and nitrogen, which might restrict their accessibility to soil microbes [67]. Moreover, changes in soil pH may influence the activity and composition of microbes, promoting the proliferation of some acid-resistant or alkaline-resistant bacteria while suppressing others [26]. Overall, the effect of Ag-NPs on soil microbial communities is a multifaceted and ever-changing process, affected by parameters including the concentration and size of NPs and the type of soil. Therefore, it is crucial to exercise caution when using Ag-NPs in agricultural activities.

5 Enironmental Implications

Using Ag-NPs in diverse applications has raised significant apprehension about potential environmental ramifications. A major worry revolves around the possible impact of Ag-NPs on non-target species. Ag-NPs may toxic many creatures, such as aquatic animals, terrestrial invertebrates, and even helpful insects [88]. Nanoparticles may infiltrate ecosystems through runoff, deposition, or direct application, endangering creatures that encounter them. Ag-NPs in aquatic settings may disturb the physiology of aquatic plants, possibly resulting in decreased populations and modified community structures [50]. In addition, advantageous insects such as pollinators can come into contact with Ag-NPs when searching for food on treated plants, potentially leading to adverse effects on crop pollination and biodiversity.

Another critical factor is the possible hazard of Ag-NPs seeping into soil and water systems. Ag-NPs have the potential to accumulate in soils when used in agricultural or other terrestrial applications [51]. Over time, these little particles may gradually seep into the underground water, possibly polluting drinking water supplies. This pollution may have detrimental effects on aquatic and terrestrial ecosystems, which can affect human well-being. Moreover, Ag-NPs in the soil might disturb the equilibrium of soil microbial communities and nutrient cycling, impacting soil health and production [85]. Stringent containment procedures and restrictions are necessary to mitigate excessive leaching and contamination while guaranteeing appropriate utilization of Ag-NPs in diverse applications.

To tackle these environmental consequences, it is crucial to use a cautious strategy when using Ag-NPs and carry out comprehensive evaluations of the risks involved. This will help evaluate the possible effects on species not intended as targets and the probability of Ag-NPs seeping into soil and water systems [21]. Furthermore, additional research is necessary to understand the behaviour, fate, and long-term effects of Ag-NPs in various ecological components. Subsequently, this understanding may be used to formulate ecologically sustainable strategies and methodologies for effectively handling the hazards linked to these nanoparticles.

6 Regulation and Safety Concerns

Ensuring responsible and sustainable deployment of Ag-NPs in agriculture requires careful attention to regulation and safety considerations. Establishing regulatory frameworks for Ag-NPs in agriculture is crucial to protecting the environment and ensuring the well-being of humans [21]. Using Ag-NPs and their applications provides a whole realm of research opportunities while also bringing out unknown potential hazards. There is a debate on whether nanoparticles should be regulated beforehand or based on evidence, either via voluntary measures or under authorities' control. Given that nanoparticles with a size of around 1 nm occupy 100% of the surface atoms, it is crucial to comprehend the surface chemistry and its significance in the toxicity risks [90]. The primary factors to be mindful of when assessing the possible hazards associated with using nanomaterials on human health are the influence on the environment and the safety of workers.

Besides, there are issues surrounding nanotechnology's ethical, legal, and social implications. Global awareness has been raised, and several nations have made efforts to control the manufacture and utilization of nanoparticles partly [17]. It is widely agreed that, shortly, the current regulatory framework designed for substances that are not at the nanoscale, together with a method of evaluating each case individually, will be used to control nanomaterials [29]. Over time, the framework will change as new information emerges that reveals the limits of the existing risk assessment paradigm. A more comprehensive awareness of the biological interactions between nanomaterials, people, and the environment will be a guiding principle in establishing a reasonable regulatory framework for nano-pesticides and nano-fertilizers.

7 Conclusion and Future Directions

The chapter results revealed that silver nanoparticles are vital in enhancing agricultural output, seed quality, and plant development. Ag-NPs have a practical use as fertilizers since they may prevent nutrient loss and effectively manage plant diseases, comparable to pesticides. This has the potential to enhance plant productivity. Furthermore, they significantly improve soil quality by increasing nutrient accessibility and reducing soil pollutants. Silver nanoparticles also enhance the water supply and efficiently provide essential nutrients to the plant. Ag-NPs are widely used in several sectors, particularly in agriculture. Furthermore, they serve as nano-fertilizers and nano-pesticides, reducing the impact of salt stress and improving plant yield. Nanotechnology is a developing field that investigates the unique physicochemical properties of nanoparticles and their many applications in areas such as agriculture. Ag-NPs are widely used in the agricultural sector due to their diverse applications. Nanoparticles are nano-fertilizers to augment plant growth and act as nano-pesticides to manage various plant-damaging insects and pests. Furthermore, they are used as nano-bacterial, nan-fungal, and nano-antiviral agents to counteract certain microorganisms that cause plant diseases. The agricultural industry has significant obstacles to inadequate crop yield and crop ailments. Consequently, researchers must prioritize the investigation of the efficacy of Ag-NPs in enhancing plant growth and optimizing agricultural methods. It is necessary to conduct tests on various plants and crops to assess the effects of environmentally friendly and bio-based Ag-NPs. Additional research is required to address the issues of poor crop productivity and the management of plant diseases by using biosynthesized Ag-NPs and other metals with lower toxicity.

References

1. Abd-Alla MH, Nafady NA, Khalaf DM (2016) Assessment of silver nanoparticles contamination on faba bean-Rhizobium leguminosarum bv. viciae-Glomus aggregatum symbiosis: implications for induction of autophagy process in root nodule. Agr Ecosyst Environ 218:163–177
2. Abdelsalam NR, Abdel-Megeed A, Ali HM, Salem MZ, Al-Hayali MF, Elshikh MS (2018) Genotoxicity effects of silver nanoparticles on wheat (Triticum aestivum L.) root tip cells. Ecotoxicol Environ Saf 155:76–85
3. Abideen Z, Hanif M, Munir N, Nielsen BL (2022) Impact of nanomaterials on the regulation of gene expression and metabolomics of plants under salt stress. Plants 11(5):691
4. Achari GA, Kowshik M (2018) Recent developments on nanotechnology in agriculture: plant mineral nutrition, health, and interactions with soil microflora. J Agric Food Chem 66(33):8647–8661. https://doi.org/10.1021/acs.jafc.8b00691
5. Afshinnia K (2018) Effect of environmental factors on the aggregation behavior of silver nanoparticles in the aquatic environment. PhD Thesis, University of South Carolina. https://search.proquest.com/openview/2bb48f8ee24af8dd5ed8e064fd63a44c/1?pq-origsite=gscholar&cbl=18750
6. Ahmad N, Sharma S, Alam MK, Singh VN, Shamsi SF, Mehta BR, Fatma A (2010) Rapid synthesis of silver nanoparticles using dried medicinal plant of basil. Colloids Surf, B 81(1):81–86
7. Ahmad Z, Tahseen S, Wasi A, Ganie IB, Shahzad A, Emamverdian A, Ramakrishnan M, Ding Y (2022) Nanotechnological interventions in agriculture. Nanomaterials 12(15):2667
8. Ahmadov IS, Ramazanov MA, Gasimov EK, Rzayev FH, Veliyeva SB (2020) The migration study of nanoparticles from soil to the leaves of plants. Biointerf Res Appl Chem 10:6101–6111
9. Ahmed M, Rauf M, Akhtar M, Mukhtar Z, Saeed NA (2020) Hazards of nitrogen fertilizers and ways to reduce nitrate accumulation in crop plants. Environ Sci Pollut Res 27(15):17661–17670. https://doi.org/10.1007/s11356-020-08236-y
10. Alabdallah NM, Hasan MM (2021) Plant-based green synthesis of silver nanoparticles and its effective role in abiotic stress tolerance in crop plants. Saudi J Biol Sci 28(10):5631–5639
11. Alghuthaymi M, Mostafa M, Abd-Elsalam KA (2020) Carbon nanotubes: an efficient sorbent for herbicide sensing and remediation. In: Carbon nanomaterials for agri-food and environmental applications. Elsevier, pp 429-457
12. An C, Sun C, Li N, Huang B, Jiang J, Shen Y, Wang C, Zhao X, Cui B, Wang C, Li X, Zhan S, Gao F, Zeng Z, Cui H, Wang Y (2022) Nanomaterials and nanotechnology for the delivery of agrochemicals: strategies towards sustainable agriculture. J Nanobiotechnol 20(1):11. https://doi.org/10.1186/s12951-021-01214-7

13. Anand R, Bhagat M (2022) Nanosilver and smart delivery in agricultural system. In: Nanotechnology in agriculture and environmental science. CRC Press, pp 156–165. https://books.google.com/books?hl=fr&lr=&id=pUOMEAAAQBAJ&oi=fnd&pg=PR145&dq=Nano-silver+exhibits+enhanced+nutrient+absorption+from+the+soil+compared+to+its+bulk+counterpart&ots=fCCF3skIwA&sig=-4GAGCWv5lGyTTCMECadX7KGjII
14. Aqeel U, Aftab T, Khan MMA, Naeem M, Khan MN (2022) A comprehensive review of impacts of diverse nanoparticles on growth, development and physiological adjustments in plants under changing environment. Chemosphere 291:132672
15. Azameti MK, Imoro A-WM (2023) Nanotechnology: a promising field in enhancing abiotic stress tolerance in plants. Crop Design 100037
16. Azimzada A, Tufenkji N, Wilkinson KJ (2017) Transformations of silver nanoparticles in wastewater effluents: links to Ag bioavailability. Environ Sci Nano 4(6):1339–1349
17. Babatunde DE, Denwigwe IH, Babatunde OM, Gbadamosi SL, Babalola IP, Agboola O (2019) Environmental and societal impact of nanotechnology. IEEE Access 8:4640–4667
18. Bachheti RK, Bachheti A, Husen A (2024) Metal and metal-oxide based nanomaterials (Synthesis, Agricultural, Biomedical and Environmental Interventions). Springer Nature Singapore Pte Ltd., 152 Beach Road, #21-01/04 Gateway East, Singapore 189721, Singapore
19. Bamal D, Singh A, Chaudhary G, Kumar M, Singh M, Rani N, Mundlia P, Sehrawat AR (2021) Silver nanoparticles biosynthesis, characterization, antimicrobial activities, applications, cytotoxicity and safety issues: an updated review. Nanomaterials 11(8):2086
20. Banerjee K, Pramanik P, Maity A, Joshi DC, Wani SH, Krishnan P (2019) Methods of using nanomaterials to plant systems and their delivery to plants (Mode of entry, uptake, translocation, accumulation, biotransformation and barriers). In: Advances in Phytonanotechnology. Elsevier, pp 123–152. https://www.sciencedirect.com/science/article/pii/B9780128153222000055
21. Chaachouay N, Zidane L (2024) Role of carbon nanotubes for herbicide detection and remediation. In: Bachheti A, Bachheti RK, Husen A (eds) Carbon-based nanomaterials. Smart nanomaterials technology. Springer, Singapore. https://doi.org/10.1007/978-981-97-0240-4_8
22. Chaachouay N, Azeroual A, Benkhnigue O, Zidane L (2023) Alkaloids: a suitable precursor for nanomaterials synthesis, and their various applications. In: Husen A (ed) Secondary metabolites based green synthesis of nanomaterials and their applications. Smart nanomaterials technology. Springer, Singapore. https://doi.org/10.1007/978-981-99-0927-8_2
23. Chaachouay N, Azeroual A, Bencherki B, Douira A, Zidane L (2023) Various metabolites and or bioactive compounds from vegetables, and their use nanoparticles synthesis, and applications. In: Husen A (ed) Nanomaterials from agricultural and horticultural products. Smart nanomaterials technology. Springer, Singapore. https://doi.org/10.1007/978-981-99-3435-5_10
24. Cocozza C, Perone A, Giordano C, Salvatici MC, Pignattelli S, Raio A, Schaub M, Sever K, Innes JL, Tognetti R (2019) Silver nanoparticles enter the tree stem faster through leaves than through roots. Tree Physiol 39(7):1251–1261
25. Denisse V-GJ, Roberto S-CC, Hermes P-H, Patricio T-GA, Andrea P-M, Fabián F-L (2021) Chapter 8—Influence of nanoparticles on the physical, chemical, and biological properties of soils. In: Amrane A, Mohan D, Nguyen TA, Assadi AA, Yasin G (eds) Nanomaterials for soil remediation. Elsevier, pp 151–182. https://doi.org/10.1016/B978-0-12-822891-3.00008-6
26. Ding J, Wu B, Chen L (2022) Application of marine microbial natural products in cosmetics. Front Microbiol 13:892505
27. Doolette CL (2015) Behaviour of silver and silver sulfide nanoparticles in the environment: effects on wastewater treatment processes and soil organisms. PhD Thesis. https://digital.library.adelaide.edu.au/dspace/handle/2440/100742
28. Dubey NK, Srivastava B, Kumar A (2008) Current status of plant products as botanical pesticides in storage pest management. J Biopest 1(2):182–186
29. Dupuy J-P, Mills K (2004) Benefits, risks, ethical, legal and social aspects of nanotechnology. https://www.nanowerk.com/nanotechnology/reports/reportpdf/report3.pdf
30. El-Saadony MT, Saad AM, Soliman SM, Salem HM, Desoky E-SM, Babalghith AO, El-Tahan AM, Ibrahim OM, Ebrahim AA, El-Mageed A (2022) Role of nanoparticles in enhancing crop tolerance to abiotic stress: a comprehensive review. Front Plant Sci 13:946717

31. Elsayh SA, Arafa RN, Ali GA, Abdelaal WB, Sidky RA, Ragab TI (2022) Impact of silver nanoparticles on multiplication, rooting of shoots and biochemical analyses of date palm Hayani cv. By in vitro. Biocatal Agric Biotechnol 43:102400
32. Farooq MA, Hannan F, Islam F, Ayyaz A, Zhang N, Chen W, Zhang K, Huang Q, Xu L, Zhou W (2022) The potential of nanomaterials for sustainable modern agriculture: present findings and future perspectives. Environ Sci Nano 9(6):1926–1951
33. Fu Q-L, Zhong C-J, Qing T, Du Z-Y, Fei J-J, Peijnenburg WJ (2021) Effects of extracellular polymeric substances on silver nanoparticle bioaccumulation and toxicity to Triticum aestivum L. Chemosphere 280:130863
34. Hamad A, Khashan KS, Hadi A (2020) Silver nanoparticles and silver ions as potential antibacterial agents. J Inorg Organomet Polym Mater 30(12):4811–4828. https://doi.org/10.1007/s10904-020-01744-x
35. Hanks NA (2015) Silver nanoparticle and silver ion water contamination: assessment of phytoremediation and point-of-use filtration media. PhD Thesis, University of Cincinnati. https://rave.ohiolink.edu/etdc/view?acc_num=ucin1446547022
36. He D, Garg S, Wang Z, Li L, Rong H, Ma X, Li G, An T, Waite TD (2019) Silver sulfide nanoparticles in aqueous environments: formation, transformation and toxicity. Environ Sci Nano 6(6):1674–1687
37. Huang D, Dang F, Huang Y, Chen N, Zhou D (2022) Uptake, translocation, and transformation of silver nanoparticles in plants. Environ Sci Nano 9(1):12–39
38. Husen A (2022) Engineered nanomaterials for sustainable agricultural production, soil improvement and stress management. Elsevier Inc., 50 Hampshire St., 5th Floor, Cambridge, MA 02139, USA. https://doi.org/10.1016/C2021-0-00054-7
39. Husen A (2023) Nanomaterials and nanocomposites exposures to plants (Response, Interaction, Phytotoxicity and Defense Mechanisms). Springer Nature Singapore Pte Ltd., 152 Beach Road, #21-01/04 Gateway East, Singapore 189721, Singapore. https://doi.org/10.1007/978-981-99-2419-6
40. Husen A, Siddiqi KS (2023) Advances in smart nanomaterials and their applications. Elsevier Inc., 50 Hampshire St., 5th Floor, Cambridge, MA 02139, USA. https://doi.org/10.1016/C2021-0-02202-1
41. Husen A (2020) Interactions of metal and metal-oxide nanomaterials with agricultural crops: an overview. In: Husen A, Jawaid M (eds) Nanomaterials for agriculture and forestry applications. Elsevier Inc. 50 Hampshire St., 5th Floor, Cambridge, MA 02139, USA, pp 167–197
42. Iqbal P, Preece JA, Mendes PM (2012) Nanotechnology: the "top-down" and "bottom-up" approaches. In: Supramolecular chemistry: from molecules to nanomaterials
43. Juan W, Kunhui SHU, Zhang LI, Youbin SI (2017) Effects of silver nanoparticles on soil microbial communities and bacterial nitrification in suburban vegetable soils. Pedosphere 27(3):482–490
44. Kale SK, Parishwad GV, Patil ASHAS (2021) Emerging agriculture applications of silver nanoparticles. ES Food Agroforest 3:17–22
45. Khan S, Zahoor M, Khan RS, Ikram M, Islam NU (2023) The impact of silver nanoparticles on the growth of plants: the agriculture applications. Heliyon. https://www.cell.com/heliyon/pdf/S2405-8440(23)04135-X.pdf
46. Kim D, Kadam A, Shinde S, Saratale RG, Patra J, Ghodake G (2018) Recent developments in nanotechnology transforming the agricultural sector: a transition replete with opportunities. J Sci Food Agric 98(3):849–864. https://doi.org/10.1002/jsfa.8749
47. Kulikova NA, Volkov DS, Volikov AB, Abroskin DP, Krepak AI, Perminova IV (2020) Silver nanoparticles stabilized by humic substances adversely affect wheat plants and soil. J Nanopart Res 22(5):100. https://doi.org/10.1007/s11051-020-04788-9
48. Lahuta LB, Szablińska-Piernik J, Glowacka K, Stalanowska K, Railean-Plugaru V, Horbowicz M, Pomastowski P, Buszewski B (2022) The effect of bio-synthesized silver nanoparticles on germination, early seedling development, and metabolome of wheat (Triticum aestivum L.). Molecules 27(7):2303

49. Lv J, Christie P, Zhang S (2019) Uptake, translocation, and transformation of metal-based nanoparticles in plants: recent advances and methodological challenges. Environ Sci Nano 6(1):41–59
50. Mahanta M, Kaushik KK (2023) Nanoparticles in aquatic environment: an overview with special reference to their ecotoxicity. In Malik JA, Sadiq Mohamed MJ (eds) Modern nanotechnology. Springer Nature Switzerland, pp 385–404. https://doi.org/10.1007/978-3-031-31104-8_17
51. Makselon J (2018) Transport and retention behavior of silver nanoparticles in soil. PhD Thesis, Dissertation, RWTH Aachen University. http://publications.rwth-aachen.de/record/730343/files/730343.pdf
52. Malandrakis AA, Kavroulakis N, Chrysikopoulos CV (2020) Use of silver nanoparticles to counter fungicide-resistance in Monilinia fructicola. Sci Total Environ 747:141287
53. Mansour WA, Abdelsalam NR, Tanekhy M, Khaled AA, Mansour AT (2021) Toxicity, inflammatory and antioxidant genes expression, and physiological changes of green synthesis silver nanoparticles on Nile tilapia (Oreochromis niloticus) fingerlings. Comp Biochem Physiol C: Toxicol Pharmacol 247:109068
54. Mittal AK, Chisti Y, Banerjee UC (2013) Synthesis of metallic nanoparticles using plant extracts. Biotechnol Adv 31(2):346–356
55. Morais LDO, Macedo EV, Granjeiro JM, Delgado IF (2020) Critical evaluation of migration studies of silver nanoparticles present in food packaging: a systematic review. Crit Rev Food Sci Nutr 60(18):3083–3102. https://doi.org/10.1080/10408398.2019.1676699
56. Mousavi SR, Rezaei M (2011) Nanotechnology in agriculture and food production. J Appl Environ Biol Sci 1(10):414–419
57. Nayeri S, Dolatyari M, Mouladoost N, Nayeri S, Zarghami A, Mirtagioglu H, Rostami A (2023) Ag/ZnO core–shell NPs boost photosynthesis and growth rate in wheat seedlings under simulated full sun spectrum. Sci Rep 13(1):14385
58. Niu Z, Xu M, Guo X, Yan J, Liu M, Yang Y (2023) Uptake of silver-containing nanoparticles in an estuarine plant: speciation and bioaccumulation. Environ Sci Technol 57(42):16075–16085. https://doi.org/10.1021/acs.est.3c04774
59. Peixoto S, Khodaparast Z, Cornelis G, Lahive E, Etxabe AG, Baccaro M, Papadiamantis AG, Gonçalves SF, Lynch I, Busquets-Fité M (2020) Impact of Ag2S NPs on soil bacterial community—a terrestrial mesocosm approach. Ecotoxicol Environ Saf 206:111405
60. Polash SA, Nadaf NY, Rahman MA, Shohael AM (2018) Green synthesis of silver nanoparticles (AgNPs): agricultural applications and future vision. J Biodiversity Environ Sci (JBES) 13(2):35–57
61. Pradas Del Real AE, Vidal V, Carrière M, Castillo-Michel H, Levard C, Chaurand P, Sarret G (2017) Silver nanoparticles and wheat roots: a complex interplay. Environ Sci Technol 51(10):5774–5782. https://doi.org/10.1021/acs.est.7b00422
62. Pradhan S, Mailapalli DR (2017) Interaction of engineered nanoparticles with the agri-environment. J Agric Food Chem 65(38):8279–8294. https://doi.org/10.1021/acs.jafc.7b02528
63. Prasad R, Kumar V, Prasad KS (2014) Nanotechnology in sustainable agriculture: present concerns and future aspects. Afr J Biotech 13(6):705–713
64. Pryshchepa O, Pomastowski P, Buszewski B (2020) Silver nanoparticles: synthesis, investigation techniques, and properties. Adv Coll Interf Sci 284:102246
65. Rai M, Yadav A, Gade A (2009) Silver nanoparticles as a new generation of antimicrobials. Biotechnol Adv 27(1):76–83
66. Rajput VD, Minkina T, Kumari A, Harish, Singh VK, Verma KK, Mandzhieva S, Sushkova S, Srivastava S, Keswani C (2021) Coping with the challenges of abiotic stress in plants: new dimensions in the field application of nanoparticles. Plants 10(6):1221
67. Rajput VD, Singh A, Singh VK, Minkina TM, Sushkova S (2021) Impact of nanoparticles on soil resource. In: Nanomaterials for soil remediation. Elsevier, pp 65–85. https://www.sciencedirect.com/science/article/pii/B9780128228913000049

68. Rajwade JM, Chikte RG, Paknikar KM (2020) Nanomaterials: new weapons in a crusade against phytopathogens. Appl Microbiol Biotechnol 104(4):1437–1461. https://doi.org/10.1007/s00253-019-10334-y
69. Rezvani E, Rafferty A, McGuinness C, Kennedy J (2019) Adverse effects of nanosilver on human health and the environment. Acta Biomater 94:145–159
70. Rienzie R, Adassooriya NM (2018) Toxicity of nanomaterials in agriculture and food. In: Rai M, Biswas JK (eds) Nanomaterials: ecotoxicity, safety, and public perception. Springer International Publishing, pp 207–234. https://doi.org/10.1007/978-3-030-05144-0_11
71. Rigby DN (2021) Uptake and accumulation of silver nanoparticles by wheat (Triticum aestivum L.). https://spiral.imperial.ac.uk/handle/10044/1/105174
72. Roco MC, Mirkin CA, Hersam MC (2011) Nanotechnology research directions for societal needs in 2020: retrospective and outlook. https://books.google.com/books?hl=fr&lr=&id=1jcOFUMLVsQC&oi=fnd&pg=PR3&dq=Nanotechnology+is+a+burgeoning+scientific+discipline+that+primarily+uses+nanoparticles+to+surpass+size+constraints+and+revolutionize+the+scientific+paradigm&ots=dBkG9o4CRB&sig=RUa62eaWtbWgfVC6PqiYDTUZDj4
73. Roy RN, Finck A, Blair GJ, Tandon HLS (2006) Plant nutrition for food security. A guide for integrated nutrient management. FAO Fertilizer and Plant Nutrition Bulletin 16:368
74. Saleh MM, Alnaddaf LM, Almuhammady AK, Salem KFM, Alloosh MT, Al-Khayri JM (2021) Applications of plant-derived nanomaterials in mitigation of crop abiotic stress. In: Al-Khayri JM, Ansari MI, Singh AK (eds) Nanobiotechnology. Springer International Publishing, pp 201–238. https://doi.org/10.1007/978-3-030-73606-4_9
75. Sarmast MK, Salehi H (2022) Sub-lethal concentrations of silver nanoparticles mediate a phytostimulatory response in tobacco via the suppression of ethylene biosynthetic genes and the ethylene signaling pathway. In Vitro Cellular Dev Biol—Plant 58(1):1–14. https://doi.org/10.1007/s11627-021-10193-1
76. Savassa SM, Castillo-Michel H, del Real AEP, Reyes-Herrera J, Marques JPR, de Carvalho HW (2021) Ag nanoparticles enhancing Phaseolus vulgaris seedling development: understanding nanoparticle migration and chemical transformation across the seed coat. Environ Sci Nano 8(2):493–501
77. Schwab F, Zhai G, Kern M, Turner A, Schnoor JL, Wiesner MR (2016) Barriers, pathways and processes for uptake, translocation and accumulation of nanomaterials in plants—critical review. Nanotoxicology 10(3):257–278. https://doi.org/10.3109/17435390.2015.1048326
78. Shang Y, Hasan MK, Ahammed GJ, Li M, Yin H, Zhou J (2019) Applications of nanotechnology in plant growth and crop protection: a review. Molecules 24(14):2558
79. Shibli R, Mohusaien R, Abu-Zurayk R, Qudah T, Tahtamouni R (2022) Silver nanoparticles (Ag NPs) boost mitigation powers of chenopodium quinoa (Q6 Line) grown under in vitro salt-stressing conditions. Water 14(19):3099
80. Shiferaw B, Prasanna BM, Hellin J, Bänziger M (2011) Crops that feed the world 6. Past successes and future challenges to the role played by maize in global food security. Food Secur 3(3):307–327
81. Shoala T (2018) Positive impacts of nanoparticles in plant resistance against different Stimuli. In: Abd-Elsalam KA, Prasad R (eds) Nanobiotechnology applications in plant protection. Springer International Publishing, pp 267–279. https://doi.org/10.1007/978-3-319-91161-8_10
82. Shukla PK, Misra P, Kole C (2016) Uptake, translocation, accumulation, transformation, and generational transmission of nanoparticles in plants. In: Kole C, Kumar DS, Khodakovskaya MV (eds) Plant nanotechnology. Springer International Publishing, pp 183–218. https://doi.org/10.1007/978-3-319-42154-4_8
83. Siddiqi KS, Husen A, Rao RAK (2018) A review on biosynthesis of silver nanoparticles and their biocidal properties. J Nanobiotechnol 16(1):14. https://doi.org/10.1186/s12951-018-0334-5
84. Siddiqi KS, Husen A (2022) Plant response to silver nanoparticles: a critical review. Crit Rev Biotechnol 42(7):773–990. https://doi.org/10.1080/07388551.2021.1975091

85. Simonin M, Richaume A (2015) Impact of engineered nanoparticles on the activity, abundance, and diversity of soil microbial communities: a review. Environ Sci Pollut Res 22(18):13710–13723. https://doi.org/10.1007/s11356-015-4171-x
86. Singh RP, Kumar S, Sainger M, Sainger PA, Barnawal D (2017) Eco-friendly nitrogen fertilizers for sustainable agriculture. In: Rakshit A, Abhilash PC, Singh HB, Ghosh S (eds) Adaptive soil management: from theory to practices. Springer, Singapore, pp 227–246. https://doi.org/10.1007/978-981-10-3638-5_11
87. Sun F, Zhang W, Hu H, Li B, Wang Y, Zhao Y, Li K, Liu M, Li X (2008) Salt modulates gravity signaling pathway to regulate growth direction of primary roots in Arabidopsis. Plant Physiol 146(1):178–188
88. Tortella GR, Rubilar O, Durán N, Diez MC, Martínez M, Parada J, Seabra AB (2020) Silver nanoparticles: toxicity in model organisms as an overview of its hazard for human health and the environment. J Hazard Mater 390:121974
89. Ullah S, Ullah R (2019) Beneficial effects of several nanoparticles on the growth of different plants species. Curr Nanosci 15(5):460–470
90. Xu L, Liang H-W, Yang Y, Yu S-H (2018) Stability and reactivity: positive and negative aspects for nanoparticle processing. Chem Rev 118(7):3209–3250. https://doi.org/10.1021/acs.chemrev.7b00208
91. Yan A, Chen Z (2019) Impacts of silver nanoparticles on plants: a focus on the phytotoxicity and underlying mechanism. Int J Mol Sci 20(5):1003
92. Yang J, Cao W, Rui Y (2017) Interactions between nanoparticles and plants: phytotoxicity and defense mechanisms. J Plant Interact 12(1):158–169. https://doi.org/10.1080/17429145.2017.1310944
93. Zhang N, Sun J, Yin L, Liu J, Chen C (2023) Silver nanoparticles : from in vitro green synthesis to in vivo biological effects in plants. Advanced Agrochem. https://www.sciencedirect.com/science/article/pii/S277323712300059X
94. Zhao Z, Xu L, Wang Y, Li B, Zhang W, Li X (2021) Toxicity mechanism of silver nanoparticles to Chlamydomonas reinhardtii: photosynthesis, oxidative stress, membrane permeability, and ultrastructure analysis. Environ Sci Pollut Res 28(12):15032–15042. https://doi.org/10.1007/s11356-020-11714-y

Beneficial and Adverse Effects of Silver Nanoparticles on Rhizosphere Biology

Pankaj Sharma, Mayur Mukut Murlidhar Sharma, Divya Kapoor, Atul Loyal, and Azamal Husen

Abstract Contemporary agricultural methods face numerous challenges, including declining soil carbon levels, shifting climatic patterns, and escalating pest infestations. Conventional agricultural techniques appear inadequate in addressing these challenges, particularly in ensuring food security. Consequently, nanotechnology emerges as a promising frontier with the potential to revolutionize the agricultural industry due to the diverse and multifarious applications of nanoparticles. Among these, silver nanoparticles stand out for their unique agricultural capabilities, offering a wide array of roles in shaping the plant rhizosphere, such as enhancing nutrient uptake, suppressing soil-borne pathogens, promoting plant growth, and mitigating environmental stressors. Additionally, their antimicrobial properties make them valuable tools for combatting plant diseases, while their ability to improve soil structure and fertility contributes to sustainable agricultural practices. However, the deliberate application of silver nanoparticles in agricultural settings inevitably leads to their accumulation within the environment, raising concerns about potential nanoparticle-related toxicities. These toxicities can manifest in various ways, including adverse effects on soil microbial communities, disruptions in ecosystem dynamics, and potential risks to human health through food chain contamination. Therefore, while silver

P. Sharma (✉) · D. Kapoor
Department of Microbiology, CCS Haryana Agricultural University, Hisar, Haryana 125004, India
e-mail: sharmap@uic.edu

P. Sharma
Department of Ophthalmology and Visual Sciences, University of Illinois Chicago, Chicago 60612, USA

M. M. M. Sharma
Department of Agriculture and Industry, Kangwon National University, Chuncheon, Korea

A. Loyal
Department of Genetics and Plant Breeding, CCS Haryana Agricultural University, Hisar, Haryana 125004, India

A. Husen
Wolaita Sodo University, PO Box 138, Wolaita, Ethiopia

Department of Biotechnology, Graphic Era (Deemed to be University), Dehradun, Uttarakhand 248002, India

A. Husen (ed.), *Plant Response to Silver Nanoparticles*, Smart Nanomaterials Technology, https://doi.org/10.1007/978-981-97-7352-7_3

nanoparticles offer promising agricultural benefits, careful consideration of their environmental impacts and management strategies is essential to ensure sustainable and safe agricultural practices.

Keywords Antifungal AgNPs · Antibacterial AgNPs · Antiviral AgNPs · PGPR Toxicity · Phytotoxicity

1 Introduction

Nanotechnology, characterized by its capacity to control materials at the nanoscale, has risen as a revolutionary discipline with a wide range of applications spanning multiple sectors. Nanoparticles, falling within the size range of 1 to 100 nm, have found extensive use in diverse fields of bioscience and biomedicine, with a growing presence in commercial applications in various sectors [3, 25, 26, 45, 73–75, 4, 8, 17–24, 27, 29, 39, 42, 71]. In addition to these applications, nanotechnology is exerting its influence in the realm of agriculture. It is emerging as a burgeoning field of science, driven by its potential to revolutionize agricultural practices and improve conditions for underserved populations [30, 36, 37, 44, 54]. Nanotechnology exhibits tremendous potential in the realm, promising advancements ranging from the optimization of crop yield and nutrient absorption to the resolution of environmental challenges. The agricultural sector's nanotechnological applications are predominantly embodied by the utilization of nano pesticides, nano fertilizers, and nano herbicides [34, 57–61]. Over the past decade, a significant scientific breakthrough has been the synthesis of lab-engineered nanoparticles (NPs). These nanoparticles are emerging as a pivotal factor across various scientific domains, notably in agriculture and related fields [12, 62–66, 69]. Among the myriad of nanoparticles receiving recognition in agriculture, silver nanoparticles persist as one of the most extensively studied classes. This enduring focus is attributed to their notable agricultural characteristics and ongoing research endeavors. In the current scenario, the production of plant-based silver nanoparticles is considered a commendable method, given their cost-effectiveness, non-pathogenic nature, presence of non-toxic phytochemical constituents, adaptability to reaction parameters, and the biochemical diversity inherent in plant extracts. Silver nanoparticles (AgNPs) possess unique characteristics and find a wide array of applications, especially in the fields of agriculture and plant biotechnology. These nanoparticles have demonstrated the ability to improve seed germination, promote plant growth, and enhance photosynthetic efficiency. Additionally, they serve as safe and effective nano-pesticides and fertilizers [31, 81–83]. Various methods employed in the agricultural application of nanoparticles, ranging from foliar spray to root treatment, ultimately result in the accumulation of nanoparticles in soil systems. The repetitive application results in continuous accrual of these NPs in the plant rhizosphere.

The rhizosphere stands out as the most intricate microbial environment on Earth, comprising a cohesive system involving plant roots, soil particles, and a diverse microbial community consisting of archaea, bacteria, virus particles, and micro-eukaryotes. This delicate interface between soil particles and plant roots forms the primary habitat influenced by plants encountered by soil microbiota [51, 52, 55–59, 67, 68]. The rhizosphere, being a highly dynamic environment around the plant roots, is highly influenced by the presence of nanoparticles in a concentration and size-dependent manner. At lower concentrations, the silver nanoparticles have been reported to enhance the plant–microbe interactions and the microbial outcome from plant–microbe symbiosis [48]. However, it might also lead to totally opposite and unexpected results when the nanoparticles are supplemented at a higher concentration. This chapter highlights the complex interplay between silver nanoparticles and rhizosphere biology, investigating the positive as well as negative outcomes stemming from this interaction (Fig. 1). Grasping these effects is essential for unlocking the potential of silver nanoparticles to improve agricultural practices, all the while being mindful of the environmental and biological consequences.

Fig. 1 A portrayal depicting positive and negative effects of silver nanoparticles on seed germination

2 Rhizosphere and Its Importance in Plant Growth

The rhizosphere is the region of soil that surrounds and is influenced by the roots of plants. It is a dynamic and complex environment where various interactions occur between plant roots, soil particles, microorganisms, and other factors. The rhizosphere plays a crucial role in plant growth and overall soil health. It is a dynamic and densely populated soil zone that supports a diverse range of both inter- and intraspecies communications. Furthermore, it serves as a dynamic location for ongoing interactions within the food web, exerting a notable impact on the flow and transformation of carbon [6, 53]. The evaluation of the rhizosphere reveals that it provides a habitat for a variety of microorganism classes. Plant metabolism significantly shapes the rhizosphere by releasing carbon dioxide (CO_2) and secreting photosynthate in the form of diverse root exudates, primarily originating from the rhizoplane and ectorhizosphere. These root exudates, containing phytohormones, play a vital role in fostering interactions within the rhizosphere by serving as energy sources for microorganisms and functioning as chemical attractants and repellents [1]. Although, different classes of microorganisms get attracted toward the rhizosphere in response to the exudates secreted by plant systems. Microorganisms inhabiting the rhizosphere, which have garnered attention in studies focusing on rhizosphere and rhizospheric engineering, are those that exert beneficial effects on plant systems. These include nitrogen-fixing microorganisms, mycorrhiza, plant growth-promoting rhizobacteria (PGPR), and microbes with antagonistic activity against plant pathogens.

Rhizosphere acts as a unique environment for the plant–microbe interactions. The multitude of interactions occurring within the rhizosphere and rhizospheric microbiome play a crucial role in regulating both plant growth and yield in their natural environments. The molecular events unfolding in the plant rhizosphere intricately shape the composition of the plant rhizospheric microbiome, also known as the rhizobium [15]. A significant portion of microbiological populations residing in the rhizospheric zone plays a crucial role in enhancing the structure and productivity of natural plant systems by ensuring persistence and tolerance against various biotic and abiotic stresses. These contributions are made through various mechanisms, including biofertilization, promotion of root growth, stress management, rhizoremediation, and disease suppression. A substantial portion of rhizospheric microbiomes acts synergistically to promote plant growth and development, increase nutrient acquisition, boost tolerance, and trigger various defense mechanisms in plant systems [47]. Bacterial members of the rhizosphere that actively participate in activities aimed at enhancing plant health are referred to as plant growth-promoting rhizobacteria. Rhizosphere-residing bacteria contribute to the promotion of plant health and growth by ensuring an active supply of various essential nutrients that may otherwise be inaccessible or limited for plant systems, such as nitrogen, iron, phosphorus, and zinc. The mechanisms facilitating superior nutrient availability include phosphate solubilization, nitrogen fixation, zinc solubilization, and iron chelation through the production of siderophores. Moreover, plant growth-promoting rhizobacteria (PGPR) also produce several plant hormones, including indole acetic acid, cytokinin, and gibberellins. Additionally, other mechanisms may involve the possession of ACC deaminase activity, the formation of biofilm, and the production of various exopolysaccharides [32, 46, 70]. The active participation of rhizospheric inhabitants in various nutrient cycles leads to the recovery of essential nutrients such as nitrogen (N), phosphorus (P), potassium (K), zinc (Zn), and iron (Fe). This process enhances the bioavailability of these nutrients to plant systems. The second category of microorganisms is tasked with promoting plant growth through various indirect means, including protection against different plant pathogenic attacks. These organisms are referred to as biocontrol agents [72]. Other than that, the rhizosphere inhabitants are also known to aid the plant systems during different biotic and abiotic stresses. PGPRs are recognized for secreting various compounds that act as osmolytes. Examples of these osmolytes include glycine-betaine, proline, ectoine, trehalose, polyols, and sucrose. PGPR actions in coordination with the compounds secreted by roots respond to various biotic and abiotic stresses. The beneficial microflora in the rhizosphere also aids plant systems in eliminating various recalcitrant and xenobiotic compounds that have accumulated in soil systems due to the rapid pace of anthropogenic activities. This, in turn, helps prevent soil degradation and sterility [5].

3 Beneficial Effects of AgNPs in Plant Rhizosphere

3.1 Seed Germination

In commercial agriculture, achieving rapid and uniform seed germination and seedling emergence is crucial for the successful establishment of crops. Germination is a process that initiates with the absorption of water by the mature dry seed, a phenomenon known as imbibition. It concludes with the elongation of the embryonic axis, typically the radicle, breaking through the seed coat. This process leads to the emergence of the root and subsequently the shoot [50]. Thus, seed industries invest a good amount of money for improving the seed vigor. Seed priming is a pre-sowing technique that involves partially hydrating seeds before planting. Priming is fundamentally a process that entails the partial imbibition, as opposed to full hydration, of seeds using various strategies. These strategies may include shortening the imbibition duration (hydropriming) or subjecting seeds to relatively low external water potential (osmopriming, matric priming, or drum priming). As a result of priming, seeds attain a 'germinating state' without the protrusion of the radicle. This 'head-start' enables primed seeds to commonly display an enhanced germination rate and greater uniformity [9]. Engineered nanoparticles (NPs) have garnered significant attention for their potential to enhance seed germination. The precise and site-specific delivery of various macromolecules, coupled with controlled release of agrochemicals, is employed to augment plant growth and bolster disease resistance. This approach enables plants to utilize nutrients more efficiently, contributing to improved agricultural outcomes. The utilization of engineered nanoparticles in agriculture represents a promising avenue for advancing crop productivity and sustainability. These unique attributes of engineered AgNPs have proposed their usage as a very important tool to enhance seed germination. Once applied to seeds, silver nanoparticles can enter plant tissues through the apoplastic pathway. This involves the nanoparticles moving through the intercellular spaces and reaching the cell membrane by passing through plasmodesmata. This pathway allows silver nanoparticles to access the interior of plant cells and exert their effects on various cellular processes. One reported effect is their ability to help in loosening the seed testa (seed coat), thereby facilitating easier germination. This property may be attributed to the interaction of silver nanoparticles with the seed coat, leading to changes in its structure or properties that promote the germination process [40]. Priming with silver nanoparticles has also been reported to enhance seed germination even in the aged seeds. For instance, Rice seeds that were aged and subsequently primed with phytosynthesized silver nanoparticles (AgNPs) at concentrations of 5 and 10 ppm exhibited a significant improvement in germination performance and seedling vigor when compared to the unprimed control, $AgNO_3$ priming, and conventional hydropriming. Nanopriming has the potential to enhance α-amylase activity, leading to an increase in soluble sugar content that supports the growth of seedlings. Additionally, nanopriming stimulates the up-regulation of aquaporin genes in germinating seeds. Meanwhile, there is an observed increase in reactive oxygen species (ROS) production in germinating seeds subjected to nanopriming compared to the unprimed control and other priming treatments. This suggests that both ROS and aquaporins play crucial roles in enhancing seed germination [40]. Similarly, Rahman et al. [49] also tested the effect of AgNPs on the seed germination of *Pisum sativum.* Pea seeds underwent individual treatments with different concentrations of an aqueous solution containing silver nanoparticles (AgNPs) for a duration of 3 h. After a period of 2 weeks, the impact of these treatments on various parameters, including seed germination, leaf number, shoot length, root length, shoot dry weight, root dry weight, fresh biomass, and dry biomass weight, was investigated. The treatment involving 2.5 mM AgNPs resulted in a notable improvement, with a 31.57% increase in fruit weight and a 9.09% increase in seed weight. Importantly, this improvement was achieved without causing any harm to the symbiotic microbes. The authors also observed a significant improvement in the colonization by AMF and there was no harm to Rhizobium. This indicates that the treatment of silver nanoparticles at an optimal concentration improves the seed germination along with a significant increase in the population of beneficial rhizospheric inhabitants. Similarly, Singh et al. [68] also a significant enhancement in seed germination and early seedling growth of wheat on treatment with at AgNPs at a lower concentration of 25 mg L^{-1} when compared to the control on both the 4th and 8th day.

3.2 Enhancement in Nutrient Uptake

Silver nanoparticles have been reported to enhance nutrient uptake in the plant rhizosphere. Their presence and interaction with the root zone can positively influence the availability and absorption of essential nutrients by plants. This phenomenon may contribute to improved nutrient utilization, ultimately promoting better plant growth and development. However, it's essential to carefully assess the potential environmental and health implications associated with the use of nanoparticles in agriculture. The judicial use of silver nanoparticles has been reported to enhance the concentrations of several key nutrients in the plant systems. For instance, Guzmán-Báez et al. [14] reported that the application of silver nanoparticles triggered a hormetic effect by boosting nitrogen, phosphorus, and potassium concentrations in leaves, and they also stimulated root length and the number of roots in tomato seedlings. In a similar way, Jhanzab et al. [80] also observed a significant increase in the NPK uptake and nutrient use efficiency of wheat on treatment with silver nanoparticles. The improved nutrient absorption observed after applying silver nanoparticles can be attributed to an increase in root surface area, which enables better interaction between soil nutrients and plant roots. Silver nanoparticles are recognized for their ability to modify soil properties and dynamics, potentially enhancing the mobility of specific nutrients within the soil system.

3.3 Disease Suppression

Silver nanoparticles possess a broad spectrum of antimicrobial activity against different bacteria, virus, and fungi. Silver nanoparticles possess the capability to penetrate bacterial cell walls, altering the structure of cell membranes, and ultimately leading to cell death [77]. They have been reported to possess a diverse antimicrobial spectrum against different bacterial and fungal plant pathogens. Guzman et al. [13] reported remarkable antimicrobial efficacy of silver nanoparticles against methicillin-resistant gram-positive bacteria, including *E. coli, P. aureginosa,* and *S. aureus.* In a similar way, Bibi et al. (2023) also evaluated the antibacterial activity of biogenic AgNPs against plant pathogenic bacteria *Erwinia carotovora* subsp. *carotovora, Erwinia carotovora* subsp. *atroseptica,* and *Ralstonia solanacearum.* The NPs exhibited excellent antiviral activity against all the bacteria under test. AgNPs have also largely been reported for their antifungal attributes. For instance, Haroon et al. [16] reported that green synthesized AgNPs using *Azadirachta indica* leaf extract exhibited excellent antimicrobial activity against different plant pathogens like *Penicillium* sp., *Aspergillus* sp., *Fusarium* sp., and *Ralstonia solanacearum.* Similarly Kim et al. [33] reported a broad spectrum antiviral activity of AgNPs against 18 different fungi at different levels. Wen et al. [76] also reported that AgNPs exhibited a dose dependent inhibition of fungal pathogen *Ustilaginoidea virens.*

Silver nanoparticles (AgNPs) show promising potential as a tool for effectively combating viral plant diseases. El Gamal et al. [10] reported that the foliar spray of AgNPs could potentially protect faba bean from bean yellow mosaic virus. Disease occurrence was halted with an AgNP concentration as low as 100 mg L^{-1}, and complete inhibition of virus infection was achieved when plants were preemptively treated with AgNPs at a concentration of 200 mg L^{-1}, administered 24 h prior to virus inoculation. Similarly, Al-Askar et al. [2] also reported the biocidal activity of silver nanoparticles against Tobacco mosaic virus in tomato plants. The authors claimed that treating plants with AgNPs led to decreased virus accumulation, delayed viral replication in all treatments, and markedly increased the expression of the CHS gene linked to flavonoid biosynthesis. In a similar way, Mahfouze et al. [41] reported the antiviral properties of AgNPs against *Banana bunchy top virus*. The authors noted that administering AgNPs at 50 ppm following virus inoculation resulted in significant changes in carotenoids, relative to both healthy plants and nano-controls. Conversely, phenol, proline, and oxidative enzymes showed significant increases in all plants treated with 50 ppm AgNPs after virus inoculation compared to the healthy control.

They have a broad spectrum mechanism of action against different plant pathogens. They release the silver ions once they come in contact with the microorganisms that disrupt multiple cellular processes essential for microbial survival. The interaction of these ions with bacterial cell membranes results in structural damage and heightened permeability, ultimately undermining membrane integrity. Moreover, silver ions have the ability to infiltrate microbial cells, where they disrupt essential cellular components like proteins and DNA, consequently disturbing cellular metabolism and triggering oxidative stress. Moreover, the elevated surface area-to-volume ratio of silver nanoparticles enables them to engage extensively with microbial surfaces, thereby amplifying their antimicrobial effectiveness. In summary, the intricate antimicrobial mechanism of silver nanoparticles renders them potent agents against a wide range of microorganisms, encompassing bacteria, fungi, and viruses.

4 Negative Effects of Silver Nanoparticles

One of the major concerns regarding the continuous application of silver nanoparticles (AgNPs) in agriculture is their accumulation in agricultural systems. With each crop cycle and repeated applications to plants, the soil serves as the ultimate sink for AgNPs, leading to their accumulation in soil systems over time. This accumulation can raise concerns about the potential long-term environmental impact of AgNPs, including their effects on soil microbial communities, ecosystem functioning, and the health of subsequent crops [35]. Additionally, the persistence of AgNPs in soil systems may increase the risk of unintended exposure to organisms in terrestrial ecosystems, highlighting the need for careful consideration of their use in agricultural practices. The presence of AgNPs in soil may disrupt microbial populations and alter microbial diversity, leading to shifts in soil microbial dynamics and functions. This, in turn, can affect soil fertility, organic matter decomposition, and the overall health of the soil ecosystem. Moreover, the buildup of AgNPs in soil systems prompts inquiries regarding their potential absorption by plants and subsequent integration into the food chain [43]. Although plants can absorb AgNPs from the soil, their presence in edible plant parts may pose health hazards to humans through the ingestion of tainted crops. Furthermore, the enduring presence of AgNPs in soil systems might heighten the likelihood of inadvertent exposure to non-target organisms, encompassing soil-dwelling species, beneficial insects, and wildlife [11].

Silver NPs have been known for their different applications in the agricultural sector. They are known to affect the plant productivity in a positive way. However, their explicit usage can also result in the nanoparticle triggered toxicity in plant systems. Different studies have shown that exposure to AgNPs can inhibit seed germination and root growth, followed by reduction in plant biomass and leaf area. Jiang et al. [28] discovered that AgNPs led to significant decreases in plant biomass, inhibited shoot growth, and induced root abscission in *Spirodela polyrrhiza*. However, the impact of AgNPs on seed germination can differ based on plant species and environmental factors. Some plant species may display heightened susceptibility to AgNPs, whereas others might demonstrate greater resilience. Moreover, environmental variables like soil composition, moisture levels, and pH levels can further modulate the degree of AgNP-induced toxicity during seed germination. The negative effects of AgNPs are persistent owing to their ability to infiltrate seed coats and interfere with vital cellular mechanisms crucial for germination, such as water absorption, enzyme function, and hormonal signaling pathways. Moreover, AgNPs might trigger oxidative stress and inflict cellular harm on sprouting seeds, ultimately resulting in compromised growth and viability [7].

Silver nanoparticles exhibit a broad spectrum antimicrobial activity against bacteria fungi and different viruses. Their accumulation in the rhizosphere can disrupt the microbial network of plant rhizosphere. The decrease in microbial diversity can disturb crucial ecological relationships and weaken the robustness of the soil microbiome. AgNPs exposure may trigger alterations in the composition and population levels of rhizospheric microbial communities. Specific microbial groups might exhibit higher susceptibility to AgNP toxicity, resulting in alterations to the structure and dynamics of the community. In addition to their biocidal nature, AgNPs can also affect the activity of rhizosphere microbiota. Zhang et al. [79] reported that the application of silver nanoparticles inhibited the growth of nitrogen-fixing bacterium *Azotobacter vinelandiii in a size dependent*. The correlation between smaller-sized AgNPs and increased toxicity was apparent through a decrease in cell counts. The existence of AgNPs resulted in the generation of intracellular reactive oxygen species (ROS), as confirmed by electron spin resonance. This further illustrated that AgNPs produced hydroxyl radicals, indicating their potential to induce oxidative harm to bacteria. Similarly, Yuan et al. [78] also reported the toxic effects of AgNPs on nitrifying bacteria *Nitrosomonas europaea* ATCC 19,718. The authors reported that treatment with silver nanoparticles resulted in a size-dependent reduction in NH_3 oxidation, cell wall damage, and nuclear disintegration in the bacteria. Similarly, Luo et al. [38] reported a significant inhibition of bacterial ammoxidation in the aquatic environment in the presence of silver nanoparticles. Thus, silver nanoparticles at higher concentrations can significantly affect the number of beneficial plant microbes in the rhizosphere zone which will ultimately lead to a decline in plant productivity.

5 Conclusion

The nano-technological innovations and interventions have revolutionized the agricultural sector. The introduction of nanofertilizers, nanopesticides, and nanoherbicides has significantly contributed toward the dose reduction of these recalcitrant xenobiotic chemicals. Alongside, the application of different metal and metal nanoparticles was also introduced to the agricultural sector owing to their positive effects on plant growth and plant health. However, the explicit usage of nanoparticles leads to their accrual ultimately resulting in nanoparticle triggered toxicity in plants as well as rhizospheric inhabitants. Thus, the application of silver nanoparticles (AgNPs) in the plant rhizosphere presents a nuanced situation, encompassing both advantageous and detrimental outcomes. While AgNPs demonstrate antimicrobial capabilities that can combat soil-borne pathogens and bolster plant health, they also hold the potential for enhancing nutrient uptake and stress resilience in plants. However, the persistent use of AgNPs raises concerns regarding their negative repercussions on soil microbial diversity, plant–microbe relationships, and overall ecosystem functionality. AgNPs have the potential to disrupt rhizospheric microbiota, leading to shifts in microbial community dynamics and compromised soil fertility. Moving forward, research efforts should prioritize elucidating the mechanisms underlying AgNP interactions with rhizospheric microbiota, while also devising sustainable approaches to mitigate their adverse effects while maximizing their agricultural benefits. This involves exploring innovative nanomaterial formulations, refining application techniques, and conducting thorough risk assessments to ensure the safe and efficient utilization of AgNPs in plant rhizospheres.

References

1. Ahkami AH, White RA III, Handakumbura PP, Jansson C (2017) Rhizosphere engineering: enhancing sustainable plant ecosystem productivity. Rhizosphere 1(3):233–243
2. Al-Askar AA, Aseel DG, El-Gendi H, Sobhy S, Samy MA, Hamdy E, El-Messeiry S, Behiry SI, Elbeaino T, Abdelkhalek A. Antiviral activity of biosynthesized silver nanoparticles from pomegranate (Punica granatum L.) peel extract against tobacco mosaic virus. Plants 12(11):2103
3. Anamika, Patra A, Shehzad S, Rani A, Sharma P, Mohammad KF, Mehta S (2023) Microbial elicitors for priming plant defense mechanisms. In: Sustainable agriculture reviews 60: microbial processes in agriculture. Springer Nature Switzerland, Cham, pp 175–196

4. Bachheti RK, Bachheti A, Husen A (2024) Metal and metal-oxide based nanomaterials (Synthesis, Agricultural, Biomedical and Environmental Interventions). Springer Nature Singapore Pte Ltd., 152 Beach Road, #21-01/04 Gateway East, Singapore 189721, Singapore
5. Bansal K, Raturi A, Katiyar U, Mishra A, Tewari S (2024) Microbial rhizoremediation as a strategy for decontaminating polluted sites and augmenting plant growth. In: Microbiome drivers of ecosystem function. Academic Press, pp 181–227
6. Besset-Manzoni Y, Rieusset L, Joly P, Comte G, Prigent-Combaret C (2018) Exploiting rhizosphere microbial cooperation for developing sustainable agriculture strategies. Environ Sci Pollut Res 25(30):29953–29970
7. Budhani S, Egboluche NP, Arslan Z, Yu H, Deng H (2019) Phytotoxic effect of silver nanoparticles on seed germination and growth of terrestrial plants. J Environ Sci Health C 37(4):330–355
8. Chawla S, Singh S, Husen A (2023) Smart nanomaterials targeting pathological Hypoxia. Springer Nature Singapore Pte Ltd., 152 Beach Road, #21-01/04 Gateway East, Singapore 189721, Singapore ISBN: 978-981-99-1717-4. https://doi.org/10.1007/978-981-99-1718-1
9. Chen K, Arora R (2013) Priming memory invokes seed stress-tolerance. Environ Exp Bot 1(94):33–45
10. El Gamal AY, Tohamy MR, Abou-Zaid MI, Atia MM, El Sayed T, Farroh KY (2022) Silver nanoparticles as a viricidal agent to inhibit plant-infecting viruses and disrupt their acquisition and transmission by their aphid vector. Adv Virol 1:1–3
11. Ferdous Z, Nemmar A (2020) Health impact of silver nanoparticles: a review of the biodistribution and toxicity following various routes of exposure. Int J Mol Sci 21(7):2375
12. Fraceto LF, Grillo R, de Medeiros GA, Scognamiglio V, Rea G, Bartolucci C (2016) Nanotechnology in agriculture: which innovation potential does it have? Front Environ Sci 4:20
13. Guzmán MG, Dille J, Godet S (2009) Synthesis of silver nanoparticles by chemical reduction method and their antibacterial activity. Int J Chem Biomol Eng 2(3):104–111
14. Guzmán-Báez GA, Trejo-Téllez LI, Ramírez-Olvera SM, Salinas-Ruíz J, Bello-Bello JJ, Alcántar-González G, Hidalgo-Contreras JV, Gómez-Merino FC (2021) Silver nanoparticles increase nitrogen, phosphorus, and potassium concentrations in leaves and stimulate root length and number of roots in tomato seedlings in a hormetic manner. Dose-Response 19(4):15593258211044576
15. Hakim S, Naqqash T, Nawaz MS, Laraib I, Siddique MJ, Zia R, Mirza MS, Imran A (2021) Rhizosphere engineering with plant growth-promoting microorganisms for agriculture and ecological sustainability. Front Sustain Food Syst 15(5):617157
16. Haroon M, Zaidi A, Ahmed B, Rizvi A, Khan MS, Musarrat J (2019) Effective inhibition of phytopathogenic microbes by eco-friendly leaf extract mediated silver nanoparticles (AgNPs). Indian J Microbiol 1(59):273–287
17. Husen A (2019) Natural product-based fabrication of zinc oxide nanoparticles and their application. In: Husen A, Iqbal M (eds) N*anomaterials* and plant potential. Springer International Publishing AG, Gewerbestrasse 11, 6330 Cham, pp 193–291. https://doi.org/10.1007/978-3-030-05569-1_7
18. Husen A (2020) Introduction and techniques in nanomaterials formulation. In: Husen, A, Jawaid M (eds) Nanomaterials for agriculture and forestry applications. Elsevier Inc. 50 Hampshire St., 5th Floor, Cambridge, MA 02139, USA, pp 1–14. https://doi.org/10.1016/B978-0-12-817 852-2.00001-9
19. Husen A (2022) Engineered nanomaterials for sustainable agricultural production, soil improvement and stress management. Elsevier Inc., 50 Hampshire St., 5th Floor, Cambridge, MA 02139, USA. https://doi.org/10.1016/C2021-0-00054-7
20. Husen A (2023a) Secondary metabolites based green synthesis of nanomaterials and their applications. Springer Nature Singapore Pte Ltd., 152 Beach Road, #21-01/04 Gateway East, Singapore 189721, Singapore. https://doi.org/10.1007/978-981-99-0927-8
21. Husen A (2023b) Nanomaterials from agricultural and horticultural products. Springer Nature Singapore Pte Ltd., 152 Beach Road, #21-01/04 Gateway East, Singapore 189721, Singapore. https://link.springer.com/book/9789819934348

22. Husen A (2023c) Nanomaterials and nanocomposites exposures to plants (Response, Interaction, Phytotoxicity and Defense Mechanisms). Springer Nature Singapore Pte Ltd., 152 Beach Road, #21-01/04 Gateway East, Singapore 189721, Singapore. https://doi.org/10.1007/978-981-99-2419-6
23. Husen A, Bachheti RK, Bachheti A (2023) Current trends in green nano-emulsions (Food, Agriculture and Biomedical Sectors). Springer Nature Singapore Pte Ltd., 152 Beach Road, #21-01/04 Gateway East, Singapore 189721, Singapore
24. Husen A, Iqbal M (2019) Nanomaterials and plant potential. Springer International Publishing AG, Gewerbestrasse 11, 6330 Cham, Switzerland. https://doi.org/10.1007/978-3-030-05569-1
25. Husen A, Jawaid M (2020) Nanomaterials for agriculture and forestry Applications. Elsevier Inc., 50 Hampshire St., 5th Floor, Cambridge, MA 02139, USA. https://doi.org/10.1016/C2018-0-02349-X
26. Husen A, Siddiqi KS (2014) Phytosynthesis of nanoparticles: Concept, controversy and application. Nanoscale Res Lett 9(229):1–24. https://doi.org/10.1186/1556-276X-9-229
27. Husen A, Siddiqi KS (2023) Advances in smart nanomaterials and their applications. Elsevier Inc., 50 Hampshire St., 5th Floor, Cambridge, MA 02139, USA. https://doi.org/10.1016/C2021-0-02202-1
28. Jiang HS, Li M, Chang FY, Li W, Yin LY (2012) Physiological analysis of silver nanoparticles and AgNO3 toxicity to Spirodela polyrhiza. Environ Toxicol Chem 31(8):1880–1886
29. Jin-Chul K, Madhusudhan A, Husen A (2021) Smart nanomaterials in biomedical applications. Springer Nature Switzerland AG, Gewerbestrasse 11, 6330 Cham, Switzerland. https://doi.org/10.1007/978-3-030-84262-8
30. Kapoor D, Yadav S, Sharma MM, Sharma P (2023) Interaction between metal nanoparticles and PGPR on the plant growth and development. In: Nanomaterials and nanocomposites exposures to plants: response, interaction, phytotoxicity and defense mechanisms. Springer Nature Singapore, Singapore, pp 327–351
31. Khan S, Zahoor M, Khan RS, Ikram M, Islam NU (2023) The impact of silver nanoparticles on the growth of plants: The agriculture applications. Heliyon, Jun 2
32. Khoso MA, Wagan S, Alam I, Hussain A, Ali Q, Saha S, Poudel TR, Manghwar H, Liu F (2023) Impact of plant growth-promoting rhizobacteria (PGPR) on plant nutrition and root characteristics: current perspective. Plant Stress 28:100341
33. Kim SW, Jung JH, Lamsal K, Kim YS, Min JS, Lee YS (2012) Antifungal effects of silver nanoparticles (AgNPs) against various plant pathogenic fungi. Mycobiology 40(1):53–58
34. Krishna SBN, Golla N (2024) Nanofungicides: a new frontier in agriculture. In: Nanofungicides. Elsevier, pp 123–137
35. Li P, Su M, Wang X, Zou X, Sun X, Shi J, Zhang H (2020) Environmental fate and behavior of silver nanoparticles in natural estuarine systems. J Environ Sci 88:248–259
36. Loyal A, Pahuja SK, Rani N, Pooja, Srivastava RK, Sharma P (2023a) Titanium oxide nanoparticles: plant response, interaction, phytotoxicity, and defence mechanisms. In: Nanomaterials and nanocomposites exposures to plants: response, interaction, phytotoxicity and defense mechanisms. Springer Nature Singapore, Singapore, pp 263–284
37. Loyal A, Pahuja SK, Sharma P, Malik A, Srivastava RK, Mehta S (2023b) Potential environmental and human health implications of nanomaterials used in sustainable agriculture and soil improvement. In: Engineered nanomaterials for sustainable agricultural production, soil improvement and stress management. Academic Press, pp 387–412
38. Luo Z, Chen Z, Qiu Z, Li Y, Du Laing G, Liu A, Yan C (2015) Gold and silver nanoparticle effects on ammonia-oxidizing bacteria cultures under ammoxidation. Chemosphere 1(120):737–742
39. Madhusudhan A, Purohit SD, Prasad R, Husen A (2024) Functional smart nanomaterials and their theranostics approaches. Springer Nature Singapore Pte Ltd., 152 Beach Road, #21-01/04 Gateway East, Singapore 189721, Singapore
40. Mahakham W, Sarmah AK, Maensiri S, Theerakulpisut P (2017) Nanopriming technology for enhancing germination and starch metabolism of aged rice seeds using phytosynthesized silver nanoparticles. Sci Rep 7(1):8263

41. Mahfouze HA, El-Dougdoug NK, Mahfouze SA (2020) Virucidal activity of silver nanoparticles against Banana bunchy top virus (BBTV) in banana plants. Bull Natl Res Centre 44(1):1–1
42. Mandal AK, Ghorai S, Husen A (2023) Functionalized smart nanomaterials for point-of-care testing. Springer Nature Singapore Pte Ltd., 152 Beach Road, #21-01/04 Gateway East, Singapore 189721, Singapore
43. McGillicuddy E, Murray I, Kavanagh S, Morrison L, Fogarty A, Cormican M, Dockery P, Prendergast M, Rowan N, Morris D (2017) Silver nanoparticles in the environment: Sources, detection and ecotoxicology. Sci Total Environ 1(575):231–246
44. Mehmood A (2018) Brief overview of the application of silver nanoparticles to improve growth of crop plants. IET Nanobiotechnol 12(6):701–705
45. Nayana AR, Joseph BJ, Jose A, Radhakrishnan EK (2020) Nanotechnological advances with PGPR applications. In: Sustainable agriculture reviews 41: nanotechnology for plant growth and development, pp 163–180
46. Nwachukwu BC, Babalola OO, Hassen AI (2024) Rhizosphere competence and applications of plant growth-promoting rhizobacteria in food production-a review. Sci African 11:e02081
47. Nwachukwu BC, Babalola OO (2021) Perspectives for sustainable agriculture from the microbiome in plant rhizosphere. Plant Biotechnol Reports 15(3):259–278
48. Przemieniecki SW, Ruraż K, Kosewska O, Oćwieja M, Gorczyca A (2024) The impact of various forms of silver nanoparticles on the rhizosphere of wheat (Triticum aestivum L.)—Shifts in microbiome structure and predicted microbial metabolic functions. Sci Total Environ:169824
49. Rahman MS, Chakraborty A, Kibria A, Hossain MJ (2023) Effects of silver nanoparticles on seed germination and growth performance of pea (Pisum sativum). Plant Nano Biol 1(5):100042
50. Rajjou L, Duval M, Gallardo K, Catusse J, Bally J, Job C, Job D (2012) Seed germination and vigor. Annu Rev Plant Biol 2(63):507–533
51. Rani K, Rani A, Sharma P, Dahiya A, Punia H, Kumar S, Sheoran S, Banerjee A (2022) Legumes for agroecosystem services and sustainability. In: Advances in legumes for sustainable intensification. Academic Press, pp 363–380
52. Rani K, Sharma P, Kumar S, Wati L, Kumar R, Gurjar DS, Kumar D, Kumar R (2019) Legumes for sustainable soil and crop management. In: Sustainable management of soil and environment, pp 193–215
53. Ryan PR, Dessaux Y, Thomashow LS, Weller DM (2009) Rhizosphere engineering and management for sustainable agriculture 321:363–383
54. Sangwan S, Sharma P, Wati L, Mehta S (2023) Effect of chitosan nanoparticles on growth and physiology of crop plants. In: Engineered nanomaterials for sustainable agricultural production, soil improvement and stress management. Academic Press, pp 99–123
55. Sharma P, Sharma MM, Patra A, Vashisth M, Mehta S, Singh B, Tiwari M, Pandey V (2020a) The role of key transcription factors for cold tolerance in plants. In: Transcription factors for abiotic stress tolerance in plants. Academic Press, pp 123–152
56. Sharma P, Sharma MM, Anamika, Kapoor D, Rani K, Singh D, Barkodia M (2020b) Role of microbes for attaining enhanced food crop production. In: Microbial biotechnology: basic research and applications, pp 55–78
57. Sharma P, Pandey V, Sharma MM, Patra A, Singh B, Mehta S, Husen A (2021a) A review on biosensors and nanosensors application in agroecosystems. Nanoscale Res Lett 16:1–24
58. Sharma P, Sharma MM, Malik A, Vashisth M, Singh D, Kumar R, Singh B, Patra A, Mehta S, Pandey V (2021b) Rhizosphere, rhizosphere biology, and rhizospheric engineering. In: Plant growth-promoting microbes for sustainable biotic and abiotic stress management. Springer International Publishing, Cham, pp 577-624
59. Sharma MM, Sharma P, Kapoor D, Beniwal P, Mehta S (2021c) Phytomicrobiome community: an Agrarian perspective towards resilient agriculture. In: Plant performance under environmental stress: hormones, biostimulants and sustainable plant growth management, pp 493–534
60. Sharma P, Meyyazhagan A, Easwaran M, Sharma MM, Mehta S, Pandey V, Liu WC, Kamyab H, Balasubramanian B, Baskaran R, Klemeš JJ (2022a) Hydrogen Sulfide: a new warrior

in assisting seed germination during adverse environmental conditions. Plant Growth Regul 98(3):401–420
61. Sharma P, Sangwan S, Kumari A, Singh S, Kaur H (2022b) Impact of climate change on soil microorganisms regulating nutrient transformation. In: Plant stress mitigators: action and application. Singapore: Springer Nature Singapore, pp 145–172
62. Sharma MM, Kapoor D, Rohilla R, Sharma P (2023a) Nanomaterials and their toxicity to beneficial soil microbiota and fungi associated plants rhizosphere. In: Nanomaterials and nanocomposites exposures to plants: response, interaction, phytotoxicity and defense mechanisms. Springer Nature Singapore, Singapore, pp 353–380
63. Sharma P, Sangwan S, Mehta S (2023b) Emerging role of phosphate nanoparticles in agriculture practices. In: Engineered nanomaterials for sustainable agricultural production, soil improvement and stress management. Academic Press, pp 71–97
64. Sharma P, Sangwan S, Kaur H, Patra A, Mehta S (2023c) Microbial remediation of agricultural residues. In: Sustainable agriculture reviews 60: microbial processes in agriculture. Springer Nature Switzerland, Cham, pp 325–358
65. Sharma P, Sangwan S, Kaur H, Patra A, Anamika, Mehta S (2023d) Diversity and evolution of nitrogen fixing bacteria. In: Sustainable agriculture reviews 60: microbial processes in agriculture, pp 95–120
66. Sharma P, Patra A, Singh B, Mehta S (2023e) Microbial rejuvenation of soils for sustainable agriculture. In: Sustainable agriculture reviews 60: microbial processes in agriculture. Springer Nature Switzerland, Cham, pp 293–323
67. Singh A, Sharma P, Kumari A, Kumar R, Pathak DV (2019) Management of root-knot nematode in different crops using microorganisms. In: Plant biotic interactions: state of the art, pp 85–99
68. Singh Y, Kaushal S, Sodhi RS (2020) Biogenic synthesis of silver nanoparticles using cyanobacterium Leptolyngbya sp. WUC 59 cell-free extract and their effects on bacterial growth and seed germination. Nanoscale Adv 2(9):3972–3982.
69. Singh S, Sangwan S, Sharma P, Devi P, Moond M (2021) Nanotechnology for sustainable agriculture: an emerging perspective. J Nanosci Nanotechnol 21(6):3453–3465
70. Solomon W, Janda T, Molnár Z (2023) Unveiling the significance of rhizosphere: implications for plant growth, stress response, and sustainable agriculture. Plant Physiol Biochem 22:108290
71. Taghiyari HR, Morrell JF, Husen A (2023) Emerging nanomaterials (Opportunities and Challenges in Forestry Sectors). Springer Nature Switzerland AG, Gewerbestrasse 11, 6330 Cham, Switzerland. https://doi.org/10.1007/978-3-031-17378-3
72. Wang C, Kuzyakov Y (2023) Rhizosphere engineering for soil carbon sequestration. Trends Plant Sci
73. Wassie AT, Ahmed IN, Bachheti A, Husen A, Bachheti RK (2024a) Plant extract-assisted green synthesis and characterization of copper nanocomposite. In: Bachheti RK, Bachheti A, Husen A (eds) Metal and metal-oxide based nanomaterials. Smart Nanomaterials Technology. Springer, Singapore. pp 43–66 https://doi.org/10.1007/978-981-99-7673-7_3
74. Wassie AT, Ahmed, IN, Bachheti A, Husen A, Bachheti RK (2024b) Role of zinc oxide nanomaterials for photocatalytic degradation of environmental pollutants. In: Bachheti RK, Bachheti A, Husen A (eds) Metal and metal-oxide based nanomaterials. smart nanomaterials technology. Springer, Singapore. pp 287–311. https://doi.org/10.1007/978-981-99-7673-7_14
75. Wassie AT, Pandey DP, Bachheti A, Husen A, Deepti Singh Vashishth, Bachheti RK (2024c) Use of green synthesized platinum nanoparticles for biomedical applications. In: Bachheti RK, Bachheti A, Husen A (eds) Metal and metal-oxide based nanomaterials. Smart Nanomaterials Technology. Springer, Singapore, pp 271–285. https://doi.org/10.1007/978-981-99-7673-7_13
76. Wen H, Shi H, Jiang N, Qiu J, Lin F, Kou Y (2023) Antifungal mechanisms of silver nanoparticles on mycotoxin producing rice false smut fungus. Iscience 26(1)
77. Yin IX, Zhang J, Zhao IS, Mei ML, Li Q, Chu CH (2020) The antibacterial mechanism of silver nanoparticles and its application in dentistry. Int J Nanomed 17:2555–2562
78. Yuan Z, Li J, Cui L, Xu B, Zhang H, Yu CP (2013) Interaction of silver nanoparticles with pure nitrifying bacteria. Chemosphere 90(4):1404–1411

79. Zhang L, Wu L, Si Y, Shu K (2018) Size-dependent cytotoxicity of silver nanoparticles to Azotobacter vinelandii: growth inhibition, cell injury, oxidative stress and internalization. PLoS ONE 13(12):e0209020
80. Jhanzab HM, Razzaq A, Jilani G, Rehman A, Hafeez A, Yasmeen F (2015) Silver nano-particles enhance the growth, yield and nutrient use efficiency of wheat. Int J Agron Agri Res 7(1):15–22
81. Kapoor D, Sharma P, Sharma MM, Yadav S, Husen A (2024 Sep 11) Exploring soil microbiota and their role in plant growth, stress tolerance, disease control and nutrient immobilizer. Biocatal Agric Biotechnol 103358
82. Sharma MM, Kapoor D, Sharma P, Husen A (2024 May 1) Role of carbon nanomaterials in the prevention of plant disease. In: Carbon-based nanomaterials: synthesis, agricultural, biomedical, and environmental interventions. Springer Nature Singapore, Singapore, pp 105–121
83. Sharma MM, Kapoor D, Loyal A, Kumar R, Sharma P, Husen A (2024 May 1) Environmental toxicity of engineered carbon nanoparticles. In: Carbon-Based nanomaterials: Synthesis, agricultural, biomedical, and environmental interventions. Springer Nature Singapore, Singapore, pp 337–353

Genotoxicity of Silver Nanoparticles in Plants and Underlying Mechanism

Tanisha, Ujwala Jadhav, and Hina Alim

Abstract In recent years, nanotechnology has made significant strides, finding widespread applications in various industries. However, this progress has resulted in the unintentional release of nanomaterials into the environment and ecosystems. One commonly used nanomaterial in this regard is silver nanoparticles (AgNPs), especially in agriculture. As plants are the foundation of ecosystems and the primary source of sustenance for humans, it is crucial to understand the impact of AgNPs on plant growth and development to assess potential risks to the environment and food safety. This chapter summarizes important findings regarding the uptake, movement, and accumulation of AgNPs in plants, elucidating their harmful effects on various levels such as morphological, physiological, cellular, and molecular. It also explores the mechanisms through which AgNPs induce toxicity in plants and discusses the strategies plants employ to tolerate and counteract these adverse effects. Despite the extensive use of AgNPs, particularly in agriculture, it is essential for the scientific community to acknowledge their potential harm to crop growth and plant development. The continued use of AgNPs in agriculture could disrupt plant biochemical and physiological processes, emphasizing the need for a deeper understanding of their toxic effects on crop plants to make informed decisions. This chapter provides the necessary knowledge to comprehend the complex interactions between AgNPs and crop plants, contributing to the assessment of environmental risks to plant, human, and animal health.

Keywords Genotoxicity · Silver nanoparticles · Mechanism · Reactive oxygen species

Tanisha · U. Jadhav · H. Alim (✉)
Department of Life Sciences, University of Mumbai, Vidyanagari Campus, Santacruz East, Mumbai 400 098, India
e-mail: hina.alim@lifesciences.mu.ac.in

A. Husen (ed.), *Plant Response to Silver Nanoparticles*, Smart Nanomaterials Technology, https://doi.org/10.1007/978-981-97-7352-7_4

1 Introduction

Engineered nanomaterials (ENMs), renowned for their minuscule size typically ranging from 1 to 100 nm and distinct physical and chemical properties, have experienced rapid advancements and integration into a variety of industrial sectors and daily life [10, 17, 44–51, 54, 57, 76, 77, 119, 122]. Among these ENMs, silver nanoparticles (AgNPs) have emerged as one of the most widely utilized nanomaterials [87, 89, 107–110]. They are featured in nearly a quarter of consumer products incorporating nanotechnology due to their established antibacterial and antifungal characteristics [2]. These versatile AgNPs are incorporated into a wide array of products, encompassing household goods, food packaging, textiles, medical equipment, healthcare items, electronics, and solutions for wastewater treatment [3, 28, 98, 101, 102, 138, 139] . In agriculture, AgNPs have found applications as additives to boost plant growth, fungicides, and agents for hastening fruit ripening [4, 90, 104, 115, 124]. Despite their extensive usage, the widespread deployment of AgNPs has unintentionally led to their release into the environment. This release occurs during AgNP manufacturing, their inclusion in products, handling, recycling, or product disposal [41, 65, 72]. As a result, AgNPs have become pervasive pollutants in surface waters like lakes, streams, and rivers, primarily introduced through biosolids from wastewater treatment [35, 92, 117]. The concentrations of AgNPs in both surface water and sewage treatment systems have been steadily on the rise [14, 30, 36]. In agriculture, water contaminated with AgNPs can infiltrate fields through fertilization and irrigation, increasing the potential exposure of crops to these nanoparticles [58]. The widespread dispersion of AgNPs into various environmental media ultimately results in their presence in the plant rhizosphere, where they are absorbed by crops [25, 67]. This sets the stage for AgNPs to enter the food chain, potentially affecting food production, quality, and, crucially, human health [13, 21, 22, 75, 80]. It's noteworthy that silver is ranked as the second most toxic metal to aquatic organisms after mercury [91]. Furthermore, AgNPs can release silver ions (Ag^+), known for their persistence, potential for bioaccumulation, and high toxicity to organisms [103]. Consequently, the ecological implications of AgNPs in natural ecosystems and their safety have received considerable attention [91, 103]. Since plants occupy a pivotal role in ecosystems as the primary trophic level and the cornerstone of the food chain [74, 79], comprehending the effects of AgNPs on plants is of paramount importance for evaluating their potential toxicity [31]. Hence, this review offers a comprehensive exploration of AgNPs, encompassing their uptake, intra-plant movement, and their impacts. It clarifies the mechanisms through which AgNPs influence plant functions and delves into the protective mechanisms employed by plants to mitigate the adverse effects of AgNPs [31]. These effects span from diminished seed germination to alterations in root growth, biomass, and leaf count, with outcomes varying based on factors like AgNP size and concentration [112]. These adverse effects manifest as disruptions in plant metabolism, changes in antioxidant activities, and variations in proteomic expression, leading to anatomical and physiological irregularities, including disturbances in photosynthesis and transpiration [83, 84, 121]. While there are instances of neutral or beneficial effects, especially in certain plant species, the majority of studies underscore the detrimental impacts of AgNPs [59, 84, 121]. With the continuous expansion of AgNP production and their increasing utilization in agricultural practices, the likelihood of their introduction into the soil is escalating [59]. Furthermore, research into the potential health risks of AgNPs for humans has uncovered various concerns, including oxidative stress and allergic responses [9]. Given these concerns, it is imperative to conduct a thorough assessment of the environmental and health effects of AgNPs before their widespread commercialization [12]. This study serves as a comprehensive exploration of AgNPs' interactions with plants, shedding light on their physiological effects, uptake mechanisms, and protective responses. Such knowledge is essential for making well-informed decisions regarding the regulation and application of AgNPs across various sectors, thereby ensuring the safety of ecosystems and human well-being.

AgNPs have emerged as versatile materials with diverse applications in biology and medicine, but their potential genotoxic and cytotoxic effects on plant systems cannot be underestimated. This discussion explores the multifaceted dimensions of AgNP-induced genotoxicity and cytotoxicity in plants, offering insights into their repercussions at molecular, cellular, and physiological levels. AgNPs exhibit genotoxic effects on plant systems, notably affecting root tips. Studies on onion root tips have revealed disturbances in cell division stages and cell disintegration as a consequence of AgNP exposure [19]. Furthermore, AgNPs can lead to chromosomal damage, characterized by bond breaks, spindle fiber dysfunction, fragmentations, and disrupted metaphase orientation in root tip cells [1]. The overproduction of reactive oxygen species (ROS) induced by AgNPs plays a pivotal role in causing DNA damage in plant cells, adding to their genotoxic potential [85]. Additionally, exposure to AgNPs has been associated with the formation of micronuclei, signifying DNA damage and chromosomal aberrations in plant cells [64]. Alterations in gene expression linked to stress tolerance and DNA repair mechanisms in plants further highlight the genotoxicity of AgNPs [18].

Cytotoxic effects of AgNPs on plant systems encompass various facets, including cellular morphological changes like poor root hair growth, collapse of cortical cells, deformations in root cap and epidermal cells, and alterations in cell turgidity [136]. AgNPs also disrupt cell division processes, resulting in abnormal cell structures and sizes in plant tissues [81, 82]. These nanoparticles impact cell membrane integrity by altering fluidity and permeability, affecting water and nutrient uptake by plants [114]. Overproduction of ROS due to AgNP exposure contributes to oxidative stress, cell membrane damage, and cell death in plants [106]. Furthermore, AgNPs-induced cytotoxicity leads to metabolic disturbances, affecting protein synthesis and various cellular processes, thereby compromising overall plant health and productivity [43]. At the molecular level, AgNPs exert profound effects on plant genetic and biochemical processes. They influence the expression of genes related to stress response, DNA repair, and cell cycle regulation, underscoring their cytotoxic and genotoxic potential [26]. AgNPs also trigger the accumulation of specific proteins in plant cells, including those involved in stress response and cell cycle regulation, reflecting the plant's efforts to combat AgNP-induced stress [111]. In response to AgNP-induced cytotoxicity, plants activate antioxidant defense mechanisms, upregulating enzymes such as superoxide dismutase (SOD), catalase (CAT), and ascorbate peroxidase (APX) [140]. This defense mechanism plays a critical role in mitigating oxidative stress triggered by AgNPs. Moreover, AgNPs can cause DNA and chromosomal disintegration, affecting various cellular processes and contributing to cytotoxicity [55]. Reactive oxygen species (ROS) generated by AgNPs serve as key mediators of damage, impacting multiple cellular components [74]. In summary, silver nanoparticles (AgNPs) have been shown to exert genotoxic and cytotoxic effects on plant systems, influencing DNA, chromosomal integrity, cellular morphology, gene expression, and metabolism. These findings underscore the importance of further research to better understand the environmental implications of AgNPs and to develop strategies for mitigating their adverse effects on plants. As research in this field continues to advance, there is growing evidence of the intricate and interconnected nature of AgNP-induced genotoxicity and cytotoxicity in plant systems. For instance, studies have indicated that AgNPs' cytotoxic effects, such as disruptions in cell division and alterations in cell morphology, can be directly linked to their genotoxic potential, as these cytological changes often result from DNA damage and chromosomal aberrations induced by AgNPs [1, 136], [81, 82]. Furthermore, the molecular-level effects of AgNPs on plants are increasingly recognized. The impact on gene expression, particularly the upregulation of genes associated with stress responses and DNA repair mechanisms, underscores the complex interplay between genotoxic and cytotoxic pathways [74, 111]. Additionally, AgNPs' induction of antioxidant defense mechanisms within plants, involving enzymes like superoxide dismutase (SOD), catalase (CAT), and ascorbate peroxidase (APX), demonstrates the intricate balance between cytotoxicity and the plant's efforts to mitigate oxidative stress [140]. The role of ROS as central mediators of both genotoxic and cytotoxic effects further emphasizes the interconnectedness of these processes [74]. ROS not only contributes to DNA damage and chromosomal instability but also disrupts cell membranes and metabolic pathways, thus amplifying the cytotoxic impact of AgNPs [55, 85, 106]. The genotoxic and cytotoxic effects of silver nanoparticles on plant systems are multifaceted and intricately linked, involving disruptions at the molecular, cellular, and physiological levels. A comprehensive understanding of these interactions is crucial for assessing the environmental implications of AgNP exposure and for devising effective strategies to mitigate their adverse effects on plants and ecosystems.

2 Synthesis of AgNPs

In general, AgNPs are produced using three distinct methods: physical, chemical, and biological approaches [24]. Physical methods, such as pyrolysis and spark discharge, have been applied for AgNP synthesis [132]. However, physical methods that involve processes like evaporation–condensation in a tube furnace at high atmospheric pressure are less commonly utilized for AgNP synthesis [20]. Chemical methods rely on essential chemicals like silver nitrate and tri-sodium citrate to synthesize AgNPs

[88]. These chemicals act as solutes when dissolved in a universal solvent like water or organic solvents, and the chemical reduction reaction plays a critical role in the synthesis process. Chemical methods include bulk girding and electrochemical techniques as popular approaches for AgNP synthesis. Both physical and chemical methods have been employed for AgNP synthesis, with chemical methods often preferred due to their higher potential for yield [138, 139]. Recent studies have emphasized the benefits of biologically mediated methods for nanoparticle synthesis, which are considered straightforward, cost-effective, predictable, and environmentally friendly, yielding AgNPs of specific sizes. Various biological systems, including bacteria, fungi, and plant extracts, have been harnessed for AgNP production. Bacterial species such as *Escherichia coli* [61], Lactobacillus strains [32], *Pseudomonas stutzeri* [131], and *Bacillus licheniformis* [60] have been utilized in AgNP synthesis. Fungi like *Fusarium oxysporum* and *Ganoderma neo-japonicum Imazeki* are commonly used for this purpose [39]. Additionally, plant extracts from *Allophylus cobbe, Artemisia princeps,* and *Typha angustifolia* have played a significant role in AgNP synthesis [130].

3 Properties of AgNPs

AgNPs show exceptional effectiveness in a wide range of applications that benefit human well-being, including cosmetics, water purification, drug delivery, textiles, imaging, biosensing, and various therapeutic uses. Notably, AgNPs act as potent antibacterial agents by disrupting cell walls and causing the formation of "pits," ultimately leading to the demise of bacterial cells like *E. coli* [130]. In the context of antifungal activity, AgNPs, when administered at doses ranging from 1 to 7 g/mL, exhibit strong effectiveness against clinical strains of *Trichophyton mentagrophytes* and Candida species [94]. Incorporating nano-Ag into polysulfone ultrafiltration membranes (nAg-PSf) enhances their antiviral potential, resulting in substantial antiviral activity against the MS2 bacteriophage, primarily due to improved membrane hydrophilicity [123]. AgNPs also possess anti-inflammatory properties by promoting apoptosis in inflammatory cells and reducing the levels of pro-inflammatory cytokines. Their ability to inhibit angiogenesis has been investigated using pigment epithelium-derived factor (PEDF), which is an anti-angiogenic agent. Additionally, AgNPs have demonstrated concentration-dependent anti-cancer properties, triggering programmed cell death as part of their molecular mechanism [138, 139].

4 Uptake and Translocation of AgNPs in Plants

AgNPs employ various routes to enter and move within plants, allowing for their distribution and buildup. These pathways are critical for comprehending the interactions between AgNPs and plant systems [33, 34, 86].

Uptake and Translocation via Vascular Tissue and Xylem:

When AgNPs enter the plant, they initially breach the cell walls and plasma membranes of the root epidermal layers. Subsequently, they move through the plant's vascular system, primarily the xylem, which plays a pivotal role in the distribution and movement of AgNPs [8]. Studies have indicated that AgNPs gradually accumulate in various regions of the roots, including border cells, root caps, epidermis, columella, and root meristem initials [34]. With prolonged exposure, AgNPs penetrate deeper into the roots, reaching lateral root primordia and root hairs. Over time, AgNPs can be detected in the vascular tissues, spreading throughout the entire plant, from the roots to the shoots [33].

Cell Wall as the Main Entry Point:

The primary point of entry for AgNPs into plant cells is through the cell wall of root cells [121]. The plant cell wall, primarily composed of cellulose, acts as a semi-permeable layer, allowing smaller AgNPs to pass while blocking larger ones, effectively acting as a natural sieve [16, 97]. Interestingly, AgNPs can induce the formation of larger pores within the cell wall, facilitating the internalization of larger AgNPs [97]. Additionally, AgNPs can navigate plant cell interiors through plasmodesmata, which are narrow pores connecting adjacent plant cells [42, 73]. In Arabidopsis, AgNPs have been observed to aggregate in plasmodesmata and the cell wall, potentially affecting intercellular communication and nutrient transport [33, 34].

Uptake via Leaves:

AgNPs can also be taken up through plant leaves. In the instance of Arabidopsis seedlings, immersion of cotyledons in AgNP-containing solutions leads to the uptake and accumulation of AgNPs in stomatal guard cells [112]. On lettuce leaves, AgNPs have demonstrated effective retention by the cuticle after foliar exposure, with penetration occurring through stomata [66]. Comparative research on AgNP uptake in soybean and rice, comparing root exposure to foliar exposure, has revealed a notably higher accumulation of Ag in plants exposed through leaves [69].

Translocation to Edible Parts:

Once AgNPs enter a plant's vascular tissues, they can be transported over long distances to various plant organs, including leaves, fruits, seeds, and other edible parts [25, 33, 75]. Consequently, there is potential for these plant components to be contaminated by AgNPs through translocation (Fig. 1).

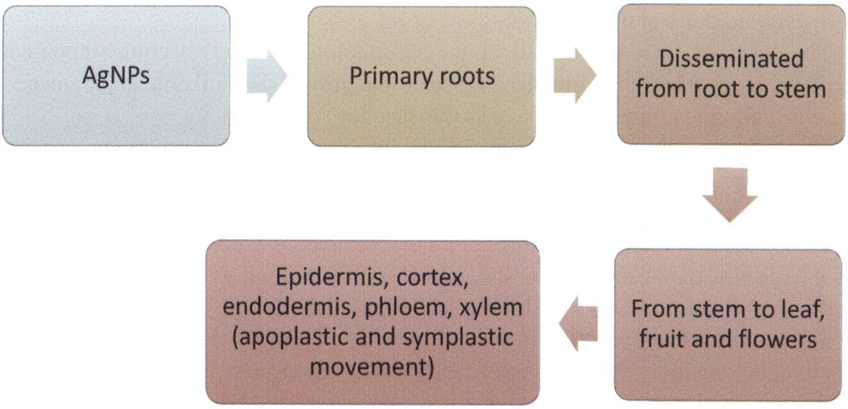

Fig. 1 A diagrammatic representation of fundamental concept regarding the absorption, transportation, and buildup of silver nanoparticles (AgNPs) within plants. Initially, these nanoparticles are taken in through the primary roots (R1). Subsequently, they are distributed from the root (R0) and travel through the stem to reach the leaves, fruits, and flowers and this process involves movement within the plant

Mechanisms and Controversies:

Numerous mechanisms govern the uptake, translocation, and accumulation of AgNPs within plant systems, including the role of aquaporins, ion channels, and interactions with the plasma membrane [118]. However, while progress has been made in understanding these processes, controversies, and issues remain, particularly regarding the characteristics of AgNPs in specific media, biodistribution, and methodologies for analyzing the release of Ag^+ ions from AgNPs [118]. Comprehending the pathways of AgNP uptake and translocation in plants is crucial for assessing their environmental impact and potential risks to human health when these plants are part of the food chain. Further research is needed to unravel the intricacies of these processes and address existing controversies and uncertainties.

5 Phytotoxicity and Genotoxicity of AgNPs: An In-Depth Examination

AgNPs, owing to their distinctive characteristics, have found widespread applications across multiple industries, including electronics, medicine, and agriculture. However, the growing utilization of AgNPs has raised environmental concerns, particularly with regard to their potential impact on plant life. This comprehensive book chapter delves into the genotoxic and phytotoxic effects of AgNPs, examining their influence on various aspects of plant biology.

1. Phytotoxicity at the Morphological Level

Phytotoxicity, when observed at the morphological level, becomes apparent through observable changes in plant structure and growth. Several crucial parameters are conventionally employed to assess this facet of AgNP toxicity.

One of the most conspicuous effects is the inhibition of seed germination and root growth when exposed to AgNPs [8, 121]. Additionally, AgNP exposure results in reduced biomass and a decrease in leaf surface area. These effects have been documented in various plant species [8, 121]. For example, *Spirodela polyrrhiza* experiences root abscission and inhibits shoot growth [56]. Arabidopsis, when subjected to higher AgNP concentrations, exhibits reduced biomass [62]. In wheat, AgNPs leads to a dose-dependent reduction in shoot and root length [26]. Rice displays diminished root elongation, as well as reduced shoot and root fresh weights [95]. Zucchini shows inhibited seed germination and decreased biomass, especially at concentrations exceeding 100 mg/L [114]. Similar results regarding toxicity concerning seed germination, biomass accumulation, and root and shoot growth have been reported for various plant species, including Arabidopsis [100], *Brassica nigra* [6], Lemna [37], *Phaseolus radiatus, Sorghum bicolor* [68], *Lolium multiflorum* [135], rice [29], wheat [134], and *Lupinus termis L.* [5].

2. Phytotoxicity at the Physiological Level

AgNPs exert their phytotoxic effects at the physiological level, leading to changes in chlorophyll content, nutrient absorption, transpiration rate, and hormonal balance within plants. These alterations can have a substantial impact on overall plant health. AgNPs disrupt chlorophyll synthesis in plant leaves, thus affecting the photosynthetic process [121]. This disruption results in a decrease in chlorophyll levels, ultimately impeding plant growth. For example, when exposed to AgNPs, Arabidopsis experiences reduced chlorophyll content, leading to inhibited growth [100]. Rice also showed a significant decline in total chlorophyll and carotenoid content following AgNP exposure [95]. Mustard (Brassica sp.) seedlings accumulate AgNPs, causing a considerable inhibition of photosynthesis [125]. AgNP exposure to Physcomitrella patens alters thylakoid structure, reduces chlorophyll b content, and disrupts the balance of essential elements in leafy gametophytes [70]. In *Lupinus termis L.* seedlings, AgNPs lead to reduced shoot and root elongation, as well as decreased total chlorophyll and total protein content after ten days of exposure [5]. Additionally, the transpiration rate in *Cucurbita pepo* is significantly reduced following AgNP exposure [40, 93, 114]. The effect of two types of negatively charged quasi-spherical AgNPs at concentrations 10. 20, 30 mg/L, and silver ions at a concentration of 30 mg/L on the growth, selected physiological aspects, and yielding of wheat (*Triticum aestivum* L.) CV, Tybalt, and on plant resistance to seedling blight were tested, and showed the result as improved leaf area compared to control [78] Fig. 2 and Table 1.

Furthermore, AgNPs can affect the fluidity and permeability of plant cell membranes, impacting water and nutrient uptake. For instance, research by Zuverza-Mena et al. [141] demonstrated that AgNP exposure in radish (*Raphanus sativus*) sprouts leads to a dose-dependent reduction in water and nutrient content (e.g., Ca, Mg, B, Cu, Mn, and Zn), suggesting that AgNPs may influence plant growth by altering water and nutrient absorption [141]. AgNPs also have the potential to

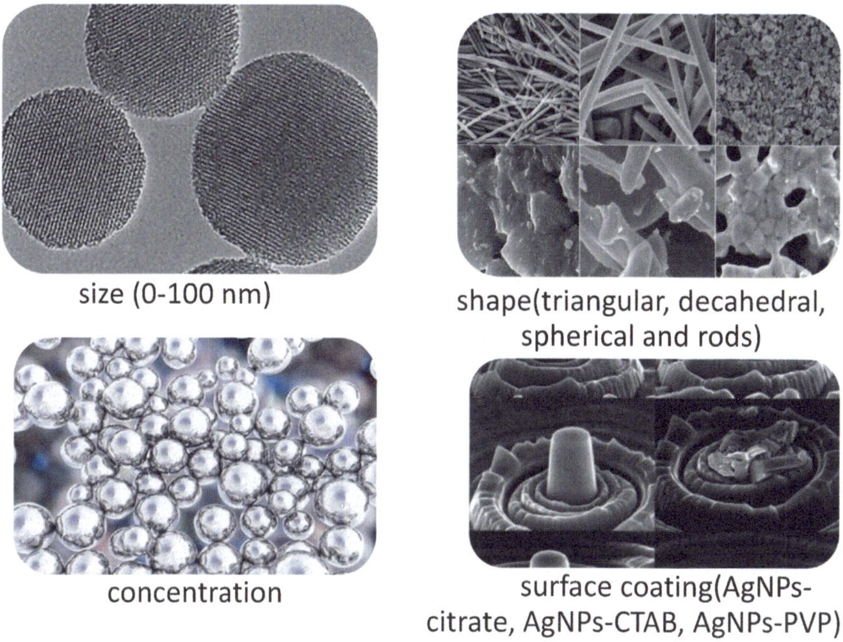

Fig. 2 A schematic illustration depicts the specific toxicity associated with silver nanoparticles (AgNPs). The harmful effects on plants caused by AgNPs depend on various characteristics of these nanoparticles, such as their size, geometry, concentration, and the type of surface coating they possess

interfere with plant hormones. For instance, the root gravitropism of Arabidopsis seedlings is inhibited in a dose-dependent manner by AgNPs. Further analysis indicates that AgNPs reduce auxin accumulation, and gene expression analysis suggests downregulation of auxin receptor-related genes upon AgNP exposure [116]. Vinkovic et al. [124] conducted hormonal analysis using ultra-high-performance liquid chromatography-electrospray and found that AgNP accumulation in pepper tissue resulted in a significant increase in total cytokinin levels, implying the importance of cytokinin in the plant's response to AgNP-induced stress [124]. In cucumber and wheat, Ag2S-NPs inhibit growth, and the expression of genes associated with the ethylene signaling pathway is significantly upregulated upon exposure to Ag2S-NPs, indicating an interaction with the ethylene signaling pathway [126].

3. Cytotoxicity and Genotoxicity

AgNPs can induce cellular and molecular toxicity in plants, which becomes evident through changes in cell structure, division, and DNA integrity. These effects are crucial components of genotoxicity.

3.1 Cellular and Nuclear Alterations

At the cellular level, exposure to AgNPs often leads to dramatic changes in plant cell morphology and nuclear structure. These alterations signify the genotoxic potential of AgNPs and highlight their ability to interfere with vital cellular processes. For

Table 1 Toxic effects of AgNPs in different plant species

Plant species	Concentration range and particle size (diameter nm)	Toxic effects of Ag-NPs	References
Glycine max L	1 mM (81.84 ± 0.6)	Decrease in stomatal conductance by 15%	Guilger-Casagrande et al. [38]
Potato *(Solanum tuberosum L.)*	0, 2, 10, 20 mg/L & (20)	Increase in total reactive oxygen species (ROS), superoxide anions, and in the activities of superoxide dismutase (SOD), catalase, ascorbate peroxidase, and glutathione reductase (GR) Higher ion leakage and finally cell death	Bagherzadeh Homaee and Ehsanpour [11]
Pisum sativum	1000, 3000 μM (20)	Reduction in photosynthetic pigments, growth, and chlorophyll fluorescence; Inhibited activities of glutathione reductase (GR) and dehydroascorbate reductase (DHAR)	Tripathi et al. [121]
Wheat (Triticum aestivum L.)	10, 20, 30 mg/L	Negative effect on roots, decreased seedling blight, and improved leaf area	Matras et al. [78]
Rice (*Oryza sativa* L.)	0, 0.2, 0.5, 1 mg/L & (20)	Decrease plant biomass, increases the amount of malondialdehyde and H_2O_2 contents in leaves	Li et al. [69]
Onion (*Allium cepa* L.)	5, 10, 20 mg/L & (20)	Causes various chromosomal aberration in mitotic as well as in meiotic cells	
Mustard (Brassica sp.)	1, 3 mM & (47)	Inhibit growth in brassica seedlings and causes oxidative stress	Vishwakarma et al. [125]
Green asparagus	> 100 mg/L & (20)	Increases ascorbic acid and total chlorophyll content at higher level	An et al. [7]
Cucurbita pepo	250, 750 mg/L & (100)	Decreases in plant biomass and transpiration	Hawthorne et al. [40]
Vicia faba	12.5, 25, 50, 100 mg/L & (60)	Reduction in mitotic index and increase in micronuclei and chromosomal aberrations	Patlolla et al. [99]
Tomato (*Lycopersicon esculentum*)	0, 100, 1000 mg/L & (10–15)	Reduction in root growth, decrease in fruit production and chlorophyll content	Song et al. [112]

(continued)

Table 1 (continued)

Plant species	Concentration range and particle size (diameter nm)	Toxic effects of Ag-NPs	References
Wolffia globosa	1, 2, 5, 8, 10 mg/L & (10)	Reduction in chlorophyll a, carotenoids, and soluble protein amount Causes oxidative damage, increases malondialdehyde (MDA) level, and upregulation of SOD activity	Zou et al. [140]
Arabidopsis thaliana	0.2, 0.5, 3 mg/L & (10)	Root growth inhibition; Disrupted the thylakoid membrane structure and decreased chlorophyll content; caused alteration of transcription of antioxidant and aquaporin-related genes	Qian et al. [100]

example, *Lolium multiflorum* seedlings exposed to AgNPs exhibit severe cellular damage, with root tip cells failing to develop root hairs, and cortical cells becoming highly vacuolated and collapsed [135].

4. **Mechanisms of Toxicity**

4.1 AgNP-Induced Oxidative Stress

The primary mechanism underlying the phytotoxicity of AgNPs is the excessive production of reactive oxygen species (ROS) induced by AgNPs. These ROS encompass singlet oxygen ($1O_2$), superoxide ($O2^-$), hydrogen peroxide (H_2O_2), and hydroxyl radical ($HO^{\cdot}$), and they accumulate within plant cells, causing oxidative stress. While ROS are typically generated as byproducts of regular metabolic processes within plant organelles under normal conditions, AgNP exposure leads to their overproduction, resulting in oxidative damage to plant biomolecules through electron transfer. This surplus production of ROS directly damages proteins, DNA, and the cell membrane, potentially leading to cell death and inhibiting plant growth [15, 105, 121, 137]. For instance, Panda et al. observed that the application of AgNP-P (AgNPs derived from *Pelargonium graveolens*) to rice plants increased ROS levels, leading to oxidative stress, which ultimately impeded the growth and development of rice plants. These findings underscore the direct correlation between ROS generation and the adverse effects of AgNPs on plants [96]. Furthermore, treating *Zea mays* seedlings with AgNPs obtained from mint (*Mentha arvensis L.*) led to elevated ROS levels, as demonstrated by a significant increase in H_2O_2 and O_2- levels in the root tip cells. This accumulation of ROS subsequently resulted in lipid peroxidation, affecting cell membrane integrity and plant growth [121].

4.2 Silver-Specific Toxicity

While AgNPs themselves contribute to ROS production, a significant portion of their phytotoxicity stems from the release of ionic silver (Ag^+) into the surrounding environment. Ag^+ ions have been implicated in various cellular processes that disrupt plant function. These ions can induce oxidative stress, interfere with the normal functioning of enzymes, and bind to cellular components, leading to structural damage and the impairment of metabolic processes [15].

Furthermore, Ag+ ions have been shown to directly impact photosynthesis by competitively substituting copper ions (Cu^{2+}) in plastocyanin (Pc), a critical electron carrier protein in the photosynthetic electron transport chain. This substitution hinders electron transfer between photosystem II and photosystem I, consequently reducing the photosynthetic efficiency of the plant and further contributing to growth inhibition and chlorosis [23, 113]. Excessive production of reactive oxygen species (ROS) is one of the outcomes of AgNPs toxicity in plants, and to cope with this stress, antioxidant enzymes such as CAT, SOD, and APX can play an important role in detoxifying and protecting plant tissues [63].

4.3 AgNP-Specific Toxicity

The toxicity of AgNPs is not solely determined by the presence of silver ions but is also influenced by various nanoparticle-specific factors, including size, surface coating, and morphology. These factors collectively regulate the interaction between AgNPs and plants.

4.3.1. Particle Size

Smaller-sized AgNPs typically possess a higher surface area per unit mass, allowing them to exert more pronounced toxic effects due to their increased potential to release Ag^+ ions and interact with cellular components. Numerous studies have reported that smaller AgNPs are often more toxic to plants. For example, Nair et al. [96] demonstrated that smaller AgNPs (10 nm) exhibited more substantial toxic effects on cucumber (*Cucumis sativus*) seedlings compared to larger particles (50 and 100 nm) at the same concentration. The smaller particles resulted in greater reductions in shoot and root elongation, as well as a more pronounced decrease in chlorophyll content, highlighting the size-dependent toxicity of AgNPs [71].

4.3.2. Surface Coating

The surface coating or functionalization of AgNPs can alter their properties and toxicity. Surface coatings can affect particle stability, dispersibility, and reactivity with plant cells. For instance, polyvinylpyrrolidone (PVP)coated AgNPs were found to be less toxic to *Arabidopsis thaliana* than uncoated AgNPs. The surface coating reduced the release of Ag^+ ions and, consequently, the oxidative stress and cytotoxicity induced by the particles [120].

4.3.3. Morphology

AgNPs can exhibit various shapes, including spheres, rods, and triangles. The morphology of AgNPs can influence their toxicity to plants. Studies have shown that AgNPs with different shapes may be taken up differently by plant cells and can result in distinct physiological responses. For example, Shah et al. investigated the effects of spherical and rod-shaped AgNPs on cucumber plants and found that rod-shaped AgNPs were more toxic, leading to greater reductions in root elongation and higher levels of oxidative stress compared to spherical AgNPs at the same concentration [27].

6 Conclusion

The interaction between plants and AgNPs is a multifaceted process influenced by a myriad of factors, including intrinsic properties of AgNPs, plant species, developmental stages, and tissue types. AgNPs can exert phytotoxicity by inhibiting seed germination, altering plant morphology, disrupting photosynthesis, affecting nutrient uptake, and inducing hormonal imbalances. Furthermore, their genotoxic effects are evident through damage to cell structure, impaired cell division, and DNA aberrations. The primary mechanisms underlying these toxic effects involve the induction of oxidative stress by excessive ROS production and the release of silver ions, which disrupt cellular processes and bind to cellular components. Additionally, nanoparticle-specific factors, such as size, surface coating, and morphology, significantly influence the toxicity of AgNPs. Understanding the genotoxic and phytotoxic effects of AgNPs is paramount to mitigating potential environmental risks associated with their use. Further research is needed to elucidate the precise mechanisms of AgNP toxicity in different plant species and under varying environmental conditions. This knowledge will aid in developing strategies to minimize the environmental impact of AgNPs while still harnessing their beneficial properties in various applications.

References

1. Abdelsalam NR, Abdel-Megeed A, Ali HM, Salem MZM, Al-Hayali MFA, Elshikh MS (2018) Genotoxicity effects of silver nanoparticles on wheat (*Triticum aestivum* L.) root tip cells. Ecotoxicol Environ Saf 155:76–85. https://doi.org/10.1016/j.ecoenv.2018.02.069
2. Acharya A, Pal PK (2020) Agriculture nanotechnology: translating research outcome to field applications by influencing environmental sustainability. Nano Impact 19:100232. https://doi.org/10.1016/j.impact.2020100232
3. Ahamed M, Posgai R, Gorey TJ, Nielsen M, Hussain SM, Rowe JJ (2010) Silver nanoparticles induced heat shock protein 70, oxidative stress and apoptosis in *Drosophila melanogaster*. Toxicol Appl Pharm 242:263–269
4. Alavi S, Dehpour A (2010) Evaluation of the nanosilver colloidal solution in comparison with the registered fungicide to control greenhouse cucumber downy mildew disease in the north of Iran. In: Proceedings of the VI international postharvest symposium, Antalya, Turkey, 1643–1646
5. Al-Huqail AA, Hatata MM, Al-Huqail AA, Ibrahim MM (2018) Preparation, characterization of silver phyto nanoparticles and their impact on growth potential of *Lupinus termis* L. seedlings. Saudi J Biol Sci 25:313–319
6. Amooaghaie R, Tabatabaei F, Ahadi AM (2015) Role of hematin and sodium nitroprusside in regulating Brassica nigra seed germination under nanosilver and silver nitrate stresses. Ecotoxicol Environ Saf 113:259–270
7. An J, Zhang M, Wang S, Tang J (2008) Physical, chemical and microbiological changes in stored green asparagus spears as affected by coating of silver nanoparticles-PVP. LWT Food Sci Technol 41:1100–1107
8. Aslani F, Bagheri S, Muhd Julkapli N, Juraimi AS, Hashemi FSG, Baghdadi A (2014) Effects of engineered nanomaterials on plants growth: an overview. Sci World J

9. Auclair J, Gagne F (2022) Shape-dependent toxicity of silver nanoparticles on freshwater cnidarians. Nanomaterials12(18):3107. https://doi.org/10.3390/nano12183107
10. Bachheti RK, Bachheti A, Husen A (2024) Metal and metal-oxide based nanomaterials (Synthesis, Agricultural, Biomedical and Environmental Interventions). Springer Nature Singapore Pte Ltd., 152 Beach Road, #21-01/04 Gateway East, Singapore 189721, Singapore
11. Bagherzadeh Homaee M, Ehsanpour AA (2016) Silver nanoparticles and silver ions: oxidative stress responses and toxicity in potato (*Solanum tuberosum* L.) grown in vitro. Hortic Environ Biotechnol 57:544–553
12. Banerjee P, Satapathy M, Mukhopahayay A, Das P (2014) Leaf extract mediated green synthesis of silver nanoparticles from widely available Indian plants: synthesis, characterization, antimicrobial property and toxicity analysis. Bioresour Bioprocess 1:3. https://doi.org/10.1186/s40643-014-0003-y
13. Beer C, Foldbjerg R, Hayashi Y, Sutherland DS, Autrup H (2012) Toxicity of silver nano particles-Nanoparticle or silver ion? Toxicol Lett 208:286–292
14. Blaser SA, Scheringer M, MacLeod M, Hungerbühler K (2008) Estimation of cumulative aquatic exposure and risk due to silver: contribution of nano-functionalized plastics and textiles. Sci Total Environ 390:396–409
15. Capaldi Arruda SC, Diniz Silva AL, Moretto Galazzi R, Antunes Azevedo R, Zezzi Arruda MA (2015) Nanoparticles applied to plant science: a review. Talanta 131:693–705
16. Carpita NC, Gibeaut DM (1993) Structural models of primary cell walls in flowering plants: consistency of molecular structure with the physical properties of the walls during growth. Plant J 3:1–30
17. Chawla S, Singh S, Husen A (2023) Smart nanomaterials targeting pathological Hypoxia. Springer Nature Singapore Pte Ltd., 152 Beach Road, #21-01/04 Gateway East, Singapore 189721, Singapore ISBN: 978-981-99-1717-4. https://doi.org/10.1007/978-981-99-1718-1
18. Chen J, Dou R, Yang Z, You T, Gao X, Wang L (2018) Phytotoxicity and bioaccumulation of zinc oxide nano particles in rice (*Oryza sativa* L.). Plant Physiol Biochem 130:604–612. https://doi.org/10.1016/j.plaphy.2018.08.019
19. Cheng Y, Yin L, Lin S, Wiesner M, Bernhardt E, Liu J (2011) Toxicity reduction of polymer-stabilized silver nanoparticles by sunlight. J Phys Chem C 115:4425–4432. https://doi.org/10.1021/jp109789j
20. Chernousova S, Epple M (2013) Silver as antibacterial agent: ion, nanoparticle, and metal. Angew Chem Int Ed 52:1636–1653. https://doi.org/10.1002/anie.201205923
21. Colman BP, Arnaout CL, Anciaux S, Gunsch CK, Hochella MFJr, Kim B, Lowry GV, McGill BM, Reinsch BC, Richardson CJ (2013) Low concentrations of silver nanoparticles in biosolids cause adverse ecosystem responses under realistic field scenario. PLoS ONE 8:e57189
22. Cvjetko P, Zovko M, Stefanic PP, Biba R, Tkalec M, Domijan AM, Vrˇcek IV, Letofsky-Papst I, Šikic S, Balen B (2018) Phytotoxic effects of silver nanoparticles in tobacco plants. Environ Sci Pollut Res 25: 5590–5602
23. De La Torre-Roche R, Hawthorne J, Musante C, Xing B, Newman LA, Ma X, White JC (2013) Impact of Ag nanoparticle exposure on p, p0-DDE bioaccumulation by Cucurbita pepo (Zucchini) and Glycine max (Soybean). Environ Sci Technol 47:718–725
24. Dhand C, Dwivedi N, Loh XJ, Ying ANJ, Verma NK, Beuerman RW, Lakshminarayanan R, Ramakrishna S (2015) Methods and strategies for the synthesis of diverse nanoparticles and their applications: a comprehensive overview. RSC Adv 5(127):105003–105037. https://doi.org/10.1039/C5RA19388E
25. Dietz KJ, Herth S (2011) Plant nanotoxicology. Trends Plant Sci 16:582–589
26. Dimkpa CO, McLean JE, Martineau N, Britt DW, Haverkamp R, Anderson AJ (2013) Silver nanoparticles disrupt wheat (*Triticum aestivum* L.) growth in a sandmatrix. Environ Sci Technol 47(2):1082–1090. https://doi.org/10.1021/es302973y
27. Dobias J, Bernier-Latmani R (2013) Silver release from silver nanoparticles in natural waters. Environ Sci Technol 47:4140–4146

28. Durán N, Marcato PD, De Souza GI, Alves OL, Esposito E (2007) Antibacterial effect of silver nanoparticles produced by fungal process on textile fabrics and their effluent treatment. J Biomed Nanotechnol 3:203–208
29. Ejaz M, Raja NI, Ahmad MS, Hussain M, Iqbal M (2018) Effect of silver nanoparticles and silver nitrate on growth of rice under biotic stress. IET Nanobiotechnol
30. Fabrega J, Luoma SN, Tyler CR, Galloway TS, Lead JR (2011) Silver nanoparticles: behaviour and effects in the aquatic environment. Environ Int 37:517–531
31. Gardea-Torresdey JL, Rico CM, White JC (2014) Trophic transfer, transformation, and impact of engineered nanomaterials in terrestrial environments. Environ Sci Technol 48:2526–2540
32. Garmasheva I, Kovalenko N, Voychuk S, Ostapchuk A, Livinska O, Oleschenko L (2016) *Lactobacillus* species mediated synthesis of silver nanoparticles and their antibacterial activity against opportunistic pathogens *in vitro*. Bioimpacts 6(4):219–223. https://doi.org/10.15171/bi.2016.29
33. Geisler-Lee J, Brooks M, Gerfen JR, Wang Q, Fotis C, Sparer A, Ma X, Berg RH, Geisler M (2014) Reproductive toxicity and life history study of silver nanoparticle effect, uptake and transport in Arabidopsis thaliana. Nanomaterials 4:301–318
34. Geisler-Lee J, Wang Q, Yao Y, Zhang W, Geisler MLIK, Huang Y, Chen Y, Kolmakov A, Ma X (2013) Phytotoxicity, accumulation and transport of silver nanoparticles by Arabidopsis thaliana. Nanotoxicology 7:323–337
35. Gottschalk F, Sonderer T, Scholz RW, Nowack B (2009) Modeled environmental concentrations of engineered nanomaterials (TiO_2, ZnO, Ag, CNT, Fullerenes) for different regions. Environ Sci Technol 43:9216–9222
36. Gottschalk F, Sonderer T, Scholz RW, Nowack B (2010) Possibilities and limitations of modeling environmental exposure to engineered nanomaterials by probabilistic material flow analysis. Environ Toxicol Chem 29:1036–1048
37. Gubbins EJ, Batty LC, Lead JR (2011) Phytotoxicity of silver nanoparticles *to Lemna minor L*. Environ Pollut 159:1551–1559
38. Guilger-Casagrande M, Bilesky-Jose N, Sousa BT (2022) Effects of biogenic silver and iron nanoparticles on soybean seedlings (*Glycine max*) BMC Plant Biol 22(1):255
39. Gurunathan S, Raman J, Abd Malek SN, John PA, Vikineswary S (2013) Green synthesis of silver nanoparticles using *Ganoderma neo-japonicum* Imazeki: a potential cytotoxic agent against breast cancer cells. Int J Nanomed 8:4399–4413. https://doi.org/10.2147/IJN.S51881
40. Hawthorne J, Musante C, Sinha SK, White JC (2012) Accumulation and phytotoxicity of engineered nanoparticles to *Cucurbita Pepo*. Int J Phytoremediat 14:429–442
41. Hedberg J, Skoglund S, Karlsson ME, Wold S, OdnevallWallinder I, Hedberg Y (2014) Sequential studies of silver released from silver nanoparticles in aqueous media simulating sweat, laundry detergent solutions and surface water. Environ Sci Technol 48:7314–7322
42. Heinlein M, Epel BL (2004) Macromolecular transport and signaling through plasmodesmata. Int Rev Cytol
43. Hossain Z, Mustafa G, Sakata K, Komatsu S (2016) Insights into the proteomic response of soybean towards Al_2O_3ZnO, and Ag nanoparticles stress. J Hazard Mater 304:291–305. https://doi.org/10.1016/j.jhazmat.2015.10.071
44. Husen A (2019) Natural product-based fabrication of zinc oxide nanoparticles and their application. In: Husen A, Iqbal M (eds) N*anomaterials* and plant potential. Springer International Publishing AG, Gewerbestrasse 11, 6330 Cham, pp 193–291. https://doi.org/10.1007/978-3-030-05569-1_7
45. Husen A (2020) Introduction and techniques in nanomaterials formulation. In: Husen, A, Jawaid M (eds) Nanomaterials for agriculture and forestry applications. Elsevier Inc. 50 Hampshire St., 5th Floor, Cambridge, MA 02139, USA, pp 1–14. https://doi.org/10.1016/B978-0-12-817852-2.00001-9
46. Husen A (2022) Engineered nanomaterials for sustainable agricultural production, soil improvement and stress management. Elsevier Inc., 50 Hampshire St., 5th Floor, Cambridge, MA 02139, USA. https://doi.org/10.1016/C2021-0-00054-7

47. Husen A (2023a) Secondary metabolites based green synthesis of nanomaterials and their applications. Springer Nature Singapore Pte Ltd., 152 Beach Road, #21-01/04 Gateway East, Singapore 189721, Singapore. https://doi.org/10.1007/978-981-99-0927-8
48. Husen A (2023b) Nanomaterials from agricultural and horticultural products. Springer Nature Singapore Pte Ltd., 152 Beach Road, #21-01/04 Gateway East, Singapore 189721, Singapore. https://link.springer.com/book/9789819934348
49. Husen A (2023c) Nanomaterials and nanocomposites exposures to plants (Response, Interaction, Phytotoxicity and Defense Mechanisms). Springer Nature Singapore Pte Ltd., 152 Beach Road, #21-01/04 Gateway East, Singapore 189721, Singapore. https://doi.org/10.1007/978-981-99-2419-6
50. Husen A, Bachheti RK, Bachheti A (2023) Current trends in green nano-emulsions (Food, Agriculture and Biomedical Sectors). Springer Nature Singapore Pte Ltd., 152 Beach Road, #21-01/04 Gateway East, Singapore 189721, Singapore
51. Husen A, Iqbal M (2019) Nanomaterials and plant potential. Springer International Publishing AG, Gewerbestrasse 11, 6330 Cham, Switzerland. https://doi.org/10.1007/978-3-030-055 69-1
54. Husen A, Siddiqi KS (2023) Advances in smart nanomaterials and their applications. Elsevier Inc., 50 Hampshire St., 5th Floor, Cambridge, MA 02139, USA. https://doi.org/10.1016/C2021-0-02202-1
55. Imlay JA, Linn S (1988) DNA damage and oxygen radical toxicity. Science 240(4857):1302–1309. https://doi.org/10.1126/science.3287616
56. Jiang HS, Li M, Chang FY, Li W, Yin LY (2012) Physiological analysis of silver nanoparticles and $AgNO_3$ toxicity to *Spirodela polyrhiza*. Environ Toxicol Chem 31:1880–1886
57. Jin-Chul K, Madhusudhan A, Husen A (2021) Smart nanomaterials in biomedical applications. Springer Nature Switzerland AG, Gewerbestrasse 11, 6330 Cham, Switzerland. https://doi.org/10.1007/978-3-030-84262-8
58. Kaegi R, Voegelin A, Sinnet B, Zuleeg S, Hagendorfer H, Burkhardt M, Siegrist H (2011) Behavior o metallic silver nanoparticles in a pilot wastewater treatment plant. Environ Sci Technol 45:3902–3908
59. Kale SK, Parishwad GV, Husainy ASN, Patil AS (2021) Emerging agriculture applications of silver nanoparticles. ES Food Agrofor 3:17–22. https://doi.org/10.30919/esfaf438
60. Kalimuthu K, Suresh Babu R, Venkataraman D, Bilal M, Gurunathan S (2008) Biosynthesis of silver nanocrystals by *Bacillus licheniformis*. Colloids Surf Biointerf 65(1):150–153. https://doi.org/10.1016/j.colsurfb.2008.02.018
61. Kathiresan K, Alikunhi NM, Pathmanaban S, Nabikhan A, Kandasamy S (2010) Analysis of antimicrobial silver nanoparticles synthesized by coastal strains of *Escherichia coli* and *Aspergillus niger*. Can J Microbiol 56(12): 1050–1059. https://doi.org/10.1139/W10-094
62. Kaveh R, Li YS, Ranjbar S, Tehrani R, Brueck CL, Van Aken B (2013) Changes in Arabidopsis thaliana gene expression in response to silver nanoparticles and silver ions. Environ Sci Technol 47:10637–10644
63. Kumar S, Masurkar P, Sravani B, Bag D, Sharma KR, Singh P, Korra T, Meena M, Swapnil P, Rajput VD, Minkina T (2023) (AgNPs) on plants: a review on phytotoxicity and defense mechanism of silver nanoparticles. J Nanopart Res 25:54. https://doi.org/10.1007/s11051-023-05708-3
64. Kumari M, Mukherjee A, Chandrasekaran N (2009) Genotoxicity of silver nanoparticles in *Allium cepa*. Sci Total Environ 407(19):5243–5246. https://doi.org/10.1016/j.scitotenv.2009.06.024
65. Kunniger T, Gerecke AC, Ulrich A, Huch A, Vonbank R, Heeb M, Wichser A, Haag R, Kunz P, Faller M (2014) Release and environmental impact of silver nanoparticles and conventional organic biocides from coated wooden façades. Environ Pollut 184:464–471
66. Larue C, Castillo-Michel H, Sobanska S, Cécillon L, Bureau S, Barthès V, Ouerdane L, Carrière M, Sarret G (2014) Foliar exposure of the crop *Lactuca sativa* to silver nanoparticles: evidence for internalization and changes in Ag speciation. J Hazard Mater 264:98–106

67. Lazareva A, Keller AA (2014) Estimating potential life cycle releases of engineered nanomaterials from wastewater treatment plants. ACS Sustain Chem Eng 2:1656–1665
68. Lee WM, Kwak JI, An YJ (2012) Effect of silver nanoparticles in crop plants *Phaseolus radiatus* and *Sorghum bicolor*: media effect on phytotoxicity. Chemosphere 86:491–499
69. Li CC, Dang F, Li M, Zhu M, Zhong H, Hintelmann H, Zhou DM (2017) Effects of exposure pathways on the accumulation and phytotoxicity of silver nanoparticles in soybean and rice. Nanotoxicology 11:699–709
70. Liang L, Tang H, Deng Z, Liu Y, Chen X, Wang H (2018) Ag nanoparticles inhibit the growth of the bryophyte, Physcomitrella patens. Ecotoxicol Environ Saf 164:739–748
71. Lin TH, Huang YL, Huang SF (1996) Lipid peroxidation in liver of rats administrated with methyl mercuric chloride. Biol Trace Elem Res 54:33–41
72. Lombi E, Donner E, Scheckel KG, Sekine R, Lorenz C, Goetz NV, Nowack B (2014) Silver speciation and release in commercial antimicrobial textiles as influenced by washing. Chemosphere 111:352–358
73. Lucas WJ, Lee JY (2004) Plasmodesmata as a supracellular control network in plants. Nat Rev Mol Cell Biol 5:712
74. Ma C, White JC, Dhankher OP, Xing B (2015) Metal based nano toxicity and detoxification pathways in higher plants. Environ Sci Technol 49(12):7109–7122. https://doi.org/10.1021/acs.est.5b00685
75. Ma X, Geiser-Lee J, Deng Y, Kolmakov A (2010) Interactions between engineered nanoparticles (ENPs) and plants: phytotoxicity, uptake and accumulation. Sci Total Environ 408:3053–3061
76. Madhusudhan A, Purohit SD, Prasad R, Husen A (2024) Functional smart nanomaterials and their theranostics approaches. Springer Nature Singapore Pte Ltd., 152 Beach Road, #21-01/04 Gateway East, Singapore 189721, Singapore
77. Mandal AK, Ghorai S, Husen A (2023) Functionalized smart nanomaterials for point-of-care testing. Springer Nature Singapore Pte Ltd., 152 Beach Road, #21-01/04 Gateway East, Singapore 189721, Singapore
78. Matras E, Gorczyca A, Pociecha E (2023) Silver nanoparticles affect wheat (*Triticum aestivum L.*) germination, seedling blight and yield Funct Plant Biol 50(5):390–406
79. Maurer-Jones MA, Gunsolus IL, Murphy CJ, Haynes CL (2013) Toxicity of engineered nanoparticles in the environment. Anal Chem 85:3036–3049
80. Maynard AD, Warheit DB, Philbert MA (2011) The new toxicology of sophisticated materials: nanotoxicology and beyond. Toxicol Sci 120:S109–S129
81. Mazumdar H (2014a) Comparative assessment of the adverse effect of silver nanoparticles to *Vigna radiata* and *Brassica campestris* crop plants. Int J Eng Res Appl 4:118–124
82. Mazumdar H (2014b) The impact of silver nano particles on plant biomass and chlorophyll content. Res Inv Int J Eng Sci 4(7):12–20. https://www.researchinventy.com/papers/v4i7/C047012020.pdf
83. Meena M, Swapnil P (2019) Regulation of *WRKY* genes in plant defense with beneficial fungus *Trichoderma*: current perspectives and future prospects. Arch Phytopathol Plant Protect 52(1–2):1–17. https://doi.org/10.1080/03235408.2019.1606490
84. Meena M, Swapnil P, Yadav G, Sonigra P (2021) Role of fungi in bio-production of nanomaterials at megascale. In: Sharma V, Shah M, Parmar S, Kumar A (eds) Fungi bio-prospects in sustainable agriculture, environment and nano-technology: Volume 3: Fungal metabolites and nano-technology. Academic Press, Elsevier Inc. Oxford, United Kingdom, pp 453–474. https://doi.org/10.1016/B978-0-12-821734-4.00006-X
85. Meriga B, Reddy BK, Rao KR, Reddy LA, Kishor PBK (2004) Aluminium-induced production of oxygen radicals, lipid peroxidation and DNA damage in seedlings of rice (*Oryza sativa*). J Plant Physiol 161(1):63–68. https://doi.org/10.1078/0176-1617-01156
86. Miralles P, Church TL, Harris AT (2012) Toxicity, uptake, and translocation of engineered nanomaterials in vascular plants. Environ Sci Technol 46:9224–9239
87. Mishra VK, Husen A, Rahman QI, Iqbal M, Sohrab SS, Yassin MO, Bachheti RK (2019) Plant-based fabrication of silver nanoparticles and their application. In: Husen A, Iqbal M (eds)

Nanomaterials and plant potential. Springer International Publishing AG, Gewerbestrasse 11, 6330 Cham, pp 135–175. https://doi.org/10.1007/978-3-030-05569-1_5
88. Mokashi VV, Gore AH, Sudarsan V, Rath MC, Han SH, Patil SR, Kolekar GB (2012) Evaluation of inter particle interaction between colloidal Ag nanoparticles coated with trisodium citrate and safranine by using FRET: spectroscopic and mechanistic approach. J Photochem Photobiol B 113:63–69. https://doi.org/10.1016/j.jphotobiol.2012.05.006
89. Mokula MR, Husen A (2023) Toxicity assessment of silver nanoparticles and silver ions on plant growth. In: Husen A (ed) Nanomaterials and nanocomposites exposures to plants. smart nanomaterials technology. Springer, Singapore, pp 157–174. https://doi.org/10.1007/978-981-99-2419-6_7
90. Monica RC, Cremonini R (2009) Nanoparticles and higher plants. Caryologia 62:161–165
91. Moreno-Garrido I, Pérez S, Blasco J (2015) Toxicity of silver and gold nanoparticles on marine microalgae. Mar Environ Res 111:60–73
92. Mueller NC, Nowack B (2008) Exposure modeling of engineered nanoparticles in the environment. Environ Sci Technol 42:4447–4453
93. Musante C, White JC (2012) Toxicity of silver and copper to Cucurbita pepo: differential effects of nano and bulk-size particles. Environ Toxicol 27:510–517
94. Nagasa GD, Belete A (2022) Review on nanomaterials and nano-scaled systems for topical and systemic delivery of antifungal drugs. J Multidiscip Healthc 15:1819–1840. https://doi.org/10.2147/JMDH.S359282
95. Nair PMG, Chung IM (2014) Physiological and molecular level effects of silver nanoparticles exposure in rice (*Oryza sativa L.*) seedlings. Chemosphere 112:105–113
96. Nair R, Varghese SH, Nair BG, Maekawa T, Yoshida Y, Kumar DS (2010) Nanoparticulate material delivery to plants. Plant Sci 179:154–163
97. Navarro E, Baun A, Behra R, Hartmann NB, Filser J, Miao AJ, Quigg A, Santschi PH, Sigg L (2008) Environmental behavior and ecotoxicity of engineered nanoparticles to algae, plants, and fungi. Ecotoxicology 17:372–386
98. Pandian AMK, Karthikeyan C, Rajasimman M, Dinesh MG (2015) Synthesis of silver nanoparticle and its application. Ecotoxicol Environ Saf 121:211–217
99. Patlolla AK, Berry A, May L, Tchounwou PB (2012) Genotoxicity of silver nanoparticles in *Vicia faba*: a pilot study on the environmental monitoring of nanoparticles. Int J Environ Res Public Health 9:1649
100. Qian H, Peng X, Han X, Ren J, Sun L, Fu Z (2013) Comparison of the toxicity of silver nanoparticles and silver ions on the growth of terrestrial plant model Arabidopsis thaliana. J Environ Sci 25:1947–1956
101. Quang Huy T, Van Quy N, Anh-Tuan L (2013) Silver nanoparticles: synthesis, properties, toxicology, applications and perspectives. Adv Nat Sci Nanosci Nanotechnol 4:033001
102. Rai M, Yadav A, Gade A (2009) Silver nanoparticles as a new generation of antimicrobials. Biotechnol Adv 27:76–83
103. Ratte HT (1999) Bioaccumulation and toxicity of silver compounds: a review. Environ Toxicol Chem 18:89–108
104. Sah S, Sorooshzadeh A, Rezazadeh H, Naghdibadi H (2011) Effect of nano silver and silver nitrate on seed yield of borage. J Med Plants Res 5:706–710
105. Sen Raychaudhuri S, Deng XW (2000) The role of superoxide dismutase in combating oxidative stress in higher plants. Bot Rev 66:89–98
106. Sharma P, Jha AB, Dubey RS, Pessarakli M (2012) Reactive oxygen species, oxidative damage, and antioxidative defense mechanism in plants under stressful conditions. J Bot 2012:217037. https://doi.org/10.1155/2012/217037
107. Siddiqi KS, Husen A (2022) Plant response to silver nanoparticles: a critical review. Crit Rev Biotechnol 42(7):773–990. https://doi.org/10.1080/07388551.2021.1975091
108. Siddiqi KS, Husen A, Rao RAK (2018a) A review on biosynthesis of silver nanoparticles and their biocidal properties. J Nanobiotechnol 16(14):1–28. https://doi.org/10.1186/s12951-018-0334-5

109. Siddiqi KS, Rashid M, Rahman A, Tajuddin HA, Rehman S (2018b) Biogenic fabrication and characterization of silver nanoparticles using aqueous-ethanolic extract of lichen (*Usnea longissima*) and their antimicrobial activity. Biomater Res 22(23):1–9. https://doi.org/10.1186/s40824-018-0135-9
110. Siddiqi KS, Rashid M, Tajuddin, Rehman S, Husen A (2019) Biofabrication of silver nanoparticles from *Diospyros montana* their characterization and activity against some clinical isolates. BioNanoScience 9:302–312. ISBN: 2191-1630
111. Siddiqui MH, Al-Whaibi MH, Firoz M, Al-Khaishany MY (2015) Role of nanoparticles in plants. In: Siddiqui MH, Al-Whaibi MH, Firoz M (eds) Nanotechnology and plant sciences: nanoparticles and their impact on plants. 1st edn. Springer International Publishing, Switzerland, pp 19–35. https://doi.org/10.1007/978-3-319-14502-0_2
112. Song U, Jun H, Waldman B, Roh J, Kim Y, Yi J, Lee EJ (2013) Functional analyses of nanoparticle toxicity: a comparative study of the effects of TiO_2 and Ag tomatoes (*Lycopersicon esculentum*). Ecotoxicol Environ Saf 93:60–67. https://doi.org/10.1016/j.ecoenv.2013.03.033
113. Speranza A, Crinelli R, Scoccianti V, Taddei AR, Iacobucci M, Bhattacharya P, Ke PC (2013) In vitro toxicity of silver nanoparticles to kiwifruit pollen exhibits peculiar traits beyond the cause of silver ion release. Environ Pollut 179:258–267
114. Stampoulis D, Sinha SK, White JC (2009) Assay dependent phytotoxicity of nanoparticles to plants. Environ Sci Technol 43(24):9473–9479. https://doi.org/10.1021/es901695c
115. Steinitz B, Bilavendran AD (2011) Thiosulfate stimulates growth and alleviates silver and copper toxicity in tomato root cultures. Plant Cell Tissue Organ Cult 107:355–363
116. Sun J, Wang L, Li S, Yin L, Huang J, Chen C (2017) Toxicity of silver nanoparticles to Arabidopsis: inhibition of root gravitropism by interfering with auxin pathway. Environ Toxicol Chem 36:2773–2780
117. Sun TY, Gottschalk F, Hungerbühler K Nowack B (2014) Comprehensive probabilistic modelling of environmental emissions of engineered nanomaterials. Environ Pollut 185: 69–76
118. Sun X, Shi J, Zou X, Wang C, Yang Y, Zhang H (2016) Silver nanoparticles interact with the cell membrane and increase endothelial permeability by promoting VE cadherin internalization. J Hazard Mater 317:570–578. https://doi.org/10.1016/j.jhazmat.2016.06.023
119. Taghiyari HR, Morrell JF, Husen A (2023) Emerging nanomaterials (Opportunities and Challenges in Forestry Sectors). Springer Nature Switzerland AG, Gewerbestrasse 11, 6330 Cham, Switzerland. https://doi.org/10.1007/978-3-031-17378-3
120. Thiruvengadam M, Gurunathan S, Chung IM (2015) Physiological, metabolic, and transcriptional effects of biologically-synthesized silver nanoparticles in turnip (*Brassica rapassp. Ra*pa L). Protoplasma 252:1031–1046
121. Tripathi DK, Singh S, Singh S, Pandey R, Singh VP, Sharma NC, Prasad SM, Dubey NK, Chauhan DK (2017) An overview on manufactured nanoparticles inplants: uptake, translocation, accumulation and phytotoxicity. Plant Physiol Biochem 110:2–12. https://doi.org/10.1016/j.plaphy.2016.07.030
122. Vance ME, Kuiken T, Vejerano EP, McGinnis SP, Hochella MFJr, Rejeski D, Hull MS (2015) Nanotechnology in the real world: redeveloping the nanomaterial consumer products inventory. Beilstein J Nanotechnol 6: 1769–1780
123. Verma P, Maheshwari SK (2019) Applications of silver nanoparticles in diverse sectors. Int J Nano Dimens 10(1):18–36
124. Vinkovic T, Novák O, Strnad M, Goessler W, Jurašin DD, Paraikovic N, Vrˇcek IV (2017) Cytokinin response in pepper plants (*Capsicum annuum L.*) exposed to silver nanoparticles. Environ Res 156:10–18
125. Vishwakarma K, Upadhyay N, Singh J, Liu S, Singh VP, Prasad SM, Chauhan DK, Tripathi DK, Sharma S (2017) Differential phytotoxic impact of plant mediated silver nanoparticles (AgNPs) and silver nitrate ($AgNO_3$) on Brassica sp. Front Plant Sci 8:1501
126. Wang P, Lombi E, Sun S, Scheckel KG, Malysheva A, McKenna BA, Menzies NW, Zhao FJ, Kopittke PM (2017) Characterizing the uptake, accumulation and toxicity of silver sulfide nanoparticles in plants. Environ Sci 4:448–460

130. Wei L, Lu J, Xu H, Patel A, Chen ZS, Chen G (2015) Silver nanoparticles: synthesis, properties, and therapeuticapplications. Drug Discov Today 20(5):595–601. https://doi.org/10.1016/j.drudis.2014.11.014
131. Wu L, Zhu G, Zhang X, Si Y (2020) Silver nano particles inhibit denitrification by altering the viability and metabolic activity of *Pseudomonas stutzeri*
132. Xu L, Wang YY, Huang J, Chen CY, Wang ZX, Xie H (2020) Silver nanoparticles: synthesis, medical applications and biosafety. Theranostics 210(20):8996–9031. https://doi.org/10.7150/thno.45413
134. Yang J, Jiang F, MaC RY, Rui M, Adeel M, Xing B (2018) Alteration of crop yield and quality of wheat upon exposure to silver nanoparticles in a lifecycle study. J Agric Food Chem 66(11):2589–2597. https://doi.org/10.1021/acs.jafc.7b04904
135. Yin L, Cheng Y, Espinasse B, Colman BP, Auffan M, Wiesner M, Rose J, Liu J, Bernhardt ES (2011) More than the Ions: the effects of silver nanoparticles on *Lolium multiflorum*. Environ Sci Technol 45:2360–2367
136. Yin L, Colman BP, McGill BM, Wright JP, Bernhardt ES (2012) Effects of silver nanoparticle exposure on germination and early growth of eleven wetland plants. Plos One 7:e47674. https://doi.org/10.1371/journal.pone.0047674
137. Yuan L, Richardson CJ, Ho M, Willis CW, Colman BP, Wiesner MR (2018) Stress responses of aquatic plants to silver nanoparticles. Environ Sci Technol 52:2558–2565
138. Zhang C, Hu Z, Li P, Gajaraj S (2016a) Governing factors affecting the impacts of silver nanoparticles on wastewater treatment. Sci Total Environ 572:852–873. https://doi.org/10.1016/j.scitotenv.2016.07.145
139. Zhang XF, Liu ZG, Shen W, Gurunathan S (2016b) Silver nanoparticles synthesis characterization properties applications and therapeutic approaches. Int J Mol Sci 17(9):1534. https://doi.org/10.3390/ijms17091534
140. Zou X, Li P, Huang Q, Zhang H (2016) The different response mechanisms of *Wolffia globosa*: light-induced silver nanoparticle toxicity. Aquat Toxicol 176:97–105. https://doi.org/10.1016/j.aquatox.2016.04.019
141. Zuverza-Mena N, Armendariz R, Peralta-Videa JR, Gardea-Torresdey JL (2016) Effects of silver nanoparticles on radish sprouts: root growth reduction and modifications in the nutritional value. Front Plant Sci 7:90

Proteomic Study on the Effects of Silver Nanoparticles Under Abiotic Stress

Ali A. Badawy, Azamal Husen, and Salem S. Salem

Abstract Numerous abiotic stressors, including drought, salt, high temperatures, floods, and heavy metal stress, can affect plants. These stressors result in financial losses and have a major detrimental impact on plant development and productivity. Consequently, novel strategies like nanotechnology are applied to lessen the detrimental consequences of these pressures on plants. Agricultural nanotechnology seeks to reduce environmental pollution and the impacts of abiotic stressors, enhance plant disease resistance, maximize water efficiency, and enhance sustainability in agriculture. By stimulating enzyme activity and bacterial adherence to the roots under abiotic stress conditions, nanoparticles help plants overcome nutritional deficits and become more resilient to environmental stressors. Proteomics is a growing discipline of research that gives an understanding of organisms at the cellular level. Moreover, proteomics examines the regulation and functions of biological systems based on analysis of the protein expression profile. This chapter explains a proteome analysis of plant organelles which generates ample information about the intrinsic mechanisms of plant stress responses against nanoparticle exposure.

Keywords Silver nanoparticles · Plants · Abiotic stress · Proteomics study · Agriculture

A. A. Badawy · S. S. Salem (✉)
Botany and Microbiology Department, Faculty of Science, Al-Azhar University, Cairo 11884, Egypt
e-mail: salemsalahsalem@azhar.edu.eg

A. Husen
Wolaita Sodo University, PO Box 138, Wolaita, Ethiopia

Department of Biotechnology, Graphic Era (Deemed to be University), Dehradun 248002, Uttarakhand, India

A. Husen (ed.), *Plant Response to Silver Nanoparticles*, Smart Nanomaterials Technology, https://doi.org/10.1007/978-981-97-7352-7_5

1 Introduction

The main necessities for optimal plant growth, development, and reproduction are light, water, carbon, and mineral nutrients. Because plants are immobile, they face a wide range of environmental stresses as well as stresses caused by living organisms from which they cannot escape [2, 15]. Extreme conditions (below or above ideal levels) limit plant growth and development. Drought, low or high temperatures, salinity, acidic conditions, heavy metal stress, and so on are the major abiotic stresses that harm plants [13, 17, 28–33, 52].

While plants suffer from nutrient deficiencies and lose normal physiological patterns of growth and development due to varying levels of early acidic conditions, exposure to salinity can cause ion toxicity [45, 53], and drought can reduce germination rates, causing suppression in photosynthesis, loss of membrane integrity and increased generation of reactive oxygen species [22, 27]. Long-term water stress can lead to decreased leaf water potential and stomata opening, reduced leaf size, inhibition of root growth, reduced seed number, size, and vigor, delayed flowering and fruiting, and limited plant growth and productivity [26, 89]. While high salinity and drought are the main causes of osmotic stress in plants. High temperatures can cause extensive denaturation and aggregation of cellular proteins, leading to cell death if not controlled. Low temperatures impair metabolic processes by altering membrane properties, altering protein structure and intermolecular interactions, and inhibiting enzymatic reactions [83]. Heavy metals can interfere with many biochemical and physiological functions, such as photosynthesis, respiration, nitrogen and protein metabolism, etc. [90]. Biological stress causes mainly plant diseases. Microorganisms can cause plant wilt, leaf spot, root rot, seed damage, and many other problems [9]. Insects can cause severe physical damage to plants, including leaves, stems, bark, and flowers. Additionally, it can serve as vectors of viruses and bacteria from infected plants to healthy plants [3, 8, 4]. Invasive plants cause severe damage to native plants through allelopathy.

The use of nanoparticles is an innovative and popular approach to address plant stress, especially abiotic stress [11, 19, 78, 79]. The application of nanomaterials in terms of nanoparticles is an essential component of nanotechnology. These nanoparticles range in size from 1 to 100 nm and have very high surface area to volume ratio and reactivity [20, 21, 60, 62–67, 72, 74]. Nanomaterials vary in size, type, composition, surface properties, and reactivity, leading to changes in the physiological and biochemical properties of the rest of the nanomaterial [10, 73, 71, 64]. Nanotechnology is an emerging field of advanced agricultural technology that enhances crop growth and development [14, 34, 35, 68, 69]. Nanoparticles play a vital role in the production of agrochemicals such as nanopesticides, nanofertilizers, and organic agriculture [58, 70, 12]. The importance of nanoparticles in agricultural products (e.g., production of edible foods, nanofoods, biosensors, nanobiocomposites) as well as in crop ecology, nanobiocomposites, biofuels, and post-harvest crop management is undeniable. Nanoparticles also play a crucial role in gene delivery to improve nutritional value and enhance resistance to different types of plant stress

[18, 23]. The application of nanoparticles in agriculture is the most important tool to improve plant growth and resistance to negative stress. To explore various links between nanoparticles and plant stress resistance, scientists have conducted several studies [7]. Nanoparticles play a vital role in improving seed germination in seedlings, including plant physiological and biochemical activities such as photosynthesis, antioxidant enzyme activity, chlorophyll quality, carbohydrate content, and production. In addition, nanoparticles also play an important role in highlighting the association and effect of stress on different plant species through physiological, biochemical, and molecular mechanisms. These particles can also improve plant morphological characteristics, biochemical properties (including enzymatic and non-enzymatic antioxidants), and expression of plant stress-related genes [76]. Silver nanoparticles (AgNPs) are one of the most important nanomaterials and their role is still under investigation [55]. AgNPs have unique properties and a wide range of applications, especially in agriculture and plant biotechnology. They increased seed germination, plant growth, and photosynthetic efficiency, while also serving as safe and effective pesticides and nanofertilizers [39]. The antimicrobial features of AgNPs have been commonly used in several industries, including healthcare, textiles, medical devices, building materials, food service, and cosmetics. AgNPs are good electric and photochemical conductors and can be used in wastewater treatment and electronic device applications. It promotes fruit ripening and plant growth and is a fungicide in agriculture [61]. Figure 1 summarizes the utilization of silver nanoparticles in some plant processes.

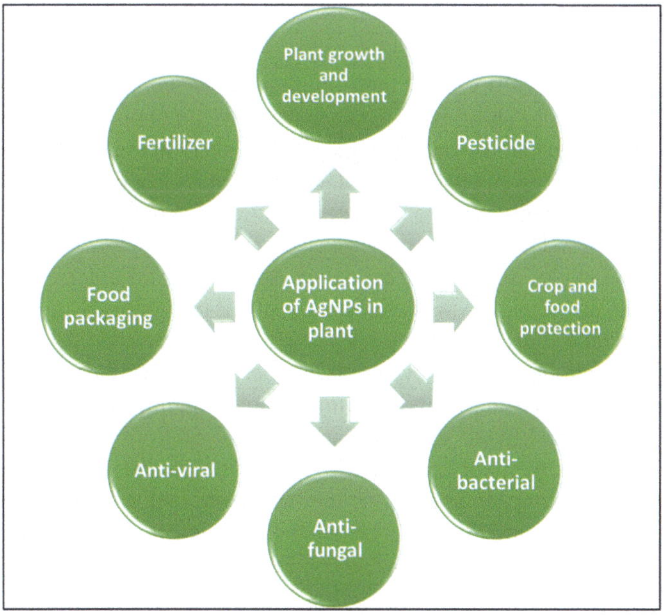

Fig. 1 Utilization of silver nanoparticles in some plant processes

2 Impact of Silver Nanoparticles on Certain Morphological Aspects

Application of AgNPs higher than 1 mg/L was detrimental to seedling growth of turnip (Brassica rapa ssp. rapa L.) plants. Seed germination rates were similar to control values while root lengths and fresh biomass were enhanced at concentrations up to 1 mg/L and continued to decline thereafter. At 10 mg/L AgNPs, seed germination rate, root lengths, and fresh biomass of turnip plants were decreased [84]. Also, the elongation of root and fresh weights of both root and shoot in rice plants were obviously suppressed in response to silver nanoparticle exposure [51].

In salinity-stressed wheat plants, priming with silver nanoparticles has markedly shifted off the inhibitory impact of salinity on the growth attributes e.g., germination percentage, lengths of plumule and radical as well as fresh and dry weights of biomass [5]. Additionally, individual application of AgNPs or chemo-blended with nicotinic acid and/or potassium nitrate caused notable enhancements in lengths of shoot and root, fresh weight of shoot and root, number of grains/spike, 100-grains weight and yield of wheat plants [36]. In a recent study on drought-stressed eggplant, the application of AgNPs on its seedlings resulted in markedly increases in plant height, leaf area, fresh and dry weights of the shoot [6].

3 Impact of Silver Nanoparticles on Certain Physiological and Metabolic Constituents

Application of AgNPs at 0.5 and 1 mg/l on *Arabidopsis thaliana* seedlings caused significant inhibition in chlorophyll content while accumulating anthocyanin, and increased lipid peroxidation [50]. In rice seedlings, A marked inhibition in the contents of sugars, chlorophylls, and carotenoids meanwhile a significant accumulation of hydrogen peroxide, proline, lipid peroxidation as a result of AgNPs (0.5 and 1 mg/l) treatments [51].

Silver nanoparticles at 1 mg/L had no marked impact on chlorophyll content, while higher levels (5 and 10 mg/L) of them resulted in significant suppressions in the contents of chlorophylls in turnip leaves. Moreover, the tested AgNPs caused a significant accumulation of anthocyanins quantities as a protective mechanism [84]. It has been documented that AgNPs priming caused significant improvements in leaf pigments in wheat plants cultivated under salt conditions [5]. Also, exposure of tobacco seedlings to silver nanoparticles at 25, 50, 75, and 100 μM resulted in elevation of reactive oxygen species, malondialdehyde (MDA) contents, enhancements in superoxide dismutase and ascorbate peroxidase activities, and suppression in pyrogallol peroxidase activity [80].

Furthermore, treating drought-stressed eggplant seedlings with silver nanoparticles significantly increased chlorophylls and protein content and augmented the activities of antioxidant enzymes while decreasing proline, malondialdehyde, and

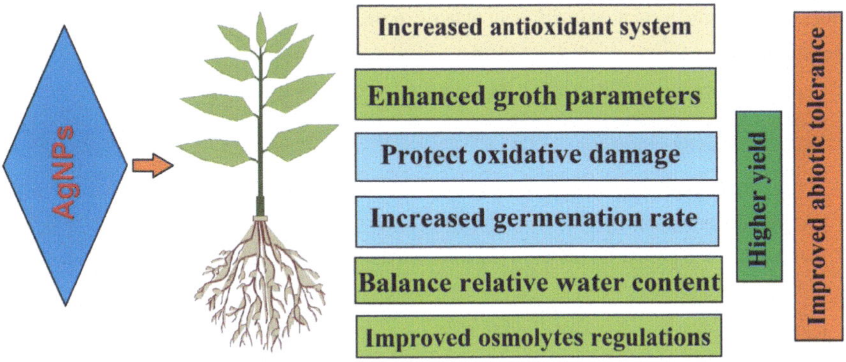

Fig. 2 The effect of silver nanoparticles on plants

hydrogen peroxide (H_2O_2) with respect to untreated seedlings [6]. Cultivating rice plants under stressful conditions and treating with AgNPs at different concentrations resulted in enhancing relative water content, membrane stability index, chlorophylls contents, activities of some antioxidant enzymes (SOD, CAT, POD), total flavonoids and phenolics contents, proline and sugars contents while contents of proteins were suppressed [81]. Recently, their application at high levels on tobacco seedlings resulted in significant boosting the contents of hydrogen peroxide and proline, induction of antioxidant enzymes activities (ascorbate peroxidase and catalase), disruption to proteins, lipids and DNA molecules, alterations in the metabolism of carbohydrates and protein [16]. Silver nanoparticles affect plants in different ways [1] (Fig. 2).

4 Impact of Silver Nanoparticles on Proteomic Level

Proteomics is an emerging discipline of research, and it pays substantially to our understanding of organisms at the cellular level. Furthermore, this is the study of functions and regulation of biological systems based on analysis of the protein expression profile [44, 57]. Proteomic responses in different plants exposed to NPs were concluded in Table 1.

Silver is found in the Earth's crust as a rare element. As a result, toxic silver ions are produced and act as bioactive molecules. They induce oxidative damage to DNA by stimulating antioxidant enzymes and reducing antioxidants [43]. Silver nanoparticles can be used in various ways in agriculture. Silver nanoparticles promote *Crocus Sativus* growth by blocking ethylene signal under stressful conditions [59]. Silver nanoparticles promote Arabidopsis growth due to inhibition of ethylene sensing and accumulation of reactive oxygen species [82]. Molecular studies show that silver nanoparticles also act as gene regulators in plants. Silver nanoparticles increased chlorophyll and protein content in beans and corn [61]. Gene expression profiling

Table 1 Proteomic responses in different plants exposed to silver nanoparticles

Protein response	Plant	References
– Improve the growth and activities of antioxidant enzymes – Improve the gene expression pattern – Improve in protein kinase signals, and transduction pathway	*Pennisetum glaucum L*	Khan et al. [38]
– Increased which proteins related to starch synthase – Decreased proteins associated with lipoxygenase – Altered proteins linked to redox and biotic stress	*Triticum aestivum*	Komatsu et al. [40], Singh et al. [77]
– Induced proteins are linked to protein synthesis and photosynthesis – Suppressed proteins attributed to signaling, glycolysis, cell wall	*Glycine max*	Hashimoto et al. [24]
– Accelerated protein degradation related proteins – Reduced protein synthesis associated proteins	*Triticum aestivum*	Jhanzab et al. [36]
– Altered abundance of root proteins involved in abiotic/biotic and oxidative stress responses – Altered leaf proteins attributed to photosynthesis	*Nicotiana tabacum*	Peharec Štefanić et al. [56]
– Decreased proteins associated with secondary metabolism, cell organization, and hormone metabolism	*Glycine max*	Hossain et al. [25]
– Altered endoplasmic reticulum and vacuolar proteins involved in sulfur metabolism	*Eruca sativa*	Vannini et al. [85]
– Changed proteins associated with signaling, cell metabolism, stress	*Triticum aestivum*	Vannini et al. [86]
– Changed proteins incorporated in cell defense and metabolism	*Glycine max*	Mustafa et al. [49]
– Inhibited protein synthesis with enhanced amino acid synthesis	*Glycine max*	Mustafa and Komatsu [46]

of AgNP-treated Arabidopsis showed that genes related to metal response, oxidative stress, and thalianol biosynthesis pathways were up-regulated, while genes related to ethylene signaling pathway were down-regulated [37].

In rice, silver nanoparticles altered the number of proteins associated with oxidative stress tolerance, as well as differential regulation of proteins associated with cell division, apoptosis, and transcription/degradation [44]. Additionally, the authors conducted a 2D gel-based proteomic study to understand the stress effects of Ag-NPs on rice (Oryza sativa). The results of the radical proteomic study indicate that Ag-NPs or ions released from cellular metabolism bind to second messenger calcium receptors, preventing calcium transport through calcium channels and calcium and sodium ATP pumps. There is marked upregulation in gene expression of glutathione biosynthesis, glutathione reductase, glutathione S-transferase, and sulfur assimilation in *Arabidopsis thaliana* due to the silver nanoparticles treatments [50].

In *Eruca sativa*, proteins related to sulfur metabolism and redox regulation were altered after exposure to silver nanoparticles [85]. Furthermore, changes in some proteins associated with endoplasmic reticulum and vacuoles indicate that these two organelles are targets for AgNPs action. AgNP induces oxidative stress, which was also confirmed by the higher expression levels of SOD and PRX mRNA obtained by semi-quantitative RT-PCR analysis. In addition, it activates common and specific elements for effective detoxification. Similar investigations were documented in *Brassica juncea, Lemna gibba, Caenorhabditis elegans,* and *Drosophila* larvae [42, 54]. Moreover, Exposure to $AgNO_3$ leads to specific increases in several proteins involved in energy production, including subunits of plastid and mitochondrial ATP synthase, carbonic anhydrase, and aconitate hydratase. This regulation may help cells generate more catabolic capacity to facilitate responses to $AgNO_3$ stress [85].

Reactive oxygen species accumulate under silver nanoparticles, causing reduced plant growth. Silver nanoparticles reduced soybean growth compared to aluminum and zinc oxide nanoparticles. In soybean leaves, decreased levels of subtilase family proteins and DAHP enzyme resulted in decreased growth of silver nanoparticles [25]. Also, the authors documented that silver nanoparticles differentially regulated proteins that mainly incorporated stress and protein degradation. Furthermore, silver nanoparticles regulate soybean growth under waterlogged conditions by regulating proteins related to amino acid metabolism and wax synthesis [48]. Silver nanoparticles are mixed with nicotinic acid and potassium nitrate to enhance soybean growth by controlling the protein quality of unfolded proteins [24]. In wheat, silver nanoparticles promote growth by altering plant metabolism [36].

A study by Komatsu et al. [41] reported the regulation of genes associated with alcohol fermentation and ethylene biosynthesis in soybean seedlings under flood stress. Although ethylene biosynthesis in many plant species is inhibited by silver nanoparticle treatment, the effects of ethylene on flood-stressed soybeans are still not well understood. Trypsin and protease inhibitor proteins were increased in AgNP-treated soybean roots under flood stress. However, these proteins are generally not present in high concentrations in soybean, suggesting that they may be involved in growth inhibition in response to flood stress [47].

In addition to these proteins, we identified three $AgNO_3$-responsive proteins whose roles in mineral responses are less well known: class I glutamine amidases, tryptophan synthase alpha chain, and the Metacaspase-4 Base p subunit. Transgenic Arabidopsis plants with elevated levels of class I glutamate transferase enhanced stress protection against environmental stress conditions through activation of cytosolic SOD [88]. Overexpression of the tryptophan synthase beta chain and increased tryptophan levels in *Arabidopsis* plants reduced lipid peroxidation and enhanced Cd tolerance [75]. Arabidopsis meta-caspase 2d is a positive mediator of cell death induced by biotic and abiotic stress [87]. In a recent study, expression of proteins that related to stress response and tolerance in tobacco plants were markedly differentiated due to AgNPs exposure [16].

5 Conclusion

Various abiotic stressors can negatively impact crop development and lower crop plant yields in agricultural crops. The current methods of managing agricultural stress are inadequate to meet the food needs of the world's population, which is expected to grow to 10 billion people by the year 2050. Utilizing nanotechnology in biological domains, Nanobiotechnology has surfaced as a viable method for augmenting agricultural output by mitigating diverse plant stressors. This chapter discusses advancements in nanobiotechnology, their function in stimulating plant development, and the underlying processes that improve plant tolerance and resilience to abiotic challenges. This chapter also explained a detailed proteome analysis of plant organelles which generates comprehensive information about the intrinsic mechanisms of plant stress responses toward nanoparticles exposure.

References

1. Abasi F, Raja NI, Mashwani ZUR, Amjad MS, Ehsan M, Mustafa N, Haroon M, Proćków J (2022) Biogenic silver nanoparticles as a stress alleviator in plants: a mechanistic overview. Molecules 27(11):3378
2. Abdel Latef AAH, Omer AM, Badawy AA, et al (2021) Strategy of salt tolerance and interactive impact of Azotobacter chroococcum and/or Alcaligenes faecalis inoculation on canola (Brassica napus L.) plants grown in saline soil. Plants 10:110
3. Abdelfattah NAH, Yousef MA, Badawy AA, Salem SS (2023) Influence of biosynthesized magnesium oxide nanoparticles on growth and physiological aspects of cowpea (Vigna unguiculata L.) plant, cowpea beetle, and cytotoxicity. Biotechnol J e2300301:1–12. https://doi.org/10.1002/biot.202300301
4. Abd-Elkhalek HF, Badawy AA, Al-Askar AA, Abd Elgawad H, Hashem AH, Salem SS (2024) Biosynthesis and characterization of selenium and silver nanoparticles using Trichoderma viride filtrate and their impact on Culex pipiens. Green Process Synth 13(1):20240025. https://doi.org/10.1515/gps-2024-0025
5. Abou-Zeid HM, Ismail GSM (2018) The role of priming with biosynthesized silver nanoparticles in the response of Triticum aestivum L. to salt stress. Egypt J Bot 58:73–85. https://doi.org/10.21608/ejbo.2017.1873.1128
6. Alabdallah NM, Hasan MM, Salih AM, et al (2021) Silver nanoparticles improve growth and protect against oxidative damage in eggplant seedlings under drought stress. Plant, Soil Environ 67:617–624. https://doi.org/10.17221/323/2021-PSE
7. Ali EF, El-Shehawi AM, Ibrahim OHM et al (2021) A vital role of chitosan nanoparticles in improvisation the drought stress tolerance in Catharanthus roseus (L.) through biochemical and gene expression modulation. Plant Physiol Biochem 161:166–175

8. Amin MA, Ismail MA, Badawy AA et al (2021) The potency of fungal-fabricated selenium nanoparticles to improve the growth performance of Helianthus annuus L. and control of cutworm Agrotis ipsilon. Catalysts 11:1551
9. Attia MS, Hashem AH, Badawy AA, Abdelaziz AM (2022) Biocontrol of early blight disease of eggplant using endophytic Aspergillus terreus : improving plant immunological, physiological and antifungal activities. Bot Stud 63:26. https://doi.org/10.1186/s40529-022-00357-6
10. Attia MS, Salem SS, Elakraa AA, Abdel-Maksoud MA, Malik A, Kiani BH, Malash MN, El-Sayyad GS (2024) Promising antagonistic effect of bimetallic silver-selenium nanoparticles against Ralstonia solanacearum-causing wilt disease in eggplant (Solanum melongena L.). Physiol Mol Plant Pathol 133:102369. https://doi.org/10.1016/j.pmpp.2024.102369
11. Badawy AA, Abdelfattah NAH, Salem SS et al (2021a) Efficacy assessment of biosynthesized copper oxide nanoparticles (CuO-NPs) on stored grain insects and their impacts on morphological and physiological traits of wheat (Triticum aestivum L.) plant. Biology (Basel) 10:233
12. Badawy AA, Husen A, Salem SS (2024) Use of nanobiotechnology in augmenting soil–plant system interaction for higher plant growth and production. In: Essential minerals in plant-soil systems. Jan 1 (pp 423–443). Elsevier. https://doi.org/10.1016/B978-0-443-16082-0.00006-0
13. Bachheti A, Deepti, Bachheti RK, Husen A (2021) Medicinal plants and their pharmaceutical properties under adverse environmental conditions. In: Husen A (ed) Harsh environment and plant resilience. Springer International Publishing AG, Gewerbestrasse 11, 6330 Cham, pp 457–502. https://doi.org/10.1007/978-3-030-65912-7_19
14. Bachheti RK, Bachheti A, Husen A (2024) Metal and metal-oxide based nanomaterials (Synthesis, Agricultural, Biomedical and Environmental Interventions). Springer Nature Singapore Pte Ltd., 152 Beach Road, #21-01/04 Gateway East, Singapore 189721, Singapore
15. Badawy AA, Alotaibi MO, Abdelaziz AM et al (2021) Enhancement of seawater stress tolerance in barley by the endophytic fungus Aspergillus ochraceus. Metabolites 11:428
16. Biba R, Cvjetko P, Tkalec M et al (2022) Effects of silver nanoparticles on physiological and proteomic responses of tobacco (Nicotiana tabacum) seedlings are coating-dependent. Int J Mol Sci 23:15923. https://doi.org/10.3390/ijms232415923
17. Chaves MM, Oliveira MM (2004) Mechanisms underlying plant resilience to water deficits: prospects for water-saving agriculture. J Exp Bot 55:2365–2384
18. Cunningham FJ, Goh NS, Demirer GS et al (2018) Nanoparticle-mediated delivery towards advancing plant genetic engineering. Trends Biotechnol 36:882–897
19. de la Rosa G, Garcia-Castañeda C, Vázquez-Núñez E et al (2017) Physiological and biochemical response of plants to engineered NMs: implications on future design. Plant Physiol Biochem 110:226–235
20. Dezfuli AAZ, Abu-Elghait M, Salem SS (2023) Recent insights into nanotechnology in colorectal cancer. Appl Biochem Biotechnol. https://doi.org/10.1007/s12010-023-04696-3
21. El-Khawaga AM, Elsayed MA, Gobara M, Soliman AA, Hashem AH, Zaher AA, Mohsen M, Salem SS (2023) Green synthesized ZnO nanoparticles by *Saccharomyces cerevisiae* and their antibacterial activity and photocatalytic degradation. Biomass Convers Bioref. https://doi.org/10.1007/s13399-023-04827-0
22. Greenberg B, Huang XD, Gerwing P et al (2010) Phytoremediation of salt impacted soils: greenhouse and field trials using plant growth promoting rhizobacteria (PGPR) to improve plant growth and salt. In: Phytoaccumulation, proceedings of the 33rd AMOP technical seminar on environmental contamination and response. Environment Canada, Ottawa
23. Hashem AH, Abdelaziz AM, Attia MS, Salem SS (2022). Selenium and nano-selenium-mediated biotic stress tolerance in plants. In: Hossain MA, Ahammed GJ, Kolbert Z, El-Ramady H, Islam T, Schiavon M (eds) Selenium and nano-selenium in environmental stress management and crop quality improvement. Sustainable plant nutrition in a changing world. Springer, Cham. https://doi.org/10.1007/978-3-031-07063-1_11
24. Hashimoto T, Mustafa G, Nishiuchi T, Komatsu S (2020) Comparative analysis of the effect of inorganic and organic chemicals with silver nanoparticles on soybean under flooding stress. Int J Mol Sci 21:1300

25. Hossain Z, Mustafa G, Sakata K, Komatsu S (2016) Insights into the proteomic response of soybean towards Al2O3, ZnO, and Ag nanoparticles stress. J Hazard Mater 304:291–305
26. Hussein H-AA, Alshammari SO, Kenawy SKM et al (2022) Grain-priming with L-arginine improves the growth performance of wheat (Triticum aestivum L.) plants under drought stress. Plants 11:1219. https://doi.org/10.3390/PLANTS11091219
27. Hussein H-AA, Alshammari SO, Abd El-sadek ME et al (2023) The promotive effect of putrescine on growth, biochemical constituents, and yield of wheat (Triticum aestivum L.) plants under water stress. Agriculture 13:587
28. Husen A (2021a) The Harsh environment and resilient plants: an overview In: Husen A (ed) Harsh environment and plant resilience. Springer International Publishing AG, Gewerbestrasse 11, 6330 Cham, pp 1–23. https://doi.org/10.1007/978-3-030-65912-7_1
29. Husen A (2021b) Morpho-anatomical, physiological, biochemical and molecular responses of plants to air pollution. In: Husen A (ed) Harsh environment and plant resilience. Springer International Publishing AG, Gewerbestrasse 11, 6330 Cham, pp 203–234. https://doi.org/10.1007/978-3-030-65912-7_9
30. Husen A (2021c) Cross talk between autophagy and hormones for abiotic stress tolerance in plants. In: Husen A (ed) Plant performance under environmental stress. Springer International Publishing AG, Gewerbestrasse 11, 6330 Cham, pp 1–15. https://doi.org/10.1007/978-3-030-78521-5_1
31. Husen A (2021d) Harsh environment and plant resilience (Molecular and Functional Aspects). Springer Nature Switzerland AG, Gewerbestrasse 11, 6330 Cham, Switzerland. https://doi.org/10.1007/978-3-030-65912-7
32. Husen A (2021e) Plant performance under environmental stress (Hormones, Biostimulants and Sustainable Plant Growth Management). Springer Nature Switzerland AG, Gewerbestrasse 11, 6330 Cham, Switzerland. https://doi.org/10.1007/978-3-030-78521-5
33. Husen A, Iqbal M (2023) Medicinal plants—Their response to abiotic stress. Springer Nature Singapore Pte Ltd., 152 Beach Road, #21-01/04 Gateway East, Singapore 189721, Singapore. https://doi.org/10.1007/978-981-19-5611-9
34. Husen A (2022) Engineered nanomaterials for sustainable agricultural production, soil improvement and stress management. Elsevier Inc., 50 Hampshire St., 5th Floor, Cambridge, MA 02139, USA. https://doi.org/10.1016/C2021-0-00054-7
35. Husen A (2023) Nanomaterials and nanocomposites exposures to plants (Response, Interaction, Phytotoxicity and Defense Mechanisms). Springer Nature Singapore Pte Ltd., 152 Beach Road, #21-01/04 Gateway East, Singapore 189721, Singapore. https://doi.org/10.1007/978-981-99-2419-6
36. Jhanzab HM, Razzaq A, Bibi Y et al (2019) Proteomic analysis of the effect of inorganic and organic chemicals on silver nanoparticles in wheat. Int J Mol Sci 20:825
37. Kaveh R, Li Y-S, Ranjbar S et al (2013) Changes in Arabidopsis thaliana gene expression in response to silver nanoparticles and silver ions. Environ Sci Technol 47:10637–10644
38. Khan I, Awan SA, Rizwan M, Akram MA, Zia-ur-Rehman M, Wang X, Zhang X, Huang L (2023) Physiological and transcriptome analyses demonstrate the silver nanoparticles mediated alleviation of salt stress in pearl millet (Pennisetum glaucum L). Environ Pollut 318:120863
39. Khan S, Zahoor M, Sher Khan R et al (2023) The impact of silver nanoparticles on the growth of plants: the agriculture applications. Heliyon 9:e16928. https://doi.org/10.1016/j.heliyon.2023.e16928
40. Komatsu S, Yamaguchi H, Hitachi K, Tsuchida K (2022) Proteomic, biochemical, and morphological analyses of the effect of silver nanoparticles mixed with organic and inorganic chemicals on wheat growth. Cells 11:1–15. https://doi.org/10.3390/cells11091579
41. Komatsu S, Yamamoto R, Nanjo Y et al (2009) A comprehensive analysis of the soybean genes and proteins expressed under flooding stress using transcriptome and proteome techniques. J Proteome Res 8:4766–4778
42. Lim D, Roh J, Eom H et al (2012) Oxidative stress-related PMK-1 P38 MAPK activation as a mechanism for toxicity of silver nanoparticles to reproduction in the nematode Caenorhabditis elegans. Environ Toxicol Chem 31:585–592

43. McShan D, Ray PC, Yu H (2014) Molecular toxicity mechanism of nanosilver. J Food Drug Anal 22:116–127
44. Mirzajani F, Askari H, Hamzelou S et al (2014) Proteomics study of silver nanoparticles toxicity on Oryza sativa L. Ecotoxicol Environ Saf 108:335–339
45. Munns R, Tester M (2008) Mechanisms of salinity tolerance. Annu Rev Plant Biol 59:651–681. https://doi.org/10.1146/annurev.arplant.59.032607.092911
46. Mustafa G, Komatsu S (2016) Insights into the response of soybean mitochondrial proteins to various sizes of aluminum oxide nanoparticles under flooding stress. J Proteome Res 15:4464–4475
47. Mustafa G, Sakata K, Hossain Z, Komatsu S (2015a) Proteomic study on the effects of silver nanoparticles on soybean under flooding stress. J Proteomics 122:100–118. https://doi.org/10.1016/j.jprot.2015.03.030
48. Mustafa G, Sakata K, Komatsu S (2016) Proteomic analysis of soybean root exposed to varying sizes of silver nanoparticles under flooding stress. J Proteomics 148:113–125
49. Mustafa G, Sakata K, Komatsu S (2015b) Proteomic analysis of flooded soybean root exposed to aluminum oxide nanoparticles. Elsevier B.V.
50. Nair PMG, Chung IM (2014) Assessment of silver nanoparticle-induced physiological and molecular changes in Arabidopsis thaliana. Environ Sci Pollut Res 21:8858–8869. https://doi.org/10.1007/s11356-014-2822-y
51. Nair PMG, Chung IM (2014) Physiological and molecular level effects of silver nanoparticles exposure in rice (Oryza sativa L.) seedlings. Chemosphere 112:105–113. https://doi.org/10.1016/j.chemosphere.2014.03.056
52. Omer AM, Osman MS, Badawy AA (2022) Inoculation with Azospirillum brasilense and/or Pseudomonas geniculata reinforces flax (Linum usitatissimum) growth by improving physiological activities under saline soil conditions. Bot Stud 63:15. https://doi.org/10.1186/s40529-022-00345-w
53. Osman MS, Badawy AA, Osman AI, Abdel Latef AAH (2021) Ameliorative impact of an extract of the halophyte Arthrocnemum macrostachyum on growth and biochemical parameters of soybean under salinity stress. J Plant Growth Regul 40:1245–1256. https://doi.org/10.1007/s00344-020-10185-2
54. Oukarroum A, Barhoumi L, Pirastru L, Dewez D (2013) Silver nanoparticle toxicity effect on growth and cellular viability of the aquatic plant Lemna gibba. Environ Toxicol Chem 32:902–907
55. Parisi C, Vigani M, Rodriguez-Cerezo E (2015) Agricultural nanotechnologies: what are the current possibilities? Nano Today 10:124–127
56. Peharec Štefanić P, Jarnević M, Cvjetko P et al (2019) Comparative proteomic study of phytotoxic effects of silver nanoparticles and silver ions on tobacco plants. Environ Sci Pollut Res 26:22529–22550. https://doi.org/10.1007/s11356-019-05552-w
57. Pokhrel LR, Dubey B (2013) Evaluation of developmental responses of two crop plants exposed to silver and zinc oxide nanoparticles. Sci Total Environ 452:321–332
58. Qian X, Ma J, Weng L et al (2020) Influence of agricultural organic inputs and their aging on the transport of ferrihydrite nanoparticles: from enhancement to inhibition. Sci Total Environ 719:137440
59. Rezvani N, Sorooshzadeh A, Farhadi N (2012) Effect of nano-silver on growth of saffron in flooding stress. Int J Agric Biosyst Eng 6:11–16
60. Said A, Abu-Elghait M, Atta HM, Salem SS (2023) Antibacterial activity of green synthesized silver nanoparticles using Lawsonia inermis against common pathogens from urinary tract infection. Appl Biochem Biotechnol. https://doi.org/10.1007/s12010-023-04482-1
61. Salama HMH et al (2012) Effects of silver nanoparticles in some crop plants, common bean (Phaseolus vulgaris L.) and corn (Zea mays L.). Int Res J Biotechnol 3:190–197
62. Salem SS (2022) Baker's yeast-mediated silver nanoparticles: characterisation and antimicrobial biogenic tool for suppressing pathogenic microbes. BioNanoScience 12(4):1220–1229. https://doi.org/10.1007/s12668-022-01026-5

63. Salem SS (2022) Bio-fabrication of selenium nanoparticles using baker's yeast extract and its antimicrobial efficacy on food borne pathogens. Appl Biochem Biotechnol 194(5):1898–1910. https://doi.org/10.1007/s12010-022-03809-8
64. Salem SS (2023) A mini review on green nanotechnology and its development in biological effects. Arch Microbiol 205:128. https://doi.org/10.1007/s00203-023-03467-2
65. Salem SS (2023) Application of nano-materials. In: Raja R, Hemaiswarya S, Narayanan M, Kandasamy S, Jayappriyan K (eds) Haematococcus. Springer, Singapore. https://doi.org/10.1007/978-981-99-2901-6_10
66. Salem SS, Fouda A (2021) Green synthesis of metallic nanoparticles and their prospective biotechnological applications: an overview. Biol Trace Elem Res 199(1):344–370. https://doi.org/10.1007/s12011-020-02138-3
67. Salem SS, Hammad EN, Mohamed AA, El-Dougdoug W (2023) A comprehensive review of nanomaterials: types, synthesis, characterization, and applications. Biointerf Res Appl Chem 13(1). https://doi.org/10.33263/BRIAC131.041
68. Salem SS, Husen A (2022) Effect of engineered nanomaterials on soil microbiomes and their association with crop growth and production. In: Engineered nanomaterials for sustainable agricultural production, soil improvement and stress management. pp 311–336. https://doi.org/10.1016/B978-0-323-91933-3.00010-6
69. Salem SS, Husen A (2023) Zinc oxide nanoparticles: plant response, interaction, phytotoxicity, and defence. In: Husen A (ed) Nanomaterials and nanocomposites exposures to plants. Smart nanomaterials technology. Springer, Singapore. https://doi.org/10.1007/978-981-99-2419-6_12
70. Salem SS, Mekky AE (2024) Biogenic nanomaterials: synthesis, characterization, and applications. In: Shah MP, Bharadvaja N, Kumar L (eds) Biogenic nanomaterials for environmental sustainability: principles, practices, and opportunities. Environmental science and engineering. Springer, Cham. https://doi.org/10.1007/978-3-031-45956-6_2
71. Salem SS, Saied E, Shah MP (2024) Advanced (nano) materials. In: Development in wastewater treatment research and processes2024 Jan 1 (pp 93–115). Elsevier. https://doi.org/10.1016/B978-0-323-99278-7.00011-0
72. Soliman MKY, Abu-Elghait M, Salem SS, Azab MS (2022) Multifunctional properties of silver and gold nanoparticles synthesis by *Fusarium pseudonygamai*. Biomass Convers Biorefin. https://doi.org/10.1007/s13399-022-03507-9
73. Soliman MK, Hashem AH, Al-Askar AA, AbdElgayed G, Salem SS (2024) Green synthesis of silver nanoparticles from Bauhinia variegata and their biological applications. Green Process Synth 13(1):20240099. https://doi.org/10.1515/gps-2024-0099
74. Soliman MKY, Salem SS, Abu-Elghait M, Azab MS (2023) Biosynthesis of silver and gold nanoparticles and their efficacy towards antibacterial, antibiofilm, cytotoxicity, and antioxidant activities. Appl Biochem Biotechnol 195(2):1158–1183. https://doi.org/10.1007/s12010-022-04199-7
75. Sanjaya, Hsiao P-Y, Su R-C et al (2008) Overexpression of Arabidopsis thaliana tryptophan synthase beta 1 (AtTSB1) in Arabidopsis and tomato confers tolerance to cadmium stress. Plant, Cell Environ 31:1074–1085
76. Silveira NM, Seabra AB, Marcos FCC et al (2019) Encapsulation of S-nitrosoglutathione into chitosan nanoparticles improves drought tolerance of sugarcane plants. Nitric Oxide 84:38–44
77. Singh P, Arif Y, Miszczuk E et al (2022) Specific roles of lipoxygenases in development and responses to stress in plants. Plants 11:979
78. Singh S, Husen A (2019) Role of nanomaterials in the mitigation of abiotic stress in plants, In: Husen A, Iqbal M (eds) Nanomaterials and plant potential. Springer International Publishing AG, Gewerbestrasse 11, 6330 Cham, pp 441–471. https://doi.org/10.1007/978-3-030-05569-1_18
79. Singh S, Husen A (2020) Behavior of agricultural crops in relation to nanomaterials under adverse environmental conditions, In: Husen, A, Jawaid M (eds) Nanomaterials for agriculture and forestry applications. Elsevier Inc. 50 Hampshire St., 5th Floor, Cambridge, MA 02139, USA, pp 219–256. https://doi.org/10.1016/B978-0-12-817852-2.00009-3

80. Štefanić PP, Cvjetko P, Biba R et al (2018) Physiological, ultrastructural and proteomic responses of tobacco seedlings exposed to silver nanoparticles and silver nitrate. Chemosphere 209:640–653. https://doi.org/10.1016/j.chemosphere.2018.06.128
81. Sultana T, Javed B, Raja NI, Zia-Ur-Rehman M (2021) Silver nanoparticles elicited physiological, biochemical, and antioxidant modifications in rice plants to control Aspergillus flavus. Green Process Synth 10:314–324. https://doi.org/10.1515/gps-2021-0034
82. Syu Y, Hung J-H, Chen J-C, Chuang H (2014) Impacts of size and shape of silver nanoparticles on Arabidopsis plant growth and gene expression. Plant Physiol Biochem 83:57–64
83. Theocharis A, Clément C, Barka EA (2012) Physiological and molecular changes in plants grown at low temperatures. Planta 235:1091–1105
84. Thiruvengadam M, Gurunathan S, Chung I-M (2015) Physiological, metabolic, and transcriptional effects of biologically-synthesized silver nanoparticles in turnip (Brassica rapa ssp. rapa L.). Protoplasma 252:1031–1046. https://doi.org/10.1007/s00709-014-0738-5
85. Vannini C, Domingo G, Onelli E et al (2013) Morphological and proteomic responses of Eruca sativa exposed to silver nanoparticles or silver nitrate. PLoS ONE 8:e68752
86. Vannini C, Domingo G, Onelli E et al (2014) Phytotoxic and genotoxic effects of silver nanoparticles exposure on germinating wheat seedlings. J Plant Physiol 171:1142–1148. https://doi.org/10.1016/j.jplph.2014.05.002
87. Watanabe N, Lam E (2011) Arabidopsis metacaspase 2d is a positive mediator of cell death induced during biotic and abiotic stresses. Plant J 66:969–982
88. Xu XM, Lin H, Maple J et al (2010) The Arabidopsis DJ-1a protein confers stress protection through cytosolic SOD activation. J Cell Sci 123:1644–1651
89. Xu Z, Jiang Y, Jia B, Zhou G (2016) Elevated-CO2 response of stomata and its dependence on environmental factors. Front Plant Sci 7:657
90. Zhang F, Zhang H, Wang G et al (2009) Cadmium-induced accumulation of hydrogen peroxide in the leaf apoplast of Phaseolus aureus and Vicia sativa and the roles of different antioxidant enzymes. J Hazard Mater 168:76–84

Silver Nanocomposite for Improved Plant-Soil System and Underlying Mechanism

Mohamed A. Amin, Tarek M. Abdelghany, and Abdelatti I. Nowwar

Abstract Global agricultural systems have faced several, unforeseen, and inescapable problems over the last ten years. Because of the unique properties of nanoparticles, including their high surface area, concentrated distribution size, reactivity, flexibility in shape, superior biocompatibility, and stability of agrochemicals, nanotechnology has provided a wide range of roles in crop production. Because of their many uses, nanoparticles can be employed as fertilizers, herbicides, insecticides, or even as a miracle drug that targets particular plant cellular organelles. Silver nanocomposites are a mixture of silver nanoparticles and one or more different substances, furthermore, one or more of them have diameters in nanometers. Silver nanocomposite (Ag NCs) can have either negative or positive impacts on plant growth, depending on the concentration, and size of the particles as well as the species of plant. Plant species come into contact with Ag NCs by foliar application, irrigation, or direct soil application. Individual NMs are not as advantageous as NCs, and plants could benefit greatly from increased nutrient availability and photosynthetic pigmentation thanks to nanocomposites. These nanocomposites readily release nutrients in the form of soluble ions. Numerous scientists have clarified the effects of silver nanoparticles on the germination of seeds and the growth and development of plants in soil, but little studies focused on the use of silver nanocomposite so, the current chapter emphasized the functions of Ag NC for improved plant-soil traits, control of plant pathogens, seed germination and root system anatomy, absorption, translocation and mode of action of Ag NC in plants.

Keywords Silver · Nanocomposites · Plant · Soil system · Mechanism

M. A. Amin (✉) · T. M. Abdelghany · A. I. Nowwar
Botany and Microbiology Department, Faculty of Science, Al-Azhar University, Nasr City, Cairo 11884, Egypt
e-mail: mamin7780@azhar.edu.eg

A. Husen (ed.), *Plant Response to Silver Nanoparticles*, Smart Nanomaterials Technology, https://doi.org/10.1007/978-981-97-7352-7_6

1 Introduction

We're going to examine about silver nanocomposites (NCs) for improved plant and soil systems in this chapter, at first silver looks like the metal that aquatic life thinks is more deadly than mercury. Additionally, silver nanoparticles (AgNPs) are extremely lethal to living things and bioaccumulative. As a result, serious questions are made regarding the environmental toxicity and safety of AgNPs injected into ecosystems, since plants constitute the primary trophic level in ecosystems and a crucial component of the food chain [41], but AgNPs are less hazardous than to silver ions for many plants for an instant, tobacco plants [20]. Nanomaterials (NMs) based on their chemical compositions, have been categorized as carbon, inorganic, organic, and nanocomposite (NCs). NCs are created when two or more different substances (with various physical and chemical properties) are combined, furthermore one or more of the phases of NCs have diameters in nanometers. NCs are created by combining clays, fullerence, and metals with matrix components to display new and upgraded characteristics as well as structure, specifically, fouling resistance, electrical, ionic transport, and ultra-hydrophobicity/hydrophilicity [59]. The interfacial features, shape, and materials employed all influence the physical characteristics of NCs [74]. In another study, polymers are combined with inorganic solids (such as oxides or clays) to generate NCs known as heterogeneous/hybrid substances at the nanometric scale. It is discovered that their structures are more intricate than those of microcomposites. The structure, composition, interfacial interactions, and constituents of individual property have a significant impact on their effect. The most popular technique for producing NCs is the in situ development and polymerization of a biopolymer and inorganic matrix. These strikingly promising materials are expected to become highly in demand quickly, which will make them extremely valuable in a variety of businesses with production facilities ranging from tiny to very large [78]. Metallic, non-matellalic, or polymeric components can be utilized as matrix materials in the preparation of nanocomposite. Based on the matrix structures, three types of NCs may be distinguished: those with a metal matrix, such as Co/Cr and Fe-MgO; those with a polymer matrix, such as polymer/TiO_2 and polymer/CNT; and those with a ceramic matrix, such as Al_2O_3/TiO_2 and Al_2O_3/ CNT (ceramic/carbon nanotube)(Fig. 1) [65]. The first type, Ceramic matrix composites are materials that have one or more separate ceramic phases purposefully added to them to improve their chemical and thermal stability as well as wear resistance. Conversely, the primary drawback of ceramics is their brittleness and low toughness, which prevents them from being used in many industrial applications. The creation of ceramic-matrix NCs (CMNCs) gets around this restriction. A matrix containing energy-dissipating elements (fiber, platelets, or particles) added to the ceramic matrix to reduce brittleness and increase fracture durability is an example of CMNCs [35]. The second type, polymer matrix nanocomposites (PMNC) are often fillers, also known as nanofillers, which are categorized as 1D (linear, like carbon nanotubes), 2D (layered, like montmorillonite), and 3D (powder). For example, silver nanoparticles [63]. An impact of attraction between nanocomposites is seen as a result of the molecular-level interaction between the polymer matrix and nanofiler. As a result, changing the matrix's composition with a tiny quantity of nanofiller particles smaller than 100 nm modifies the characteristics of the composite material [5]. The third type, metal matrix nanocomposites (MMNCs) are multiphase materials made of an alloy or ductile metal matrix with a small amount of nanoscale reinforcing material implanted in it. High ductility, toughness, strength, and modulus are characteristics of MMNC. Also, nanocomposite (NCs) can be categorized based on their matrix, for example, zeolite, biochar, bimetallic and chitosan-based. Chitosan has a strong predilection for metals because it contains free amine groups throughout its polymeric structure. It can be combined with other elements to create NCs materials [73]. NCs and mixtures of NMs use to improve crop protection and production. The application of NCs can have either negative or positive impacts on plant growth, depending on the kind, concentration, and size of the particles as well as the species of plant. Plant species come into contact with NCs by foliar application, irrigation, or direct soil application [6, 29]. NCs have emerged as a crucial part of NMs for examining and promoting the life of plants cycle, because of their inherent, special thermal, mechanical, and chemical properties [22, 25].

A silver nanocomposite is created by modifying the AgNPs with graphene, metal, fiber, ceramic, and polymeric components. This enhances the material's antibacterial, electrical conductivity, chemical durability, photocatalysis, and other properties. Their applications span a wide range of industries, including medical, environmental,

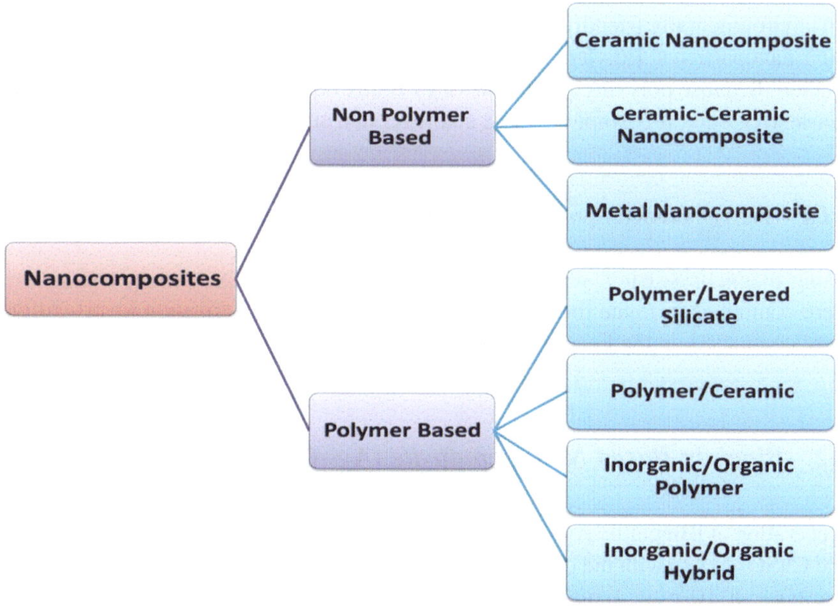

Fig. 1 Types of nanocomposites

industrial, biological, and food. They serve as a valuable resource for comprehensive studies on the characteristics of nano-silver particles and the ongoing advancement of their potential applications [75]. Individual NMs are not as advantageous as NCs, and plants could benefit greatly from increased nutrient availability thanks to nanocomposites. These nanocomposites readily release nutrients in the form of soluble ions [36]. According to McWilliams, the global market for NCs was valued at US $2.0 billion in 2017 and is projected to increase to US $7.3 billion by 2022, with a compound annual growth rate (CAGR) of 29.5% [56].

2 Preparation of Silver Nanocomposite

There are several nanocomposites of silver were created using organic and inorganic compounds. Moreover, numerous techniques were applied under different conditions for silver NCs formulation.

2.1 Metal-Based Nanocomposite

2.1.1 Preparation of Bimetallic Ag^+/Cu^{2+} Nanocomposite

An environmentally friendly polyol technique was used to create bimetallic Ag^+ and Cu^{2+} nanoparticles. Ethylene glycol serves as a reducing agent and solvent in this method. Because polyvinylpyrilidone (PVP) can effectively stop Ag^+ and Cu^{2+} from self-aggregating, it is employed as a stabilizer and surfactant [70].

2.1.2 Fe/Ag Nanocomposite Synthesis

Using magnetic stirring, where water solutions of both $AgNO_3$ and $FeSO_4{\cdot}7H_2O$ were combined to create the Fecore@Agshell NCs. Characterization investigations were conducted on the Fecore@Agshell NCs after they were purified [34].

2.2 Biochar-Based Nanocomposite (Ag–Cu/biochar NCs)

Preparation of Bimetallic Ag–Cu/biochar NCs were formulated via mixing of $AgNO_3$ and $CuSO_4.5H_2O$ with the deionized water containing biochar (biochar prepared via pyrolysis of 10 g of *Atriplex halimus* fine powder at 550 °C for 3 h) then sonicated. followed by the addition of of *Atriplex halimus* L shoot extract which is

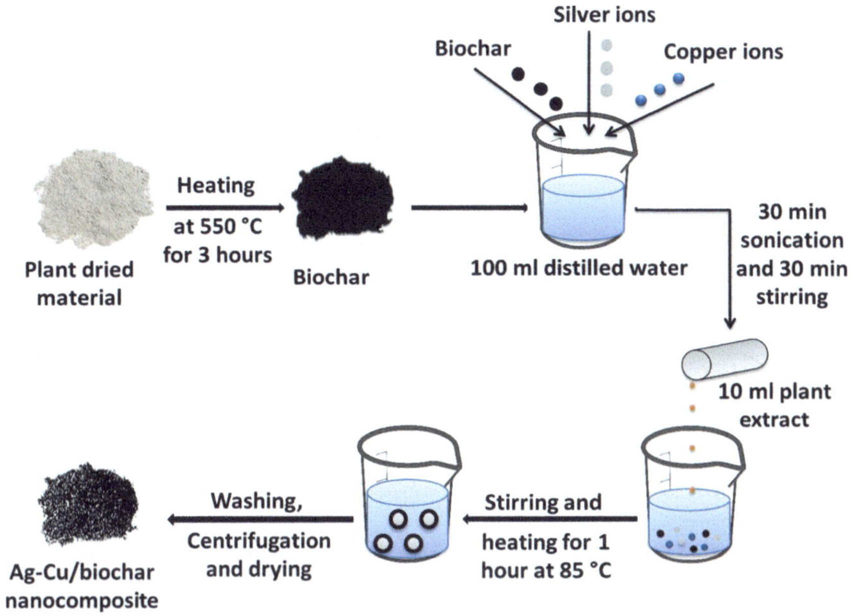

Fig. 2 Design of bimetallic Ag–Cu/biochar nanocomposite

used as a reducing and stabilizing agent to get bimetallic Ag–Cu/biochar nanocomposite (Fig. 2). According to [38]. The obtained Ag–Cu/biochar nanocomposite was centrifuged and washed repeatedly with deionized water.

2.3 *Chitosan-Based Nanocomposite*

2.3.1 Making Chitosan/Ag Nanocomposite

Chitosan-based nanocoposite can be synthesized using a variety of techniques. Common techniques for preparing nanochitosan include ionotropic gelation, emulsion crosslinking, reverse micelles, soaking method, and spray drying method [43]. These techniques provide less force and a lower utilization of organic solvents. As mentioned by Umoren and others Chitosan/Ag NCs, were formulated via several steps. In the first step, 100 mL of 0.1 M CH_3COOH acid solution was combined with 2.0 g of chitosan and swirled. Secondly, the polymer solution made in the first stage was mixed with 0.02 g of $AgNO_3$ that had been dissolved in 5 mL of distilled water. For three hours, the chitosan-$AgNO_3$ solution was agitated at 150 rpm. Thirdly, the chitosan-$AgNO_3$ was mixed with 5 mL of *Phoenix dactylifera* or *Rumex vesicarius* extracts as the reducing agent. In the end, a mixture of chitosan-$AgNO_3$ and plant

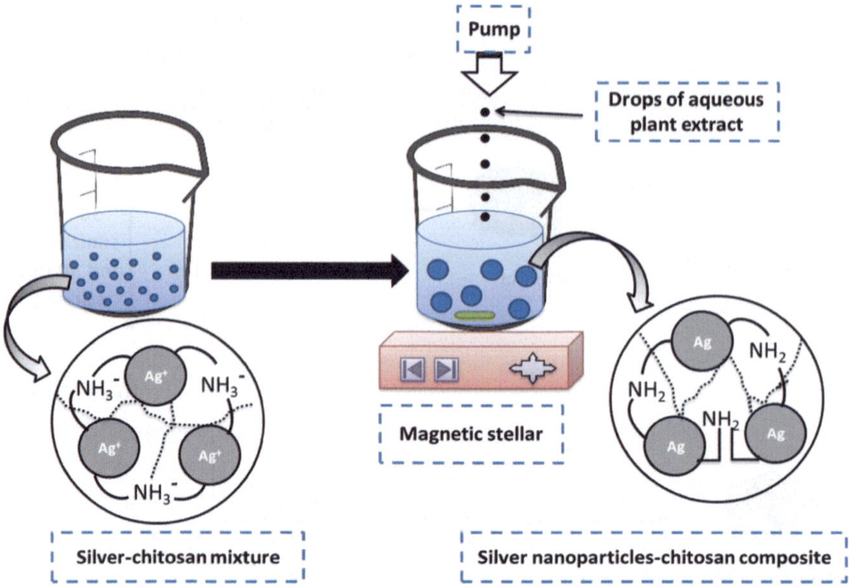

Fig. 3 Silver-chitosan nanocomposite preparation

extract solution was stirred continuously for 24 h at room temperature to get Chitosan/Ag NCs as shown in Fig. 3 [87].

2.3.2 Chitosan/Fe/Ag Nanocomposite

The NCs of **chitosan/Fe/Ag** were created in the manner mentioned by Govindan et al. [34]. Generally, 100 mL of 1% acetic acid was used to dissolve 0.5 g of pure **chitosan** powder. The mixture was then vortexed to promote total dissolution and centrifuged for two hours at 150 rpm. $FeSO_4 \cdot 7H_2O$ and $AgNO_3$ salts, each weighing 0.5 g, were added to this mixture, and it was agitated for two hours. Then, a dark-grey precipitate with gas evolution was seen when freshly made *Alchornea laxiflora* leaf extract (broth) was applied dropwise to this mixture. After another hour of stirring, the liquid was completely reduced from the metal salts. For characterization, the CS-Fe@Ag NC was dried at 100 °C for four hours and kept dry in a plastic sample vial.

2.4 *Zeolite (ZU) Based Nanocomposite*

2.4.1 Ag_2O/ZU Nanocomposite

For 48 h in the dark, 10 g of (ZU)-Na was added to 666 mL of 0.1 M $AgNO_3$ solution. After that, the samples were dried for 24 h at 60° and cleaned using ultra-purity (UP) water [16].

2.4.2 AgO/ZU Nanocomposite Synthesis

Initially, by introducing 100 g of zeolite to 250 mL of 0.04 M $AgNO_3$ solution and keeping it in the dark for 24 h, ZU-Na was transformed to Ag^+ form via ion exchange. Next, after 48 h in the dark, 13.5 g of the resulting Ag^+ form was added to 135 mL of 50 mM $AgNO_3$ solution. Ultimately, the reduction was accomplished by submerging the zeolite in a 0.25 M sodium borohydride solution (as a reducing agent) for one hour under cold, ice-assisted conditions, until the material was fully covered. After reduction, the mixture was filtered, and the zeolite was dried for eight hours at 90° in a bench oven [85].

3 Characterization of Silver Nanocomposite

Various analytical spectroscopic techniques were employed to characterize the silver chitosan nanocomposite (AgCNCs), in order to ascertain their size distribution, surface area, shape, and crystallinity index. UV–visible spectroscopy was used to confirm the synthesized AgNPs and the constructed AgCNCs in an initial manner. The presence of silver nanoparticles was unequivocally established by the greatest absorbance at 420 nm, in contrast, two distinct, strong peaks were seen in the case of the NCs at 420 and 240 nm, respectively, indicating the successful creation of the nanocomposite system with the coating of chitosan molecules on the silver nanoparticles [71]. In another study found that the absorption peak for chitosan was observed at 280 nm [14]. Scanning Electron Microscopy (SEM) was used to finish the surface topological analyses of both entities. The less than 100 nm-sized AgNPs were uniformly smooth, somewhat spherical, and shaped as seen by the electronic micrographs. In contrast, the surface of AgCNCs in the nanocomposite system had an irregular surface topology and was rough due to chitosan stacking on the silver nanoparticles. This also caused the average AgCNC particle size to grow to less than 1000 nm. For the confirmation of fundamental elements present in the synthesized AgNPs and AgCNCs, energy dispersive X-ray spectroscopic analysis was performed which appeared presence of silver, carbon, and oxygen in synthesized nanoparticles and nanocomposite system with the energy levels at 0.265 keV, 0.510 keV, 2.917 keV for the AgNPs and 0.277 keV, 0.525 keV, 2.983 keV for the AgCNCs. AgNPs had a higher mass percent of elemental silver (51.6%) than AgCNCs (2.85%), which demonstrated the successful fabrication process. Attenuated total reflection (ATR-FTIR) spectroscopy was employed to identify the functional groups in chitosan, AgNPs, and nanocomposite systems (AgCNCs) as shown in Fig. 4 [71]. To precisely ascertain the average hydrodynamic diameter of the nanomaterial and nanocomposite system, the DLS technique was utilized. This nanocomposite system's Polydispersity Index(PDI) value was found to be 0.573 [37]. AgNPs had a zeta potential of − 22.1 mV, while AgCNCs had a zeta potential of +63.2 mV. The silver nanoparticles' negative charge confirms their stability in the liquid medium with high surface energy by showing their well-dispersed behavior in the medium and indicating strong repulsion between the silver nanoparticles. The presence of the protonated amino group (NH^{3+}) of the chitosan molecule explained the good characteristics of the nanocomposite system and further supported the chitosan's successful protective coating of the AgNPs' surface [68].

4 Silver Nanocopmosites Improve Plant Growth, Seed Germination, Root System, and Soil Traits

Researchers now have a greater knowledge of the connection between the molecular structure and macroscopic characteristics of biological materials found in plants thanks to nanotechnology. Nanotechnology has completely changed several aspects of agriculture and the food industry, including processing, pathogen detection, quality control, and the creation of new tools that allow plants to absorb nutrients [23]. The application of nanocomposite enhances the plant's physiological and biochemical processes as well as the physicochemical characteristics of soils [32]. There is evidence that NMs such as NPs and NCs have the capacity to release nutrients gradually and under control. Clay-based NCs have altered qualities that facilitate the delayed release of fertilizers and increase the soil's ability to store water. Clay can bind with minerals and prolong their retention period in the soil [46]. One of the most vital roles of NCs as fertilizers, which have a gradual release, which means that they will be accessible to crops during their full life cycle over time, nanofertilizers can effectively reduce chronic problems like eutrophication (excessive development of plants and algae as a result of one or more limiting growth elements required for photosynthesis being more readily available) by up to 35–40% in agricultural production. They can also serve as an alternative to conventional fertilizers by reducing nutrient loss from the environment and increasing nutrient use efficiency [52]. Utilizing nanotechnology, such as nanoparticles and NCs, can control the release of nutrients and pesticides through regulated releases and target-specific distribution. It can also increase the efficiency with which nutrients are used, decrease nutrient loss, minimize dosage decrease toxic effects, and even help with water management [45]. There are few studies on the effect of silver NCs on plant or soil systems. One of these studies on Ag/ZnO nanocomposite at 60 ppm with nitrogenous fertilizers

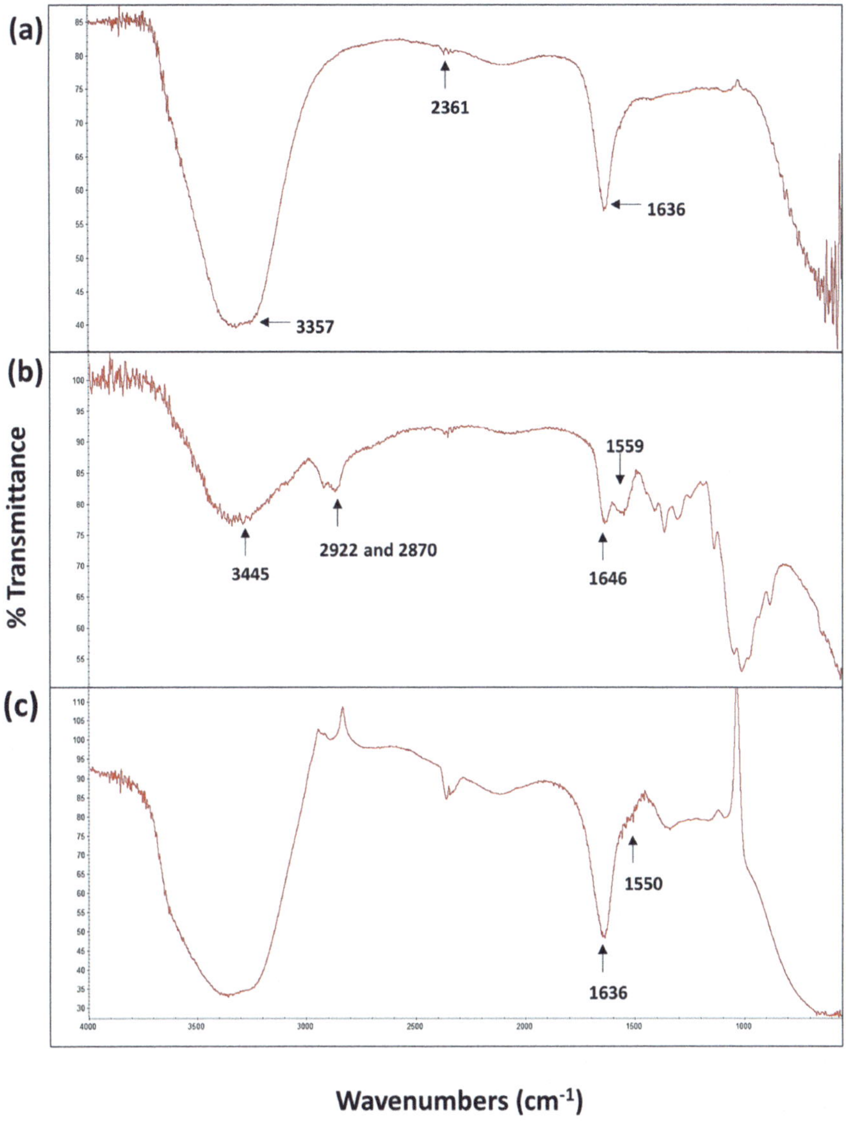

Fig. 4 Analysis of functional groups by FTIR spectroscope, **a** spectroscopic examination of AgNPs), **b** chitosan and **c** nanocomposite system (AgCNCs) [71]

(urea and ammonium sulphate) appeared significantly increased in yields traits (the number of grains per pot, spike length, 100-grain weight, grain yield per pot, and harvest index), and protein content, but showed a decrease in proline content, soluble sugar content, total phenolic, total flavonoid, malondialdehyde levels, and enzymatic antioxidants (SOD, POD, and CAT) of wheat varieties (Galaxy-13 and Pak-13) [24].

In another study, [69] revealed that *Cicer arietinum* L. plant seedling's germination, growth development, fresh and dry weights and photosynthetic pigmentation were favorably impacted by gum acacia (GA) capped Cu-Ag bimetallic nanoparticles (Cu/Ag@GA2 at 200 mg/L) as compared to the controls (to distilled water, Cu@GA NPs, Ag@GA NPs, and Cu/Ag NPs). Cu/Ag@GA2 could therefore be used as a useful nanofertilizer for the *C. arietinum* L. plant. Every seed cultivated in the presence of Cu/Ag@GA2 germinated. Cu/Ag@GA2 at 200 mg/L and distilled water treated seeds exhibited 85% and 22% germination respectively while the number of leaves, roots, shoots, and total lengths after 14 days of treatment was 85, 9.2 cm, 30.2 cm, and 39.4 cm respectively as compared to 45, 3.8 cm, 15 cm, and 18.8 cm for the plants grown in distilled water. On the other hand, Silver NCs of bimetallic Ag/Cu (1:3) showed reduced response in germination frequency, shoot and root lengths of *Eruca sativa* plants after 6 weeks of inoculation Additionally, Ag/Cu 1:3 appeared increased in antioxidant capacity, total flavonoid, total phenolic activities, total protein content, superoxide dismutase and polyphenol oxidase activities but, plantlets treated with bimetallic Ag/Au (1:3) showed the reduced phenolic content [92].

A study was conducted on silver nanocomplexes (GFAgNPs at 2.0 mg L^{-1}), which were created by capping silver ions with extracts from the seaweed *Gracilaria foliifera,* caused 100% shoot bud initiation with the maximum number of shoot buds (112.5 shoots $explant^{-1}$). Additionally, phycochemically coated silver nanocomplexes were shown to be unique growth-regulating chemicals that could serve as a substitute for commercial cytokinins in the large-scale generation of genetically homogeneous plantlets for industrial purposes through tissue culture [2]. Another study carried out on *Capsicum annuum* plants under different NPs (NPs of chitosan (CS), nanocomposite of CSAg, and CSCu). After 24-h seed priming with 1, 10, and 20 ppm concentrations of NPs (CS, CSAg, and CSCu), the seedlings displayed better physiological and biochemical traits. When applied to the leaves of *Capsicum* spp. plants at varying concentrations (0, 20, 40, 60, 80, and 100 ppm), the phytochemical characteristics increased and the plants were 70–85% protected from thrips (small, slender insects with wings that are fringed, that feed by making holes in the host tissue's epidermal (outer) layer and sucking out the contents of the cells, which causes the leaf surface to stipple, fleck, or turn silver). Additionally, it increased the levels of total flavonoids (40–125%), carotenoids (20–30%), total phenolics (20–45%), reducing sugars (15–40%), total antioxidant activity (10–82%), and ferric-reducing antioxidant power (10–100%), DPPH (76–83 mg mL^{-1}) activity and 125–142% of total capsaicinoids [55]. Studies were conducted to investigate the effectiveness of nanocomposite of Ag-chitosan nanoparticles on seeds of chickpeas. Significant growth-promoting effects on chickpea seed germination, seedling length, fresh weight, dry weight, chlorophyll contents, rise in the activity of several enzymes, including α, β-amylase, ascorbate peroxidase (APX), peroxidase (POD), catalase (CAT) and MDA content were demonstrated using this nanocomposite (0.1%, w/v) [8]. Additionally, silver nanocomposite had different effects on seed germination and root system anatomy of plants. A study on maize (*Zea mays* L. cv. NK-199) and cabbage (*Brassica oleracea* var. capitata L.), showed the anatomical changes in root system under the effect of Citrate–nanoAg, nanoZnO, $AgNO_3$, and $ZnSO_4$. Silver quantification in maize seedlings (whole seedlings evaluated) revealed that treatment with $AgNO_3$ (200 μg/mL) resulted in Ag (as total Ag) uptake more than five folds higher than the control, but in the presence of Citrate-NanoAg composite was 4.5 times greater than the control. For the $AgNO_3$ and citrate–NanoAg treatments, the biouptake concentrations were 22 and 1.8 nAg/mg dry weight of seedlings, respectively. The moisture content of maize seedlings treated with $AgNO_3$ (168.7% dry weight) or Citrate-nanoAg (163.9% dry weight) was found to be higher than that of the control group (147.3% dry weight). When compared to the control in cabbage plant, citrate-nanoAg suppressed root growth by 24.1% at the highest tested quantity (73.4 μg/mL as total Ag) [67]. It is well known that when plants experience water stress, the number of xylem vessels increases, and when sufficient water is provided, the number of xylem vessels decreases. Therefore, metaxylem number assessment in plants has been useful in understanding osmotically-induced stress caused by water stress [76]. Metaxylem counts varied within a range of 5–6 with both types of NP treatment (Citrate–NanoAg: 73.4 μg Ag/mL; NanoZnO: 1000 μg/mL). Metaxylem counts varied within a range of 5–6 with both types of NP treatment (Citrate–nanoAg: 73.4 μg Ag/mL; nanoZnO: 1000 μg/mL). The proportion of metaxylem frequency in maize primary roots varied considerably in response to treatments with citrate–nanoAg, $ZnSO_4$, and $AgNO_3$ [47]. AgNO3 treatment in cabbage resulted in a germination suppression that was larger than the effect Citrate-nanoAg at similar doses had [67]. In another study, citrate-nanoAg (particle diameter = 2–20 nm) inhibited barley and ryegrass germination, but not in flax or lettuce at as high as 100 μg/mL levels [26].

If we take about AgNPs as a vital part of silver Nanocomposite, AgNPs are among the most widely used engineered NPs in a wide range of consumer products and are expected to enter natural ecosystems including soil via diverse pathways [7]. Soil is the primary resource used to produce food and is the main source of trace elements that enter the food chain, soil has always been essential to human health. Because of the close relationship between the "soil–plant systems," any alteration in the chemical or biological characteristics of the soil would inevitably affect the growth, development, and productivity of plants. A build-up of artificial nanoparticles in the environment could significantly change the productivity and quality of food crops reliant on soil. Specifically, the bulk of AgNPs emitted from consumer goods are anticipated to enter terrestrial ecosystems via biosolids applied to the land [15]. AgNPs have been widely reported to pose a threat to many residents of the soil system, such as free-living nitrogen-fixing bacteria and symbiotic interactions between fungus and bacteria, which can ultimately impact the physico-chemical features of the soil. Hence, by adversely affecting "soil–plant systems," which provide essential ecosystem services, AgNPs may change the biogeochemistry and productivity of ecosystems [60]. AgNPs can enter soils through a variety of ways, including the wet or dry deposition of AgNPs suspended in the air, point sources like production facilities, the application of organic wastes in agriculture, the use of a spray to encourage plant growth, the recycling of sewage sludge as fertilizer for agricultural soils, waste incineration plants, landfills, and AgNPs containing consumer products [88]. AgNPs-traits (such as dissolution rate dispersibility, aggregation, dissolution rate, agglomeration, size, transformation and charge, and surface chemistry) can be altered by their interactions with soil substrates, which may regulate AgNPs stability and, in turn, their availability and transportation through the environment, as well as their eventual toxicity to species [13]. A positive charged soil may prevent the long-distance mobility of negative charge-bearing citrate-capped AgNPs. On the contrary, a negatively charged soil may cause AgNPs to be more mobile in such soils [90]. In greater soil pH values both Ag and AgNPs, were shown to increase Ag-sorption since they resulted in more cation exchange processes and more negatively charged sites, respectively [31]. AgNPs have been shown to move more easily when they are in the presence of organic materials, such as humic acid or surfactants, while higher ionic strength and divalent cations have been shown to promote AgNPs aggregation and retention [86]. Ag ions were shown to be closely bonded in complexes when organic matter was present, while there was more availability of free Ag ions in acidic soils. The most bioavailable species may not be AgNPs, Ag, or Ag complexes due to their propensity to oxidize in a soil environment [13]. The decomposition of organic materials and nitrogen fixation are processes that are greatly aided by soil microorganisms. Furthermore, legumes and soil bacteria develop symbiotic partnerships that serve as a key source of fixed nitrogen for both plants and other plants. However, any disruptions to soil microbial consortia can have a substantial impact on ecosystems, since they can impact soil fertility, water quality and ecology, as well as plant growth and productivity [89]. Additionally, AgNPs have been shown to have a deleterious effect on soil exo-enzyme activities, with urease activity showing a greater susceptibility to AgNPs [79]. It has been demonstrated that exposing soil bacteria to 100–1,000 mg L^{-1} of several AgNPs considerably reduces nitrification [9]. Regarding this matter, it has been shown that the toxicity of AgNPs to nitrification bacteria is largely contingent upon their size, with AgNPs less than 5 nm in diameter being known to substantially suppress nitrification bacterial activity [18]. By collecting NPs into their biomass, plants can exert tremendous control over the fate and movement of various NPs in the environment [54]. According to previous reports, the main methods that soil exposes terrestrial plants to NPs are through purposeful subsurface release for environmental remediation, land applications of contaminated biosolids, wastewater effluent discharge, and possible leaching from nano-enabled items [93]. It has been suggested that plants are primarily exposed to NPs through their roots, which may store a significant portion of the accumulated NPs and eventually cause chemical or physical toxicity in plants. It has been suggested that adequate hydraulic conductivity is necessary for AgNPs movement and contact with rhizobacteria and plant roots [27].

5 Control of Plant Pathogens by Silver Nanocomposite

The effective application of AgNPs as antifungal and antibacterial agents in several medical fields [11, 39, 81] has resulted in their usage in the management of phytopathogens [48–50]. AgNPs wide range of antibacterial action inhibits a number of plant illnesses brought on by fungi that produce spores, applying AgNPs well in advance of fungal spore penetration and colonization within plant tissues can increase their efficacy [53]. In order to generate chitosan metallic nanoparticles, metals such as silver, copper, zinc, iron, and nickel are extremely compatible and have been shown to be non-toxic to the agricultural plant system [19]. Nevertheless, it was also noted that pansy flowers and cucumber leaves had some chemical damage when exposed to an extremely high concentration of nanosilica-silver. Utilizing AgNPs as investigated was also conducted into alternatives to fungicides for the management of phytopathogenic fungi that develop sclerotia. Fungal hyphae exposed to AgNPs experienced significant damage as a result of the hyphal wall layers separating and the hyphae collapsing. Studies revealed that AgNPs had detrimental effects on conidial germination and were detrimental to unidentified fungus species of the genus Raffaelea that cause oak tree mortality [83]. Fabricated spheres of AgNPs conjugated with chitosan showed antifungal property against many pathogenic fungi for example *Aspergillus niger*, *Penicillium sp.*, *Candida sp.* and *Cordyceps militaris*; composite of AgNPs with chitosan might directly affect bacteria or fungi or act as catalysts to encourage plants to produce defense mechanisms [73]. *Aspergillus niger* and *Aspergillus flavus*, postharvest molds, and *Fusarium oxysporum*, *Rhizopus stolonifer*, and *Botrytis cinerea*, which infect plants and crops, were all significantly inhibited by chitosan-AgNPs [58]. Also, another study found that chitosan silver nanocomposite demonstrated a greater suppression of fungus radial development than chitosan nanoparticles alone, against *F. oxysporum* and *Aspergillus sp.* [21] Another investigation showed that AgNPs capped on graphene oxide (GO) driven by DNA was efficiently for reducing the formation of bacterial spot (a significant tomato disease caused by *Xanthomonas perforans,* resulting in a reduction of 10–50% in cell viability on plants and in culture. In a greenhouse trial on the same nanocomposite at 100 ppm, tomato transplants showed a significant reduction in the severity of bacterial spot disease when compared to untreated plants. The treatment produced results comparable to the current grower standard therapy, but it did not cause phytotoxicity [62]. In comparison to the sick plant, chitosan-hybridized biogenic AgNPs enhanced a number of morphological and physiological characteristics, including root and shoot lengths, fresh and dry weights, and levels of carotenoid and chlorophyll of tomato plants. Additionally, it reduced tomato leaf spot disease by as much as 60% [10]. Another study carried on wheat plants found that, chitosan NPs appeared to treat Fusarium head blight (FHB) (one of the most serious wheat diseases brought on by *F. graminearum* in humid and warm regions). Spraying plants at anthesis and twice or three times during the growing season can protect them from disease so, these nanocomposite can be used as biological pesticides [44]. Additionally, when infected chilies plants with thrips infestation were blasted with different nanoparticles and composite of chitosan (CS), nanocomposite of CSAg, and CSCu, they prevented the sporulation of thrips. All NPs over 60 ppm concentration significantly suppressed sporulation by 70–85% as compared to control [55].

In a different investigation, chitosan hybridized biogenic silver nanoparticles (Ch@BSNP) which are made using *Trichoderma viride* extracellular metabolites exhibited significant growth inhibition against three bacterial phytopathogens (*Pseudomonas syringae, Erwinia chrysanthemi,* and *Xanthomonas campestris* of tomato plants) and four plant pathogenic fungal strains (*Sclerotium rolfsii, Rhizoctonia solani, Alternaria alternata, and Alternaria brassicicola*). Additionally, Ch@BSNP dramatically decreased the incidence of disease in tomatoes cultivated in greenhouses, which alleviated the biotic stress caused by several pathogens and was evident in improvements to growth and health. It has been demonstrated to encourage the manufacture of carotenoid and chlorophyll in plants challenged by pathogens [33].

6 Mechanisms of Absorption, Translocation, and Mode of Action of Silver Nanocoposite in Plants

Silver-based nanocomposites are used as Nano fertilizers more than monometallic nanoparticles because they have a larger surface-area-to-volume ratio than monometallic nanomaterials and due to their synergistic properties. For instant, [3] found that using Ag/ZnO as Nano fertilizers is more advantageous than monometallic nanoparticles. This characteristic permits a gradual and regulated release, hence facilitating the crops' efficient uptake of nutrients. Consequently, this method might be considered as a starting point for novel and long-lasting nutrient distribution schemes. Also, Ehsan et al. [24] found that green synthesized Ag/ZnO (60 ppm) + urea (100 mg/L) significantly increased wheat production and protein content with decreases in the production of proline, soluble sugar content, total phenolic, total flavonoid, MDA levels and decrease in enzymatic antioxidants by reducing ROS activity.

Additionally, Prabha et al. [69] revealed that gum acacia (GA) capped Cu-Ag bimetallic nanoparticles (Cu/Ag@GA2) are more effective on gram seedling germination, growth, and photosynthetic pigmentation than monometallic nanoparticles, this is maybe as a result of the synergistic effects of the two metals makes bimetallic nanoparticles. Also, the controlled and progressive release of GA-capped bimetallic NPs enabled the crops to absorb nutrients more efficiently. The growth parameters showed a more marked drop upon increasing the concentration of Cu/Ag@GA2 to 400 mg/L. Therefore, it can be concluded that greater concentrations reverse the stimulating action of Cu/Ag@GA2 and result in a sharp reduction in root elongation. The aforementioned results made it possible to identify 400 mg/L as the potentially hazardous dose that caused gram seeds to undergo metabolic alterations. The observed decrease in growth at concentrations beyond 200 mg/L could perhaps be attributed to the increased accumulation and maximal absorption of Cu/Ag@GA2, resulting in phytotoxic effects on gram seeds. Furthermore, it has been documented that NP concentrations above a certain threshold can cause cytotoxicity, genotoxicity, and tissue damage in addition to inhibiting cell division and finally causing cell death [91].

Either apoplastic or symplastic pathways can carry the nanoparticles consumed by roots or leaves from the exposed surfaces to the plant's vascular system. The intercellular layer, or the gap between the cell wall and the plasma membrane, is where they go after exiting the cell wall pores. It is possible to suppose that the combination and synergistic action of bimetallic NPs facilitated Cu/Ag@GA2's promotion of the seed's germination. They improved the physiological parameters and supplied the nutrients for the seeds that were growing. In addition to encouraging plant growth, the antibacterial properties of Cu and Ag, the two metals that were added to the polysaccharide matrix, may shield the seedling from bacterial invasion [69].

In another study, bimetallic Ag Cu 1:3 appeared to increase in antioxidant capacity, total flavonoid, total phenolic activities, total protein content, superoxide dismutase, and polyphenol oxidase activities but, plantlets treated with bimetallic Ag Au 1:3 showed reduced contents. This may be a result of the total flavonoid and total phenolics depending upon both the surface area of the NPs used and the size of the nanoparticles; the smaller size of nanoparticle more stress will be induced flavonoids and phenolics biosynthesised. Under stress conditions induced by bimetallic nanoparticles of the smallest size, more protein was produced in response to the defence mechanism of plant cells. Superoxide dismutase (SOD) and polyphenol oxidase (POD) activities play an important role in protecting plants against oxidative damage and so it is triggered in plants treated with nanoparticles like AgNPs [12]. Cells release some of the secondary proteins to balance the oxidative damage exerted as a result of stress [92].

In maize, silver nanocomposite/citrate exhibited suppresses germination. The inhibition may result from a number of sources, but the most common ones being investigated to explain observed toxicity in many plant species are particular ion toxicity, somatically-induced stress from water loss, or hormonal imbalance [61]. After being treated with Citrate–NanoAg, appeared elongated in cell morphology at the zone of elongation. This maybe a result of NMs could cause other biological interactions and, as a result, different harmful effects from their intended ion toxicity. Furthermore, a decrease in the flexibility of the cell wall may result from metals attaching to the walls of the cells and to other subcellular structures. By using the same silver nanocomposite, Citrate–NanoAg in maize plants had little to no effect on root growth. Citrate-NanoAg, $ZnSO_4$, and $AgNO_3$ treatments resulted in a substantially different proportion of metaxylem frequency in maize primary roots. However, the proportion of metaxylem count was not statistically different ($p > 0.1$) with nanoZnO treatment. Because with either ionic salt treatment, the formation of seven metaxylem vessels was observed in maize, this is indicative of higher stress exerted by biolabile metal ions than the evaluated NPs. Based on total Ag analysis for maize seedlings; these results revealed that the examined NPs had reduced effects on the primary root anatomy. This may be because of a distinct mechanism of action between the NPs and the corresponding ionic salts, or it could be because of lower metal uptake. It is impossible to rule out the stimulation of transcription factors like VND6, VND7, or auxin or cytokinin to induce more metaxylem vessels in maize [67]. The vital role of silver Nanocomposite as antifungal of many fungal species such as *fusarium oxysporum, rhizopus stolonifer*, and *botrytis cinerea*, which infect plants and crops may be as a result of fungal cell membranes may have negative charges, and chitosan coating could increase the antifungal efficacy of bio-AgNPs [28]. Against *Curvularia lunata, Aspergillus fumigates* and *Cladosporium* sp. fungus, Chitosan-AgNPs has shown a highly substantial antifungal activity. At 50 mg/L, the largest mycelial growth reduction was seen in *C. lunata* (97%), followed by *Cladosporium sp.* (89%), and *A. fumigates* (83%). The mycelial growth of 62% of *Cladosporium sp.*, 60% of *C. lunata*, and 42% of *A. fumigates* was reduced by the lower dose (6.25 mg/L) [4]. In another study, nanosilver composite (AgNP–Na zeolite A) showed antibacterial properties towards both Gram-positive *Staphylococcus aureus* and Gram-negative *Escherichia coli* bacteria. However, they demonstrated much greater efficacy against *E. coli* in comparison to *S. aureus* [42]. In many searches discussed the crucial role of silver nanocomposite in biological controls of many bacterial and fungal diseases which could cause many diseases in plant, animal, and human for instant, Citosan-Ag Nanocomposite shows antibacterial potential against *Enterococcus faecalis*, the nanocomposite system exhibited the lowest minimal inhibitory concentration (1.56 $\mu g\ ml^{-1}$) but they did not show any antifungal action [71]. Positively charged silver nanoparticles may bind to negatively charged bacterial cell walls, altering the structure and compromising the permeability of the cell membrane as a potential mode of action. A thorough investigation showed that the bacterial cell surface had holes or "pits" that formed, which caused the nanoparticles to accumulate [82]. Chitosan is used as a stabilizer and reducing agent in the preparation of nanoparticles in two ways: electrostatically and satirically. The chitosan's NH2-group reacts with Ag^+ to increase the stability of the nanoparticles [57]. Additionally, the vital role of chitosan/Ag nanocomposite against the fungal disease of Chili pepper plant may be as a result of the positively charged NPs, which might attach to DNA/RNA, which may have an impact on transcription and translation and prevent the growth of fungi. Furthermore, chitosan may directly defend against microbes by destroying their cell surface through electrostatic contact. Conversely, chitosan may perform an indirect effect by stimulating the plant immune system through increased levels of secondary metabolites and antioxidants [55]. In another perspective, Chitosan nanoparticles' higher mobility, smaller particle size, and larger surface area all contribute to their higher efficiency. In soybeans, exposure to nanoiron oxide increased leaf, pod dry weight, and grain yield, among other agronomic traits [77]. There are other multifunctional silver nanocomposites in many fields as shown in Table 1.

7 Conclusion and Future Remarks

This chapter summarizes the different methods for preparing and characterization of silver nanocomposite. Also, the multifunctional silver nanocomposite for improving seed germination, plant growth, dynamic changes in primary and secondary metabolites of plants, and soil traits could serve as a substitute for commercial cytokinins in the large-scale generation of genetically homogeneous plantlets for industrial

Table 1 Other applications of silver nanocomposites

Silver nanocomposite	Benefits	Key references
Ag–Au	Show strong adsorption, and good electrical conducting properties, and is used in enzyme electrode design	Ren et al. [72]
Agar gum/silver	Use as optical sensors for ammonia detection	Pandey et al. [66]
Cellulose/silver	exhibited improved antimicrobial agent in the field of water treatment	Chen and Peng [17]
Graphene oxide-silver	Improved both its dye degradation and thermoluminescence properties	Kumari et al. [51]
Reducing graphene oxide/Ag	Antibacterial activity of multidrug-resistant (MDR) *Klebsiella pneumoniae* (Kp) strains at 30 μg/mL	Tan et al. [84]
Cellulose/Ag	Reduction of 4-nitrophenol and the high catalytic ability	Fujii et al. [30]
Chitosan–iron/silver	Gave higher maximum removals of BOD (48.7%), COD (15.5%), and TDS (54.9%) than pure CS with BOD (25.5%), COD (3.6%) and TDS (38.9%)	Olajire and Bamigbade [64]
Citosan-Ag	Antibacterial potential against *Enterococcus faecalis*, the lowest minimal inhibitory concentration (1.56 $\mu g\ ml^{-1}$) but they did not show any antifungal action	Raza et al. [71]
Silver-doped zinc oxide (Ag-ZnO) nanoparticles	Observed a high of 83% in radical scavenging activity	Shreema et al. [80]
Doping ZnO NPs with selenium or silver NPs	Enhanced their antimicrobial and photocatalytic properties	Ifijen et al. [40]
Ag/ZnO Nanocoposite	Comparing Ag/ZnO NPs to 4-nitrophenol (NPh), which showed a degradation rate of 81.37%, the former showed good photocatalytic degradation (88.9%) of the methyl violet	Afzal et al. [1]

purposes through tissue culture, and it can be used as biological pesticides and phytopathogen biological control. Additionally, why is silver-based nanocomposite used as nanofertilizers more than monometallic nanoparticles? Because they have a larger surface-area-to-volume ratio than monometallic nanomaterials, enhance the plant physiological and biochemical processes as well as the physicochemical characteristics of soils due to their synergistic properties. Also, appeared that the superior antibacterial capabilities, chemical durability, photocatalysis, and other qualities of the developed silver-based composite materials make them widely used in medical, environmental, biological, and food domains. Also, this chapter discussed the crucial role of silver nanocomposite in biological controls of many bacterial and fungal diseases which could cause many diseases in plant. However, research on the silver nanocomposites toxicity mechanism is still lacking, and opinions on the material's safety are divided. Thus, the purpose of this paper is to serve as a foundation for further studies on silver nanocomposites for example, using these silver nanocomposites for improving growth and metabolites of plants under biotic and biotic stress which can reduce reactive oxygen species activity.

References

1. Afzal MA, Javed M, Aroob S, Javed T, Alnoman MM, Alelwani W, Bibi I, Sharif M, Saleem M, Rizwan M (2023) The biogenic synthesis of bimetallic Ag/ZnO nanoparticles: a multifunctional approach for methyl violet photocatalytic degradation and the assessment of antibacterial, antioxidant, and cytotoxicity properties. Nanomaterials 13(14):2079
2. Ahmad A, Ahmad N, Anis M, Faisal M, Alatar AA, Abdel-Salam EM, Meena RP, Sivanesan I (2022) Biotechnological advances in pharmacognosy and in vitro manipulation of Pterocarpus marsupium Roxb. Plants 11(3):247
3. Ahmed SF, Mofijur M, Rafa N, Chowdhury AT, Chowdhury S, Muntasha ABM, Islam S, Ong HC (2022) Green approaches in synthesising nanomaterials for environmental nanobioremediation: technological advancements, applications, benefits and challenges. Environ Res 204:111967
4. Al-Zahrani SS, Al-Garni SM (2023) Fabrication of antifungal AgNPs capped with chitosan using endophytic fungus Curvularia kusanoi. Biocatal Agric Biotechnol 51:102769
5. Alexandre M, Dubois P (2000) Polymer-layered silicate nanocomposites: preparation, properties and uses of a new class of materials. Mater Sci Eng R Rep 28(1–2):1–63
6. Amin MA, Badawy AA (2017) Metabolic changes in common bean plants in response to zinc nanoparticles and zinc sulfate. Int J Innov Sci Eng Technol 4:321–335
7. Anjum NA, Gill SS, Duarte AC, Pereira E, Ahmad I (2013) Silver nanoparticles in soil–plant systems. J Nanopart Res 15:1–26
8. Anusuya S, Banu KN (2016) Silver-chitosan nanoparticles induced biochemical variations of chickpea (Cicer arietinum L.). Biocatal Agric Biotechnol 8:39–44
9. Arnaout CL, Gunsch CK (2012) Impacts of silver nanoparticle coating on the nitrification potential of Nitrosomonas europaea. Environ Sci Technol 46(10):5387–5395
10. Attia MS, Abdelaziz AM, Al-Askar AA, Arishi AA, Abdelhakim AM, Hashem AH (2022) Plant growth-promoting fungi as biocontrol tool against fusarium wilt disease of tomato plant. J Fungi 8(8):775
11. Bachheti RK, Bachheti A, Husen A (2024) Metal and metal-oxide based nanomaterials (Synthesis, Agricultural, Biomedical and Environmental Interventions). Springer Nature Singapore Pte Ltd., 152 Beach Road, #21-01/04 Gateway East, Singapore 189721, Singapore
12. Benn TM, Westerhoff P (2008) Nanoparticle silver released into water from commercially available sock fabrics. Environ Sci Technol 42(11):4133–4139
13. Benoit R, Wilkinson KJ, Sauvé S (2013) Partitioning of silver and chemical speciation of free Ag in soils amended with nanoparticles. Chem Cent J 7:1–7
14. Biao, Tan S, Wang Y, Guo X, Fu Y, Xu F, Zu Y, Liu Z (2017) Synthesis, characterization and antibacterial study on the chitosan-functionalized Ag nanoparticles. Mater Sci Eng C 76:73–80
15. Blaser SA, Scheringer M, MacLeod M, Hungerbühler K (2008) Estimation of cumulative aquatic exposure and risk due to silver: contribution of nano-functionalized plastics and textiles. Sci Total Environ 390(2–3):396–409
16. Bo A, Sarina S, Zheng Z, Yang D, Liu H, Zhu H (2013) Removal of radioactive iodine from water using Ag2O grafted titanate nanolamina as efficient adsorbent. J Hazard Mater 246:199–205
17. Chen L, Peng X (2017) Silver nanoparticle decorated cellulose nanofibrous membrane with good antibacterial ability and high water permeability. Appl Mater Today 9:130–135
18. Choi O, Hu Z (2008) Size dependent and reactive oxygen species related nanosilver toxicity to nitrifying bacteria. Environ Sci Technol 42(12):4583–4588

19. Chouhan D, Mandal P (2021) Applications of chitosan and chitosan based metallic nanoparticles in agrosciences—A review. Int J Biol Macromol 166:1554–1569
20. Cvjetko P, Zovko M, Štefanić PP, Biba R, Tkalec M, Domijan A-M, Vrček IV, Letofsky-Papst I, Šikić S, Balen B (2018) Phytotoxic effects of silver nanoparticles in tobacco plants. Environ Sci Pollut Res 25:5590–5602
21. Dananjaya SHS, Erandani WKCU, Kim C-H, Nikapitiya C, Lee J, De Zoysa M (2017) Comparative study on antifungal activities of chitosan nanoparticles and chitosan silver nano composites against Fusarium oxysporum species complex. Int J Biol Macromol 105:478–488
22. Dawadi S, Katuwal S, Gupta A, Lamichhane U, Thapa R, Jaisi S, Lamichhane G, Bhattarai DP, Parajuli N (2021) Current research on silver nanoparticles: synthesis, characterization, and applications. J Nanomater 2021:1–23
23. Duncan TV (2011) Applications of nanotechnology in food packaging and food safety: barrier materials, antimicrobials and sensors. J Colloid Interf Sci 363(1):1–24
24. Ehsan M, Iqbal Raja N, Mashwani ZUR, Zohra E, Abasi F, Ikram M, Mustafa N, Wattoo FH, Proćków J, de la Lastra JMP (2022) Effects of phytogenically synthesized bimetallic Ag/ZnO nanomaterials and nitrogen-based fertilizers on biochemical and yield attributes of two wheat varieties. Nanomaterials 12(17):2894
25. El-Batal AI, Ismail MA, Amin MA, El-Sayyad GS, Osman MS (2023) Selenium nanoparticles induce growth and physiological tolerance of wastewater-stressed carrot plants. Biologia 1–17.
26. El-Temsah YS, Joner EJ (2012) Impact of Fe and Ag nanoparticles on seed germination and differences in bioavailability during exposure in aqueous suspension and soil. Environ Toxicol 27(1):42–49
27. Fabrega J, Luoma SN, Tyler CR, Galloway TS, Lead JR (2011) Silver nanoparticles: behaviour and effects in the aquatic environment. Environ Int 37(2):517–531
28. Fan Z, Qin Y, Liu S, Xing R, Yu H, Chen X, Li K, Li P (2018) Synthesis, characterization, and antifungal evaluation of diethoxyphosphoryl polyaminoethyl chitosan derivatives. Carbohydr Polym 190:1–11
29. Farooq MA, Fatima A, Rehman S, Batool A, Gul I, Din AA, Anwer H, Arshad M (2023) Phytotoxicity response and defense mechanisms of nanocomposites/mixture of nanoparticles. In: Nanomaterials and nanocomposites exposures to plants: response, interaction, phytotoxicity and defense mechanisms. Springer, pp 43–58
30. Fujii Y, Imagawa K, Omura T, Suzuki T, Minami H (2020) Preparation of cellulose/silver composite particles having a recyclable catalytic property. ACS Omega 5(4):1919–1926
31. Geranmayeh Oromieh A (2011) Evaluating solubility, aggregation and sorption of nanosilver particles and silver ions in soils
32. Ghassemi-Golezani K, Rahimzadeh S, Farhangi-Abriz S (2023) Nanomaterials and nanocomposites exposures to plants: an overview. In: Nanomaterials and nanocomposites exposures to plants: response, interaction, phytotoxicity and defense mechanisms, pp 19–41
33. Giri VP, Pandey S, Kumari M, Tripathi A, Katiyar R, White JC, Mishra A (2022) Hybridization of chitosan and biosynthesized silver nanoparticles to enhance antimicrobial activity against phytopathogens in tomato (Solanum lycopersicum). ACS Agric Sci Technol 2(4):719–733
34. Govindan S, Nivethaa EAK, Saravanan R, Narayanan V, Stephen A (2012) Synthesis and characterization of chitosan–silver nanocomposite. Appl Nanosci 2:299–303
35. Harmer MP, Chan HM, Miller GA (1992) Unique opportunities for microstructural engineering with duplex and laminar ceramic composites. J Am Ceram Soc 75(7):1715–1728
36. He J, Li J, Gao Y, He X, Hao G (2023) Nano-based smart formulations: a potential solution to the hazardous effects of pesticide on the environment. J Hazard Mater: 131599
37. Honary S, Ghajar K, Khazaeli P, Shalchian P (2011) Preparation, characterization and antibacterial properties of silver-chitosan nanocomposites using different molecular weight grades of chitosan. Tropical J Pharmaceut Res 10(1)
38. Hosny M, Fawzy M, Eltaweil AS (2022) Phytofabrication of bimetallic silver-copper/biochar nanocomposite for environmental and medical applications. J Environ Manag 316:115238
39. Husen A, Siddiqi KS (2023) Advances in smart nanomaterials and their applications. Elsevier Inc., 50 Hampshire St., 5th Floor, Cambridge, MA 02139, USA. https://doi.org/10.1016/C2021-0-02202-1

40. Ifijen IH, Maliki M, Anegbe B (2022) Synthesis, photocatalytic degradation and antibacterial properties of selenium or silver doped zinc oxide nanoparticles: a detailed review. OpenNano:100082
41. Indraratne SP, Pierzynski GM, Baker LR, Vara Prasad PV (2021) Nano-oxides immobilize cadmium, lead, and zinc in mine spoils and contaminated soils facilitating plant growth. Can J Soil Sci 101(3):543–554
42. Jiraroj D, Tungasmita S, Tungasmita DN (2014) Silver ions and silver nanoparticles in zeolite A composites for antibacterial activity. Powder Technol 264:418–422
43. Kashyap PL, Xu X, Heiden P (2015) Chitosan nanoparticle based delivery systems for sustainable agriculture. Int J Biol Macromol 77:36–51
44. Kheiri A, Moosawi SA, Malihipour A, Saremi H, Nikkhah M (2016) Application of chitosan and chitosan nanoparticles for the control of Fusarium head blight of wheat (Fusarium graminearum) in vitro and greenhouse. Int J Biol Macromol 93:1261–1272
45. Konappa N, Udayashankar AC, Dhamodaran N, Krishnamurthy S, Jagannath S, Uzma F, Pradeep CK, De Britto S, Chowdappa S, Jogaiah S (2021) Ameliorated antibacterial and antioxidant properties by Trichoderma harzianum mediated green synthesis of silver nanoparticles. Biomolecules 11(4):535
46. Kráľová K, Jampílek J (2022) Metal-and metalloid-based nanofertilizers and nanopesticides for advanced agriculture. In: Inorganic nanopesticides and nanofertilizers: a view from the mechanisms of action to field applications. Springer, pp 295–361
47. Kubo M, Udagawa M, Nishikubo N, Horiguchi G, Yamaguchi M, Ito J, ... Demura T (2005) Transcription switches for protoxylem and metaxylem vessel formation. Genes Dev 19(16):1855–1860.
48. Kumar A, Choudhary A, Kaur H, Guha S, Mehta S, Husen A (2022) Potential applications of engineered nanoparticles in plant disease management: a critical update. Chemosphere 295(133798):1–21. https://doi.org/10.1016/j.chemosphere.2022.133798
49. Kumar A, Choudhary A, Kaur H, Mehta S, Husen A (2021) Metal-based nanoparticles, sensors and their multifaceted application in food packaging. J Nanobiotechnol 19(256):1–25. https://doi.org/10.1186/s12951-021-00996-0
50. Kumar A, Choudhary A, Kaur H, Mehta S, Husen A (2021) Smart nanomaterial and nanocomposite with advanced agrochemical activities. Nanoscale Res Lett 16(156):1–26. https://doi.org/10.1186/s11671-021-03612-0
51. Kumari S, Sharma P, Yadav S, Kumar J, Vij A, Rawat P, ... Majumder S (2020) A novel synthesis of the graphene oxide-silver (GO-Ag) nanocomposite for unique physiochemical applications. ACS Omega 5(10):5041–5047
52. Lal R (2008) Soils and Indias food security
53. Li L, Pan H, Deng L, Qian G, Wang Z, Li W, Zhong C (2022) The antifungal activity and mechanism of silver nanoparticles against four pathogens causing kiwifruit post-harvest rot. Front Microbiol 13:988633
54. Ma X, Geiser-Lee J, Deng Y, Kolmakov A (2010) Interactions between engineered nanoparticles (ENPs) and plants: phytotoxicity, uptake and accumulation. Sci Total Environ 408(16):3053–3061
55. Mawale KS, Giridhar P (2023) Chitosan nanoparticles modulate plant growth, and yield, as well as thrips infestation in Capsicum spp. Int J Biol Macromol 127682
56. McWilliams A (2018) Nanocomposites, nanoparticles, nanoclays and nanotubes: global markets to 2022. Wellesley: BCC Research LLC
57. Mohammed MA, Syeda JT, Wasan KM, Wasan EK (2017) An overview of chitosan nanoparticles and its application in non-parenteral drug delivery. Pharmaceutics 9(4):53
58. Mondéjar-López M, López-Jimenez AJ, Ahrazem O, Gómez-Gómez L, Niza E (2023) Chitosan coated-biogenic silver nanoparticles from wheat residues as green antifungal and nanoprimig in wheat seeds. Int J Biol Macromol 225:964–973
59. Naaz S, Sachdev S, Husain R, Pandey V, Israil Ansari M (2023) Nanomaterials and nanocomposites: significant uses in plant performance, production, and toxicity response. In: Nanomaterials and nanocomposites exposures to plants: response, interaction, phytotoxicity and defense mechanisms. Springer, pp 1–18

60. Navarro E, Baun A, Behra R, Hartmann NB, Filser J, Miao A-J, Quigg A, Santschi PH, Sigg L (2008) Environmental behavior and ecotoxicity of engineered nanoparticles to algae, plants, and fungi. Ecotoxicology 17:372–386
61. Nawaz K, Hussain K, Majeed A, Khan F, Afghan S, Ali K (2010) Fatality of salt stress to plants: morphological, physiological and biochemical aspects. Afr J Biotechnol 9(34)
62. Ocsoy I, Paret ML, Ocsoy MA, Kunwar S, Chen T, You M, Tan W (2013) Nanotechnology in plant disease management: DNA-directed silver nanoparticles on graphene oxide as an antibacterial against Xanthomonas perforans. ACS Nano 7(10):8972–8980
63. Ogasawara T, Ishida Y, Ishikawa T, Yokota R (2004) Characterization of multi-walled carbon nanotube/phenylethynyl terminated polyimide composites. Compos A Appl Sci Manuf 35(1):67–74
64. Olajire AA, Bamigbade LA (2021) Green synthesis of chitosan-based iron@ silver nanocomposite as adsorbent for wastewater treatment. Water Resour Ind 26:100158
65. Omanović-Mikličanin E, Badnjević A, Kazlagić A, Hajlovac M (2020) Nanocomposites: a brief review. Heal Technol 10:51–59
66. Pandey S, Goswami GK, Nanda KK (2012) Green synthesis of biopolymer–silver nanoparticle nanocomposite: an optical sensor for ammonia detection. Int J Biol Macromol 51(4):583–589
67. Pokhrel LR, Dubey B (2013) Evaluation of developmental responses of two crop plants exposed to silver and zinc oxide nanoparticles. Sci Total Environ 452:321–332
68. Potara M, Jakab E, Damert A, Popescu O, Canpean V, Astilean S (2011) Synergistic antibacterial activity of chitosan–silver nanocomposites on Staphylococcus aureus. Nanotechnology 22(13):135101
69. Prabha M, Malviya T, Kumar A, Singh V (2023) Effect of Gum acacia capped Cu-Ag bimetallic nanoparticles on germination and growth of gram seeds (Cicer arietinum L.). Mater Today: Proc
70. Rahman L-u, Qureshi R, Yasinzai MM, Shah A (2012) Synthesis and spectroscopic characterization of Ag-Cu alloy nanoparticles prepared in various ratios. C R Chim 15(6):533–538
71. Raza S, Ansari A, Siddiqui NN, Ibrahim F, Abro MI, Aman A (2021) Biosynthesis of silver nanoparticles for the fabrication of non cytotoxic and antibacterial metallic polymer based nanocomposite system. Sci Rep 11(1):10500
72. Ren X, Meng X, Chen D, Tang F, Jiao J (2005) Using silver nanoparticle to enhance current response of biosensor. Biosens Bioelectron 21(3):433–437
73. Saharan V, Sharma G, Yadav M, Choudhary MK, Sharma SS, Pal A, Raliya R, Biswas P (2015) Synthesis and in vitro antifungal efficacy of Cu–chitosan nanoparticles against pathogenic fungi of tomato. Int J Biol Macromol 75:346–353
74. Sahay R, Reddy VJ, Ramakrishna S (2014) Synthesis and applications of multifunctional composite nanomaterials. Int J Mech Mater Eng 9:1–13
75. Saravanan A, Senthil Kumar P, Hemavathy RV, Jeevanantham S, Jawahar MJ, Neshaanthini JP, Saravanan R (2022) A review on synthesis methods and recent applications of nanomaterial in wastewater treatment: challenges and future perspectives. Chemosphere:135713
76. Schneider H, Wegner LH, Haase A, Zimmermann U (2007) Long-distance water transport under controlled transpirational conditions: minimal-invasive investigations by means of pressure probes and NMR imaging. In: The Apoplast of higher plants: compartment of storage, transport and reactions: the significance of the apoplast for the mineral nutrition of higher plants, pp 251–264
77. Sekhon BS (2014) Nanotechnology in agri-food production: an overview. In: Nanotechnology, science and applications, pp 31–53
78. Sen M (2020) Nanocomposite materials. In: Nanotechnology and the environment, pp 1–12
79. Shin Y-J, Kwak JI, An Y-J (2012) Evidence for the inhibitory effects of silver nanoparticles on the activities of soil exoenzymes. Chemosphere 88(4):524–529
80. Shreema K, Mathammal R, Kalaiselvi V, Vijayakumar S, Selvakumar K, Senthil K (2021) Green synthesis of silver doped zinc oxide nanoparticles using fresh leaf extract Morinda citrifoliaand its antioxidant potential. Mater Today: Proc 47:2126–2131

81. Siddiqi KS, Husen A, Rao RAK (2018) A review on biosynthesis of silver nanoparticles and their biocidal properties. J Nanobiotechnol 16(14):1–28
82. Slavin YN, Asnis J, Hńfeli UO, Bach H (2017) Metal nanoparticles: understanding the mechanisms behind antibacterial activity. J Nanobiotechnol 15:1–20
83. Slavin YN, Bach H (2022) Mechanisms of antifungal properties of metal nanoparticles. Nanomaterials 12(24):4470
84. Tan S, Wu X, Xing Y, Lilak S, Wu M, Zhao JX (2020) Enhanced synergetic antibacterial activity by a reduce graphene oxide/Ag nanocomposite through the photothermal effect. Colloids Surf, B 185:110616
85. Tauanov Z, Tsakiridis PE, Shah D, Inglezakis VJ (2019) Synthetic sodalite doped with silver nanoparticles: characterization and mercury (II) removal from aqueous solutions. J Environ Sci Health, Part A 54(9):951–959
86. Thio BJ, Reginald MO, Montes MA, Mahmoud D-W, Zhou D, Keller AA (2012) Mobility of capped silver nanoparticles under environmentally relevant conditions. Environ Sci Technol 46(13):6985–6991
87. Umoren SA, Solomon MM, Nzila A, Obot IB (2020) Preparation of silver/chitosan nanofluids using selected plant extracts: characterization and antimicrobial studies against gram-positive and gram-negative bacteria. Materials 13(7):1629
88. VandeVoort AR, Arai Y (2012) Environmental chemistry of silver in soils: current and historic perspective. Adv Agron 114:59–90
89. Yang Y, Wang J, Xiu Z, Alvarez PJJ (2013) Impacts of silver nanoparticles on cellular and transcriptional activity of nitrogen-cycling bacteria. Environ Toxicol Chem 32(7):1488–1494
90. Yu S-J, Yin Y-G, Liu J-f (2013) Silver nanoparticles in the environment. Environ Sci Process Impacts 15(1):78–92
91. Zafar H, Ali A, Ali JS, Haq IU, Zia M (2016) Effect of ZnO nanoparticles on Brassica nigra seedlings and stem explants: growth dynamics and antioxidative response. Front Plant Sci 7:535
92. Zaka M, Abbasi BH (2017) Effects of bimetallic nanoparticles on seed germination frequency and biochemical characterisation of Eruca sativa. IET Nanobiotechnol 11(3):255–260
93. Zhang W, Yao Y, Sullivan N, Chen Y (2011) Modeling the primary size effects of citrate-coated silver nanoparticles on their ion release kinetics. Environ Sci Technol 45(10):4422–4428

Interaction of Arbuscular Mycorrhizal Fungi and Silver Nanoparticles in a Soil Matrix for Enhanced Plant Growth and Production

Hiral Desai and Megha Barot

Abstract Growing interest in the use of nanotechnology in agriculture for the purpose of enhancing plant growth and production has been seen in recent years. The unique qualities of silver nanoparticles (AgNPs), such as their capacity to improve nutrient absorption and antibacterial activity, have made them stand out as potential candidates. Different variables, such as soil characteristics, exposure time, concentration, and size of the nanoparticles, affect these interactions. AgNPs at low concentrations stimulate the growth of arbuscular mycorrhizal fungus (AMF), maybe through modifications to the soil microbial habitat. On the other hand, increased AgNP concentrations restrict the growth of AMF, which is linked to their cytotoxic effects and disruption of symbiotic relationships. Simultaneously, AMF and plants have long been known to share a symbiotic relationship that promotes nutrient uptake and stress tolerance. AMF colonization affects AgNP uptake, distribution, and toxicity in plants, emphasizing the function of mycorrhizal symbiosis in reducing stress brought on by AgNP. This chapter highlights the effect of arbuscular mycorrhizal fungi and silver nanoparticles as well as the impact of AgNPs and AMF interaction on plant development and productivity within a soil matrix.

Keywords Plant performance · Silver nanoparticles · Arbuscular mycorrhizal fungus · Soil health · Sustainable agriculture

H. Desai
Department of Biotechnology, Smt. S. S. Patel Nootan Science & Commerce College, Sankalchand Patel University, Visnagar, Gujarat 384315, India

M. Barot (✉)
Research and Development Cell & Department of Environmental Science, Parul Institute of Applied Sciences, Parul University, Vadodara, Gujarat 391760, India
e-mail: meghabarot1011@gmail.com

A. Husen (ed.), *Plant Response to Silver Nanoparticles*, Smart Nanomaterials Technology, https://doi.org/10.1007/978-981-97-7352-7_7

1 Introduction

The application of nanotechnology to improve plant development and productivity has drawn a lot of interest to the field of agriculture in recent years. By enhancing plants' ability to absorb nutrients and preventing plant-infecting diseases, nanoparticles have improved agricultural use [3]. Studies have shown that nanoparticles have a distinct impact on higher plants, with both beneficial and detrimental outcomes observed [5, 22–25, 35]. The effect of nanoparticles on the condition and structure of mycorrhizal fungus is yet largely unexplored. Arbuscular mycorrhiza has been the subject of the majority of research, with results that are ambiguous, much like those about their influence on plants. Many plant roots can develop symbiotic connections with arbuscular mycorrhizal fungus (AMF), which is advantageous [38]. Furthermore, it has been demonstrated that AgNPs have an adverse impact on AMF mycorrhizal growth [7]. In contrast, AMF mycorrhizal clover development was positively impacted by nano-Ag and FeO particles [15]. The addition of nanoparticles additionally influenced the properties of the soil, which can improve the composition and organization of AMF communities [2, 33]. A detailed study of AMF and AgNP interaction in the soil matrix is crucial for harnessing their combined potential to improve plant development and production.

2 Silver Nanoparticles

The nanoparticles are clusters of atoms ranging from 1 to 100 nm in size. The Greek word "nano" means "extremely small," and it is synonymous with dwarf. Because of their higher surface area and strong antibacterial qualities, metallic nanoparticles are the most promising. Many nanomaterials have been developed, including copper, magnesium, zinc, gold, titanium, alginate, and silver [1, 17, 36]. However, AgNPs shown to be the most successful due to their strong antimicrobial potential [16]. These characteristics render AgNPs appealing for various agricultural uses, such as crop improvement, nutrient transport, and plant disease control.

3 Arbuscular Mycorrhizal Fungi (AMF)

AMF soil fungi belong to the Glomeromycota phylum. In soil ecosystems, AMF constitutes a significant group of plant symbionts which are considerably associated with maximum higher plants (more than 80% of higher plants) [37]. Plants and AMF association is mutualistic, benefiting both partners. In addition to higher phosphorus (P) uptake to boost the production of biomass [12, 39], the host plant can benefit against metal stress [11, 26]. In addition to their well-studied functions of immobilizing metal in cell wall components or their secreted chemicals like glomain, AMF have also been proposed to regulate the genes expression implicated in plant metal transportation mechanisms [9, 26, 32, 42]. The findings demonstrated a substantial suppression of dry mass of roots in the absence of AMF following treatment with 2000 mg/kg AgNPs. At every degree of AgNP addition, the AMF inoculation improved the shoot's dry mass [10].

The benefits of these fungi association with host plants are the availability of soil nutrients that are not easily accessible to roots (particularly low mobility nutrients like P and Zn), resistance against biotic and abiotic factors such as plant pathogens, salinity, drought, and enhanced plant growth, and toxic heavy metals [19, 30, 37]. Mycorrhizal hyphae have been shown in earlier research to have a significant capacity for binding metals, with concentrations of metals higher than those found in roots [26, 41]. By altering the pH of the soil, AM fungus may indirectly reduce the bioavailability of metals stated by Hu et al. [20]. The lack of evidence by Cao et al. [8] that AMF have a major impact on soil pH suggests that this mechanism is not the cause of the lower Ag uptake by mycorrhizal plants. According to Balestrini and Lanfranco [6], AM fungus may control the expression of genes related to their hosts' metal transport. AM fungus have the ability to release chemicals rich in carbon into soil, which attracts microorganisms and promotes their growth [40, 44].

4 Interaction of AMF and AgNPs in Soil Matrix

Plant development and productivity may be significantly impacted by the way AgNPs and AMF interact in the soil matrix. AMFs have a substantial direct or indirect effect on the promotion of plant growth. It supports healthy plant aquaporin function, or the uptake of water, as well as the numerous other nutrients necessary for plant development and growth. However, the use of various nanoparticles in agriculture sector is continuously growing, although scientists have differing views on the contribution of nanoparticles to plant development. Furthermore, a few efforts have been made to use AMF to reduce the harmful effects of nanoparticles.

Previous research indicated that AMF growth in plant roots and surrounding soil may be facilitated by certain nanoparticles interacting with it. Feng et al. [15] reported that the biological effects of AgNPs, iron oxide nanoparticles (FeONPs), Fe_2O_3, and silver in bulk material form on AMF in mycorrhizal clover are dose-dependent. The findings showed a substantial enhancement in fungal growth and function in plants treated with nanoparticles. Increased AMF colonization in the soil around plant roots and related conditions will inevitably lead to better plant growth. Cao et al. [8] studied that in AgNP-treated soils, AMF inoculation increased the accumulation of plant biomass relative to non-AMF plants, demonstrating an effective symbiosis. They observed the positive effects of arbuscular mycorrhizal fungi on silver nano-particle-treated maize seedlings. Control samples (non-AM fungal inoculated) showed less uptake of phosphorous than the treated ones. In the AM-treated seedling genes involved in P uptake (Pht1) were also up-regulated. Apart from it,

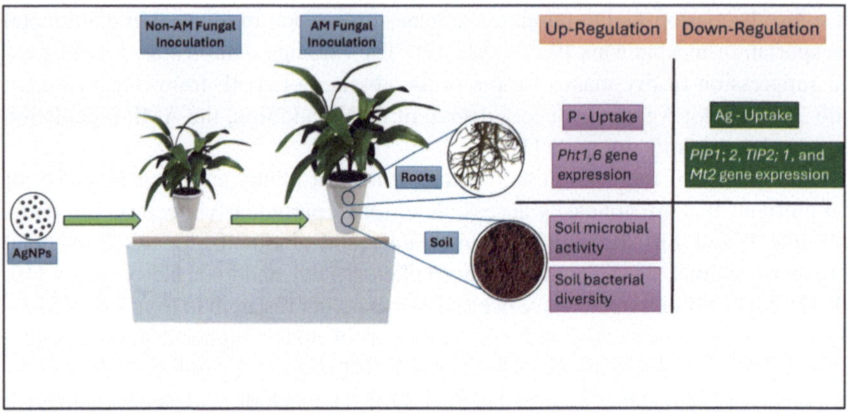

Fig. 1 Schematic diagram of effect of AgNPs and AMF on plant growth

other important genes involved in metal transport (Mt2; PIP; and TIP) were also investigated. They found that these metal transporter genes were relatively less expressed in plants treated with AMF as compared to non-AM-treated maize plants. Moreover, both soil microbial activity and diversity were found to be increased in AM-inoculated, AgNPs-treated plants. Together, their findings suggested that inoculation of AMF can alleviate the metal toxicity imparted by AgNPs (Fig. 1).

AgNPs have antibacterial capabilities, but under modest NPs treatments, the colonization of mycorrhizal maize roots infected with AMF did not reduce; in fact, it increased. The advantageous function of AMF in promoting plant development and P acquisitions in comparison to non-AMF inoculation plants under these nanoparticles stress is consistent with positive gene expression of P transporter.

Ag concentration and the activities of antioxidant enzymes in plants were reduced as a result of AMF's improved capacity to reduce AgNP stress (via increased growth and ecological behaviors) in response to the elevated AgNPs content [15]. Inhibiting plant metal uptake or preventing metal from moving from plant roots to shoots is frequently strongly linked to the reduction of AMF's effects on metal-induced phytotoxicity [4, 26]. Additionally, a number of earlier research showed that AM fungus reduced overall metal acquisitions, which in turn reduced the phytotoxicity of heavy metals [4, 43]. Moreover, the inoculation of AMF significantly altered the makeup of the native bacterial community under AgNPs treatments, indicating that the AMF had a selective effect. In order to aid plants in resisting AgNPs stress, the resultant bacterial community may collaborate with the AM fungal inoculum. The synergistic effects of combination inoculation of AMF and certain bacterial functional members on plant resistance to exotic stress provide support for this concept [9, 13].

Nitrogen and phosphorus intake by plants can be improved by AMF [34]. AgNPs have the capacity to interact with soil constituents and release nutrients in suitable forms that are soluble in plants, which may further improve nutrient availability and uptake in the soil [27].

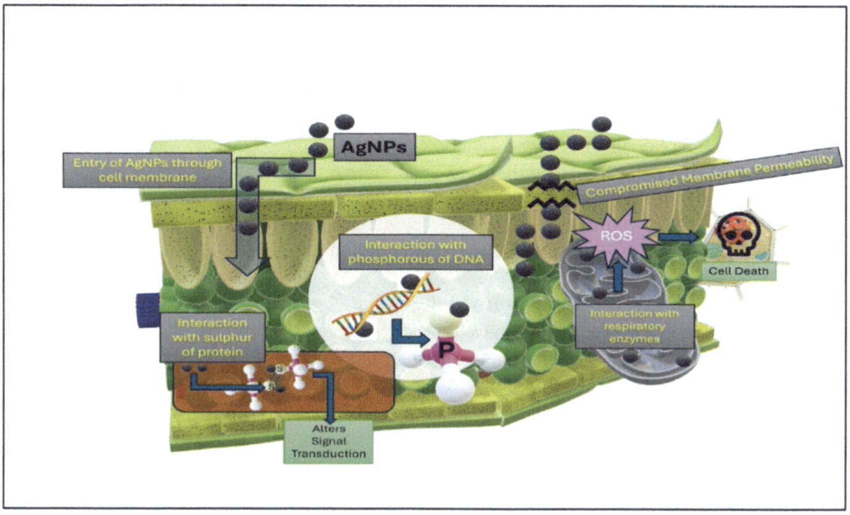

Fig. 2 Graphical representation of mechanism of cell lysis by silver nanoparticles

Ag NPs' (and heavy metals') stimulatory influence on mycorrhizal colonization may be explained by mycorrhizal fungi allocating more carbon to them in order to obtain the benefits of their protective properties [28]. The antibacterial properties of NPs may directly contribute to the decrease in rhizobial nodule growth in host plants. The mechanisms through which NPs exert their antibacterial effects can generally be attributed to oxidative stress, metal ion release or non-oxidative mechanisms like cell membrane damage [14, 18, 21, 29, 31]. These mechanisms can all take place concurrently. Figure 2 illustrates the mode of action of AgNPs against microbes (Fig. 2).

5 Conclusion

Arbuscular mycorrhizal fungi and silver nanoparticles interact in a complex and diverse way in the soil matrix, potentially affecting plant growth and productivity both positively and negatively. Utilizing the combined potential of AMF and AgNPs to improve agricultural sustainability and food security requires an understanding of the mechanisms behind this relationship. To maximize the benefits of using AgNPs and AMF together while minimizing any negative effects on ecosystem functioning and soil health, more research is required.

References

1. Ahmad Z, Pandey R, Sharma S, Khuller GK (2006) Alginate nanoparticles as antituberculosis drug carriers: formulation development, pharmacokinetics and therapeutic potential. Indian J Chest Dis Allied Sci 48(3):171
2. Ameen F, Alsamhary K, Alabdullatif JA, ALNadhari S (2021) A review on metal-based nanoparticles and their toxicity to beneficial soil bacteria and fungi. Ecotoxicol Environ Saf 213:112027
3. Anand R, Bhagat M (2019) Silver nanoparticles (AgNPs): as nanopesticides and nanofertilizers. MOJ Biol Med 4(1):19–20
4. Andrade SA, Gratão PL, Azevedo RA, Silveira AP, Schiavinato MA, Mazzafera P (2010) Biochemical and physiological changes in jack bean under mycorrhizal symbiosis growing in soil with increasing Cu concentrations. Environ Exp Bot 68(2):198–207
5. Bachheti RK, Bachheti A, Husen A (2024) Metal and metal-oxide based nanomaterials (Synthesis, Agricultural, Biomedical and Environmental Interventions). Springer Nature Singapore Pte Ltd., 152 Beach Road, #21-01/04 Gateway East, Singapore 189721, Singapore
6. Balestrini R, Lanfranco L (2006) Fungal and plant gene expression in arbuscular mycorrhizal symbiosis. Mycorrhiza 16:509–524
7. Cao JL, Feng YZ, He SY, Lin XG (2017) Silver nanoparticles deteriorate the mutual interaction between maize (Zea mays L.) and arbuscular mycorrhizal fungi: a soil microcosm study. Appl Soil Ecol 119:307–316
8. Cao J, Feng Y, Lin X, Wang J (2020) A beneficial role of arbuscular mycorrhizal fungi in influencing the effects of silver nanoparticles on plant-microbe systems in a soil matrix. Environ Sci Pollut Res 27:11782–11796
9. Chen XW, Wu L, Luo N, Hui C, Wong MH, Li H (2019) Arbuscular mycorrhizal fungi and the associated bacterial community influence the uptake of cadmium in rice. Geoderma 337:749–757
10. Chen L, Xiao Y (2023) Silver nanoparticles and arbuscular mycorrhizal fungi influence Trifolium repen root-associated AMF community structure and its co-occurrence pattern. Sci Hortic 320:112232
11. Cornejo P, Pérez-Tienda J, Meier S, Valderas A, Borie F, Azcón-Aguilar C, Ferrol N (2013) Copper compartmentalization in spores as a survival strategy of arbuscular mycorrhizal fungi in Cu-polluted environments. Soil Biol Biochem 57:925–928
12. Deguchi S, Uozumi S, Touno E, Kaneko M, Tawaraya K (2012) Arbuscular mycorrhizal colonization increases phosphorus uptake and growth of corn in a white clover living mulch system. Soil Sci Plant Nutr 58(2):169–172
13. del Mar AM, Kohler J, Caravaca F, Roldán A (2009) Differential effects of Pseudomonas mendocina and Glomus intraradices on lettuce plants physiological response and aquaporin PIP2 gene expression under elevated atmospheric CO2 and drought. Microb Ecol 58:942–951
14. Fan R, Huang YC, Grusak MA, Huang CP, Sherrier DJ (2014) Effects of nano-TiO2 on the agronomically-relevant Rhizobium–legume symbiosis. Sci Total Environ 466:503–512
15. Feng Y, Cui X, He S, Dong G, Chen M, Wang J, Lin X (2013) The role of metal nanoparticles in influencing arbuscular mycorrhizal fungi effects on plant growth. Environ Sci Technol 47(16):9496–9504
16. Gong P, Li H, He X, Wang K, Hu J, Tan W et al (2007) Preparation and antibacterial activity of Fe3O4@Ag nanoparticles. Nanotechnology 18:604–611
17. Gu H, Ho PL, Tong E, Wang L, Xu B (2003) Presenting vancomycin on nanoparticles to enhance antimicrobial activities. Nano Lett 3(9):1261–1263
18. Gurunathan S, Han JW, Dayem AA, Eppakayala V, Kim JH (2012) Oxidative stress-mediated antibacterial activity of graphene oxide and reduced graphene oxide in Pseudomonas aeruginosa. Int J Nanomed 5901–5914
19. Hildebrandt U, Regvar M, Bothe H (2007) Arbuscular mycorrhiza and heavy metal tolerance. Phytochemistry 68(1):139–146

20. Hu JL, Wang HS, Wu FY, Wu SC, Cao ZH, Lin XG, Wong MH (2014) Arbuscular mycorrhizal fungi influence the accumulation and partitioning of Cd and P in bashfulgrass (Mimosa pudica L.) grown on a moderately Cd-contaminated soil. Appl Soil Ecol 73:51–57
21. Huang YC, Fan R, Grusak MA, Sherrier JD, Huang CP (2014) Effects of nano-ZnO on the agronomically relevant Rhizobium–legume symbiosis. Sci Total Environ 497:78–90
22. Husen A (2020a) Interactions of metal and metal-oxide nanomaterials with agricultural crops: an overview, In: Husen A, Jawaid M (eds) Nanomaterials for agriculture and forestry applications. Elsevier Inc. 50 Hampshire St., 5th Floor, Cambridge, MA 02139, USA, pp 167–197
23. Husen A (2020b) Carbon-based nanomaterials and their interactions with agricultural crops. In: Husen, A, Jawaid M (eds) Nanomaterials for agriculture and forestry applications. Elsevier Inc. 50 Hampshire St., 5th Floor, Cambridge, MA 02139, USA, pp 199–218
24. Husen A (2022) Engineered nanomaterials for sustainable agricultural production, soil improvement and stress management. Elsevier Inc., 50 Hampshire St., 5th Floor, Cambridge, MA 02139, USA. https://doi.org/10.1016/C2021-0-00054-7
25. Husen A (2023) Nanomaterials and nanocomposites exposures to plants (Response, Interaction, Phytotoxicity and Defense Mechanisms). Springer Nature Singapore Pte Ltd., 152 Beach Road, #21-01/04 Gateway East, Singapore 189721, Singapore. https://doi.org/10.1007/978-981-99-2419-6
26. Joner EJ, Briones R, Leyval C (2000) Metal-binding capacity of arbuscular mycorrhizal mycelium. Plant Soil 226:227–234
27. Khan S, Zahoor M, Khan RS, Ikram M, Islam NU (2023) The impact of silver nanoparticles on the growth of plants: the agriculture applications. Heliyon
28. Kiers, ET, Duhamel M, Beesetty Y, Mensah JA, Franken O, Verbruggen E, ... Bücking H (2011) Reciprocal rewards stabilize cooperation in the mycorrhizal symbiosis. Science 333(6044):880–882
29. Leung YH, Ng AM, Xu X, Shen Z, Gethings LA, Wong MT, ... Leung FC (2014) Mechanisms of antibacterial activity of MgO: non-ROS mediated toxicity of MgO nanoparticles towards Escherichia coli. Small 10(6):1171–1183
30. Miransari M (2010) Contribution of arbuscular mycorrhizal symbiosis to plant growth under different types of soil stress. Plant Biol 12(4):563–569
31. Nagy A, Harrison A, Sabbani S, Munson Jr RS, Dutta PK, Waldman WJ (2011) Silver nanoparticles embedded in zeolite membranes: release of silver ions and mechanism of antibacterial action. Int J Nanomed 1833–1852
32. Noori A, White JC, Newman LA (2017) Mycorrhizal fungi influence on silver uptake and membrane protein gene expression following silver nanoparticle exposure. J Nanopart Res 19:1–13
33. Oehl F, Laczko E, Bogenrieder A, Stahr K, Bösch R, van der Heijden M, Sieverding E (2010) Soil type and land use intensity determine the composition of arbuscular mycorrhizal fungal communities. Soil Biol Biochem 42(5):724–738
34. Qi S, Wang J, Wan L, Dai Z, da Silva Matos DM, Du D, Egan S, Bonser SP, Thomas T, Moles AT (2022) Arbuscular mycorrhizal fungi contribute to phosphorous uptake and allocation strategies of Solidago canadensis in a phosphorous-deficient environment. Front Plant Sci 13:831654
35. Rizwan M, Ali S, Qayyum MF, Ok YS, Adrees M, Ibrahim M, Zia-ur-Rehman M, Farid M, Abbas F (2017) Effect of metal and metal oxide nanoparticles on growth and physiology of globally important food crops: a critical review. J Hazard Mater 322:2–16
36. Schabes-Retchkiman PS, Canizal G, Herrera-Becerra R, Zorrilla C, Liu HB, Ascencio JA (2006) Biosynthesis and characterization of Ti/Ni bimetallic nanoparticles. Opt Mater 29(1):95–99
37. Smith SE, Read D (2008) Mineral nutrition, toxic element accumulation and water relations of arbuscular mycorrhizal plants. In: Mycorrhizal symbiosis, 145-VI
38. Smith SE, Jakobsen I, Grønlund M, Smith FA (2011) Roles of arbuscular mycorrhizas in plant phosphorus nutrition: interactions between pathways of phosphorus uptake in arbuscular mycorrhizal roots have important implications for understanding and manipulating plant phosphorus acquisition. Plant Physiol 156(3):1050–1057

39. Smith SE, Smith FA, Jakobsen I (2003) Mycorrhizal fungi can dominate phosphate supply to plants irrespective of growth responses. Plant Physiol 133(1):16–20
40. Tisserant E, Malbreil M, Kuo A, Kohler A, Symeonidi A, Balestrini R (2013) Genome of an arbuscular mycorrhizal fungus provides insight into the oldest plant symbiosis. Proc Natl Acad Sci USA 110:20117–20122
41. Toler HD, Morton JB, Cumming JR (2005) Growth and metal accumulation of mycorrhizal sorghum exposed to elevated copper and zinc. Water Air Soil Pollut 164:155–172
42. Vodnik D, Grčman H, Maček I, Van Elteren JT, Kovačevič M (2008) The contribution of glomalin-related soil protein to Pb and Zn sequestration in polluted soil. Sci Total Environ 392(1):130–136
43. Vogel-Mikus K, Pongrac P, Kump P, Necemer M, Regvar M (2006) Colonisation of a Zn, cd and Pb hyperaccumulator Thlaspi praecox Wulfen with indigenous arbuscular mycorrhizal fungal mixture induces changes in heavy metal and nutrient uptake. Environ Pollut 139:362–371
44. Zhang L, Shi N, Fan JQ, Wang F, George TS, Feng G (2018) Arbuscular mycorrhizal fungi stimulate organic phosphate mobilization associated with changing bacterial community structure under field conditions. Environ Microbiol 20:2639–2651

Effect of Silver Nanoparticles on Seed Germination and Seedling Growth

Anshu Sharma, Mayur Mukut Murlidhar Sharma, Pankaj Sharma, and Azamal Husen

Abstract The growing interest in incorporating nanotechnology into agriculture has been notable in recent years, particularly with the emergence of silver nanoparticles (AgNPs) as potential catalysts for enhancing plant growth and overall productivity. This chapter delves into the intricate relationship between AgNPs and key stages of plant development, centering on seed germination and seedling growth. Through a thorough examination of existing literature and original research findings, the chapter sheds light on the multifaceted effects of AgNPs on a variety of plant species. The exploration begins by dissecting the mechanisms governing the interaction between AgNPs and seeds during the germination process. Factors such as nanoparticle concentration, size, and exposure duration are meticulously scrutinized to unveil their nuanced impact on germination rates, dormancy alleviation, and early root development. Subsequent sections delve into the complex cellular and molecular responses triggered by AgNPs within seedlings, revealing insights into processes like nutrient uptake, photosynthesis, and antioxidative defense mechanisms. Furthermore, the chapter critically assesses the potential environmental implications of AgNPs within the soil–plant system, considering both positive and adverse effects on the ecosystem. The chapter concludes by providing valuable insights into

---

A. Sharma
Department of Chemical Engineering, Kangwon National University, Chuncheon, Korea

M. M. M. Sharma
Department of Agriculture and Industry, Kangwon National University, Chuncheon, Korea

P. Sharma (✉)
Department of Microbiology, CCS Haryana Agricultural University, Hisar, Haryana 125004, India
e-mail: sharmap@uic.edu

Department of Ophthalmology and Visual Sciences, University of Illinois Chicago, Chicago 60612, USA

A. Husen
Wolaita Sodo University, PO Box 138, Wolaita, Ethiopia

Department of Biotechnology, Graphic Era (Deemed to be University), Dehradun, Uttarakhand 248002, India

A. Husen (ed.), *Plant Response to Silver Nanoparticles*, Smart Nanomaterials Technology, https://doi.org/10.1007/978-981-97-7352-7_8

the future prospects of deploying AgNPs in sustainable agriculture practices, emphasizing the imperative for a balanced approach that maximizes benefits while minimizing potential risks. This book chapter is positioned as an invaluable resource for researchers, agronomists, and policymakers aiming to deepen their understanding of the dynamic interplay between AgNPs and plant development, thereby contributing to the evolving realm of nano-agriculture.

Keywords AgNPs · Seed germination · Seedling growth · Sustainable Agriculture · Nanoparticles

1 Introduction

In the preceding decades, our world has confronted a multitude of challenges, including persistent demographic pressures, constantly shifting climatic conditions, and intensified competition for resources. The convergence of these factors has created a substantial threat, necessitating a crucial call to safeguard global food sustainability [50–52]. Present-day agricultural approaches demonstrate glaring inadequacies when it comes to coping with the soaring global appetite for food. The escalating anxieties regarding health and environmental considerations pose an imposing challenge, casting a formidable shadow over the ambitious endeavor to achieve sustainability in global food production [31, 32, 64]. Nano-technological interventions have unquestionably unleashed a revolutionary boon upon the agricultural sector, showcasing an awe-inspiring prowess in achieving targeted delivery, meticulous mastery over release mechanisms, unprecedented solubility heights, and the amplification of shelf life to almost mythical proportions. In the annals of time, a spectacular and unprecedented surge has manifested in the relentless pursuit and integration of nanotechnology within the agricultural domain [5, 20, 22, 23, 25, 48, 49, 2, 10, 13, 21, 78–80]. A dazzling array of nanoparticles (NPs), akin to mystical agents, has been acclaimed for their transcendental roles in agriculture, morphing into aqueous reservoirs, orchestrating the celestial precision of nutrient and fertilizer dispensation. Beyond this agricultural alchemy, the metallic and metal oxide NPs stand as titans, wielding biocidal powers of cosmic proportions, as they valiantly combat an entire cosmos of bacterial and fungal adversaries, painting a narrative of triumph in the agro-scientific saga. Natural or synthetic polymers can be combined to form spherical, polymeric particles known as nanoparticles (NP). They can range from 10 to 500 nm in size. NP technology is rapidly advancing, providing novel applications but because of its size, careful consideration is required for assessments and environmental implications of NPs on seedlings [53, 54]. Nanotechnology in agriculture has emerged as a revolutionary approach, offering precision and efficiency to enhance various aspects of farming. The application of nanomaterials, such as NPs and nanosensors, allows for targeted delivery of nutrients, pesticides, and fertilizers, minimizing wastage and environmental impact (Fig. 1). Nanotechnology enables the development of smart delivery systems that respond to specific environmental conditions, ensuring optimal resource utilization. Additionally, nanoscale sensors facilitate real-time monitoring of crop health, soil conditions, and water quality, providing farmers with valuable data for informed decision-making. The use of nanotechnology in agriculture holds the promise of increased crop yields, improved resource efficiency, and sustainable farming practices, contributing to global food security while minimizing the ecological footprint of agricultural activities [55, 56]. Additionally, Nanotechnology has demonstrated its potential to significantly impact seed germination in agriculture. Researchers are exploring the use of nanomaterials, such as nanoparticles and nanocoatings, to enhance the germination process and improve seedling vigor. NPs can be engineered to deliver nutrients and growth-promoting substances directly to the seeds, providing a controlled and targeted release during the critical germination phase [57–59]. Additionally, nanocoatings can be applied to seeds to improve their water absorption and retention abilities, promoting better hydration and ensuring optimal conditions for germination. Nanotechnology also offers the development of nanoscale sensors to monitor and regulate environmental factors affecting germination, such as temperature and moisture levels. These advancements in nanotechnology contribute to more efficient and sustainable agricultural practices by increasing seedling success rates and overall crop productivity. Silver nanoparticles (AgNPs), are the metal NPs that are attracting the most attention due to their capacity to boost crop growth and output. They have special qualities with a wide range of uses, including antibacterial and anticancer effects. However, the study for its interactions with seeds is still in its infancy. Due to the rapidly growing population and the resulting increased need for basic necessities, one of our top priorities is to invest in the study of information that will advance agricultural technology to produce crops more efficiently [3 77, 41–43, 66]. Some plants are more sensitive to the NP, while others showed relatively higher resistance. In developing nations, high-yield crop varieties, herbicides, and artificial fertilizers are employed to achieve widespread positive crop output [71]. However, the extensive use of chemical pesticides and fertilizers—which are necessary to enable green evolution—has unintended consequences, including a reduction in soil diversity. Since they retain a steady and continuous release of agrochemicals and nutrients, nano-encapsulated traditional fertilizers, herbicides, and pesticides can be used to precisely dose fertilizers into crops. One of the main causes for concern in the environment has been abiotic stress. In terms of plant development and productivity, the main abiotic stresses include drought, salt, and extremely high and low temperatures. Recently, a number of metallic NP (AgNPs, AuNPs, FeNPs, TiO_2NPs, and ZnNPs) have been used to various crop plants to enhance stress tolerance, plant growth, and seed germination. In addition, using AgNPs to promote plant development and seed germination can reduce the negative effects of using chemical pesticides and fertilizers, speeding up crop output without having to deal with the drawbacks of the traditional approach. Among the cutting-edge technologies that must be investigated to solve the aforementioned issues is nanotechnology, which can be used to create strategies for increasing crop output that also ensures environmental safety and increased consumption efficiency. These days, plant growth, photosynthetic quantum efficiency, and seed germination are all improved by using AgNPs, which are also

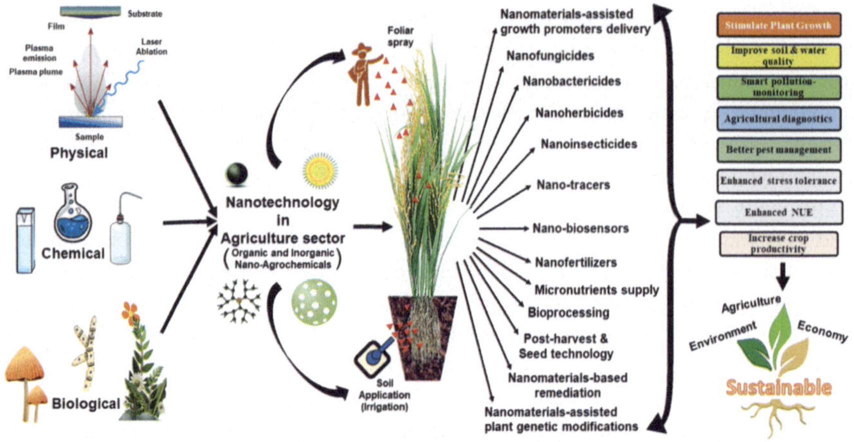

Fig. 1 Various applications of nanotechnology in the agriculture sector

employed as antimicrobial agents to prevent plant disease. The dosage of AgNPs affects a plant's ability to respond to them by either enhancing or inhibiting growth.

2 Synthesis of Silver Nanoparticles

2.1 *(a) Physical*

The two most significant physical methods are laser ablation and evaporation–condensation. Source materials were evaporated with the help of the little ceramic heater. Because of the extremely steep temperature gradient at the heater surface compared to a tube furnace, the evaporated vapor may cool at a suitably quick rate. This enables the production of tiny NPs at high concentrations. The heater surface's temperature does not change over time; therefore, nanoparticle production is extremely steady. This physical approach can be helpful as a calibration tool for nanoparticle measuring equipment as well as a nanoparticle producer for long-term investigations on the toxicity of inhalation [24]. By using a laser to ablate metallic bulk materials in solution, AgNPs can be produced [33].

2.2 *(b) Chemical*

AgNPs may be biosynthesized in a number of methods, such as chemical reduction, ultrasonic-aided reduction, microwave-assisted synthesis, and more. These are only a handful of the numerous. The most popular method for creating AgNPs is chemical reduction using both inorganic and organic reducing agents. Silver ions (Ag^+) in aqueous or non-aqueous solutions are generally reduced using a variety of reducing agents, including sodium citrate, ascorbate, sodium borohydride (NaBH4), elemental hydrogen, polyol process, Tollens reagent, N, N-dimethylformamide (DMF), and poly (ethylene glycol)-block copolymers. These reducing chemicals decrease Ag^+, resulting in the agglomeration of silver into oligomeric clusters and the production of metallic silver (Ag0). As a result of these clusters, metallic colloidal silver particles eventually develop [28]. Using microemulsion methods, it is possible to synthesize silver NPs that are uniform and controlled in size. The initial spatial separation of reactants (metal precursor and reducing agent) in two immiscible phases is the foundation for the synthesis of NPs in two-phase aqueous organic systems. The pace of interactions between metal precursors and reducing agents is dependent on the interface between the two liquids and the intensity of inter-phase transport between two phases, which is mediated by a quaternary alkyl-ammonium salt. Metallic colloidal silver particles are eventually formed when metal clusters formed at the interface are transferred to the organic medium by the inter-phase transporters and stabilized as a result of their surface being coated with stabilizer molecules that occur in the non-polar aqueous medium [29].

2.3 (c) Biological

Simple prokaryotic bacterial cells to eukaryotic fungus and plants have the capacity to synthesize NPs [35, 60–63]. By changing a few crucial parameters, such as substrate concentration, pH, light, temperature, buffer strength, electron donor (such as glucose or fructose), biomass and substrate concentration, mixing speed, and exposure duration, the sizes and morphologies of the NPs may be regulated. The most commonly explored bacterial species for the synthesis of AgNPs include *Escherichia coli, Lactobacillus sp., Bacillus cereus, Acinetobacter sp., Pseudomonas sp., Corynebacterium sp.* and *Klebsiella pneumonia* [44, 45, 65]. However, among the different bacteria used, the supernatant from Enterobacter cloacae (*Enterobacteriaceae*) was found to be most potent for the speedy fabrication of AgNPs [47]. *Lactobacillus* strains produce NPs in their bacterial cells when exposed to silver ions. According to reports, AgNPs may be grown by subjecting lactic acid bacteria found in buttermilk whey to various silver ion combinations. Due to the action of sugars and enzymes in the cell wall, AgNPs were first nucleated on the cell surface. Subsequently, the metal nuclei were carried into the cell, where they expanded and aggregated to bigger particles [27, 36]. High-stability AgNPs (40 nm) can be produced by bioreduction of aqueous silver ions with a culture supernatant of *Bacillus licheniformis*, a nonpathogenic bacteria [28].

3 Properties of Silver Nanoparticles

AgNPs are thought to be the best choice for combating pathogenic microbial activity because of their special qualities. AgNPs' physical and chemical characteristics, such as their surface chemistry, size, distribution, shape, morphology, composition, coating/capping, agglomeration, rate of dissolution, reactivity of the particles in solution, efficiency of ion release, cell type, and, lastly, the kind of reducing agents used during synthesis, are essential for determining their cytotoxicity, Shape plays an equivalent role in determining toxicity [12, 75]. Their toxicity is mostly determined by the presence of biological or chemical coatings on the surface of the NPs [68]. AgNP surface charges may have an impact on a cell's toxicity. For example, these NPs' positive surface charge makes them more appropriate and prolongs their half-life compared to negatively charged NPs [69]. Many variables, including pH, temperature, and incubation duration, among others, affect physiochemical characteristics of AgNPs [1]. The surface area to mass ratio of nanomaterials is significantly higher than that of micrometric particles, with a thousand-fold increase in substance.

4 Method of Applying Nanoparticles to Plants

For the past ten years, nanoscience has gained popularity in the field of plant science. It employs different metal NP as fertilizers or carriers of agrochemicals [73].

Under stressful and unstressed conditions, these NPs have been shown to have an impact on seedling development, plant growth, germination, root growth [72], enhanced glucose metabolism [34], reactive oxygen species [74], and nutrient transfer [11]. Different techniques, such as hydroponic culture for sustainable agriculture, foliar spraying, mixing with soil, and seed priming, can be used to apply NPs to plants. Among the several application techniques, seed priming technology is the most widely used technique for encouraging NP penetration via seeds by passive diffusion with water, and it has demonstrated beneficial effects. Previous research confirmed the idea that because smaller particles have a bigger surface area than larger ones, they may be more dangerous [67]. The main mechanisms by which these NPs are absorbed by leaves during foliar application are stomatal permeation, epidermal absorption, and internalization. Since NPs treated to seeds did not reach soil, seed priming may be recommended as a potential method for their commercial application. It also reduces the amount of NPs that disperse into ecosystems [7]. Among its numerous benefits is its ability to combat plant diseases and pathogens and to supply vital micronutrients through leaves that are seldom found in nutrition-deficient soil [6, 8, 30].

5 Effects on Root and Shoot Length

Studies have shown that crop productivity has greatly benefited from the use of nanomaterials in many ways. In the agricultural sector, AgNPs have been found to have effects on gas exchange rate, plant growth and development, and seed germination under a variety of abiotic conditions. AgNP causes damage to the cell wall and allows vacuoles to enter during their penetration of the root cell. Toxic symptoms tend to manifest more strongly in roots than in shoots because roots are the first target tissue to come into contact with contaminants [16]. This led researchers to hypothesize that the damage may have been caused by larger particles penetrating through the tiny pores in the cell walls. According to reports, AgNPs treatment primarily affected shoot and root growth and also seed germination. A significant reduction in root and shoot length of wheat, tomato, radish, flax, onion, ryegrass, two-rowed barley, pearl millet, Chinese cabbage, mung bean, bishop bean, corn has been seen when used with AgNP.

6 Positive Effects of Silver Nanoparticles on Seed Germination and Seedling Growth

AgNPs are effective nanomaterials that may be used in agriculture to increase plant growth and development under stressful environmental circumstances, including seed germination. Recently, plant biologists have shown sufficient interest in using AgNPs because of its environmentally benign qualities. AgNPs are mostly given to plants by foliar spray, soil mixing techniques, or seed priming technologies [7]. Benefits of low AgNP concentrations (10 and 15 mg/L) with urea have been demonstrated in terms of decreased diseased condition in seeds, increased germination rate, increased chlorophyll contents, increased stomatal conductance, and higher seedling masses in oilseed rape and cucumber under heat stress [1].

Although there are studies attributed to both negative and positive impacts of AgNPs on seed germination and plant growth (Table 1). They can even substitute toxic pesticides and highly concentrated mineral fertilizers. In drought-stricken eggplant, foliar spraying AgNPs increased growth metrics, photosynthetic pigments, proline, hydrogen peroxide (H_2O_2), and antioxidant activities. These results have been obtained through seed priming and foliar application. In order to assess if biocompatible AgNPs might enhance rice seed germination and starch metabolism following seed priming application using 5 and 10 mg/L AgNPs under standard circumstances, kafr lime leaf extract was used in the synthesis process. They penetrate the seed coat which enhances water intake, increases ROS and H_2O_2, and improves seed germination as well as starch metabolism [34]. Higher AgNP concentrations were shown to have no significant influence on the mean germination time or corn germination percentage, but they did cause a substantial rise in the germination rate of corn seeds. When 1.5 mg/ml of AgNPs were exposed to maize seeds, the greatest germination rate of 6.5 seeds per day was noted. Also, watermelon and zucchini plants when exposed to AgNPs exhibited substantially higher germination percentages and rates. At 2 mg/ml AgNPs, watermelon had the greatest germination rate (1.59 seeds/day) and germination percentage (73.33%). The maximum germination rates (1.68 and 1.66 seeds/day) and germination percentages (86.67% and 90%) for zucchini plants were noted at 0.5 and 2.5 mg/ml of AgNPs, respectively [74]. In a study conducted on fenugreek, the application of five distinct concentrations of AgNPs (0, 10, 20, 30, and 40 $\mu g\ mL^{-1}$) was executed. Throughout the 12-day plant growth period post-germination, a daily administration of 15 ml from each concentration was executed. Examination of seed germination outcomes revealed that lower concentrations of AgNPs positively influenced seed germination and the initial growth of seedlings in Fenugreek, whereas higher concentrations demonstrated mild adverse effects. Control plants exhibited the lowest levels of these parameters, with an escalating application of AgNPs correlating with a diminishing trend in these compounds. Furthermore, a significant positive impact on metrics such as root length, root fresh weight, root dry weight, and root elongation was discerned for all seeds compared to the unexposed control germination. The findings highlighted a statistically significant effect of AgNPs on germination percentage at $P \leq 0.05$, underscoring the noteworthy influence of AgNPs on the germination process [19]. In another study, it was concluded that the concentration range from 0.001 to 0.5 mg/L exhibited no discernible impact on seed germination, while heightened concentrations resulted in inhibitory effects. AgNPs were noted for fostering an augmentation in both shoot and root fresh mass within the ranges of 0.06–1 mg/L and 0.03–0.1 mg/L, respectively. However, concentrations surpassing 3 and 5 mg/L induced inhibition in shoot and root fresh mass, respectively. The application of the examined NPs within the 0.005–0.5 mg/L range stimulated an increase in wheat root length, whereas concentrations exceeding 5 mg/L led to inhibition in root length [4]. In a distinct research endeavor, the study delved into the phytotoxic impact of AgNPs on *Pennisetum glaucum*, an essential crop plant. Notably, seeds subjected to treatment with synthesized AgNPs displayed enhanced germination rates, indicating a positive influence on the initial stages of seed development. However, a more detailed examination revealed that the subsequent seedling growth of the tested species was significantly affected when exposed to different concentrations of AgNPs. The nuanced influence of AgNPs on seedling growth underscores the complexity of their interaction with the specific crop plant. This finding suggests that while AgNPs may contribute positively to certain aspects of seed germination, their effects on subsequent seedling development require careful consideration [39].

7 Phytotoxicity Profile of Silver Nanoparticles

Most studies that have been done on AgNP reveal negative effects on seed germination and seedling growth, however some positive features are also reported. When rice seeds were exposed to AgNPs, the germination rate decreases. Additionally, the

Table 1 Effects of silver NPs on seed germination and seedling growth

Plant attribute	Impact	Species	Key references
Seed germination	Positive	*Phaseolus vulgaris*	Pražak et al. [40]
Seed germination and seedling growth	Negative	Tobacco	Biba et al. [9]
Seed germination	Positive	*Lycopersicon esculentum*	Karami et al. [26]
Seedling growth	Positive	*Oryzya sativa*	Gupta et al. [17]
Seed germination	Positive	*Lens culinaris Medikus*	Hojjat et al. [18]
Seed germination	Positive	*Boswellia Ovalifoliolata*	Savithramma et al. [46]
Seedling growth	Negative	Radish	Zuverza-Mena et al. [76]
Seed germination	Negative	*Vigna radiata*	Nair et al. [37]
Seedling growth	Positive	*Solanum lycopersicum*	Noshad et al. [38]
Seed germination	Negative	*Lactuca sativa*	de Paiva Pinheiro et al. [15]

length of both the roots and shoots of the seedlings were also negatively affected because NP may disrupt key physiological processes, such as water absorption and nutrient uptake, leading to impaired seed germination and growth. The negative impact of AgNPs on seed germination and seedling growth is dose-dependent. The prevention of phytotoxic effects in rice seedlings requires careful consideration of the ideal sizes and concentrations. Higher concentrations of AgNPs lead to more significant reductions in germination and growth parameters [70]. Their presence also negatively affects seed germination traits of *Thymus kotschyanus* seeds, where germination percentage was reduced and the germination rate was retarded. Also, when 200 mg/L of AgNPs were applied in vitro to cucumber and wheat plants, phytotoxic effects were seen during the vegetative development stage as opposed to germination [14].

8 Conclusion

With less space available for farming and more people living in the world than ever before, nanoparticle-mediated enhanced agriculture undoubtedly looks like a great approach to go toward a better future. The AgNPs can cause all sorts of impacts toward crops and plants, including mean germination time, root length, fresh and dry weight, germination percentage, and overall germination time. It is important to note that the effect AgNPs have on plants is highly dependable on the sizes, shapes, and concentrations of the nanomaterial. In addition to lowering production costs, the low nanoparticle concentration used in seed priming also lessens the likelihood that NPs would spread widely across the ecosystem. When applied properly, AgNPs can positively control a plant's ability to grow and function in both stressful and unstressed environments. AgNPs have important implications for agriculture, as the use of AgNPs in various industries, such as electronics and textiles, could potentially contaminate agricultural soils and water sources. The negative effects of AgNPs are not consistent across all crops. Different crop species exhibit varying levels of sensitivity to AgNPs, with some being more affected than others. Plant germination can be accelerated and plant yield can be increased by nanomaterial exposure. But it is also necessary to investigate the effective control over the morphology and monodispersity of AgNPs. Evaluating the effects of AgNPs on seed, and different germination traits for a better understanding of their potential impacts on plant growth and development is still required. Further research is required to assess the long-term consequences of NP exposure on plant populations and ecosystems. Future trials and tests would also determine the precise concentration, best application manner, and optimal application time. This information can be an aid in the development of more safe and sustainable NP for their applications in agriculture and other fields.

References

1. Alabdallah NM, Hasan MM (2021) Plant-based green synthesis of silver nanoparticles and its effective role in abiotic stress tolerance in crop plants. Saudi J Biol Sci 28(10):5631–5639
2. Almutairi ZM, Alharbi A (2015) Effect of silver nanoparticles on seed germination of crop plants. Int J Nuclear Quantum Eng 9(6):689–693
3. Anamika, Patra A, Shehzad S, Rani A, Sharma P, Mohammad KF, Mehta S (2023) Microbial elicitors for priming plant defense mechanisms. In: Sustainable agriculture reviews 60: microbial processes in agriculture. Springer Nature Switzerland, Cham, pp 175–196
4. Asanova AA, Yashin SE, Trofimova TV, Polonskiy VI (2019) Application of silver nanoparticles to improve wheat seedlings growth. In: IOP conference series: earth and environmental science, vol 315, no 5. IOP Publishing, p 052041
5. Bachheti RK, Bachheti A, Husen A (2024) Metal and metal-oxide based nanomaterials (Synthesis, Agricultural, Biomedical and Environmental Interventions). Springer Nature Singapore Pte Ltd., 152 Beach Road, #21-01/04 Gateway East, Singapore 189721, Singapore
6. Bakhtiari M, Moaveni P, Sani B (eds) The effect of iron nanoparticles spraying time and concentration on wheat. biological forum; 2015: Research Trend
7. Balusamy SR, Joshi AS, Perumalsamy H, Mijakovic I, Singh P (2023) Advancing sustainable agriculture: a critical review of smart and eco-friendly nanomaterial applications. J Nanobiotechnol 21(1):372
8. Batool SU, Javed B, Sohail, Zehra SS, Mashwani Z-u-R, Raja NI et al (2021) Exogenous applications of bio-fabricated silver nanoparticles to improve biochemical, antioxidant, fatty acid and secondary metabolite contents of sunflower. Nanomaterials 11(7):1750
9. Biba R, Matić D, Lyons DM, Štefanić PP, Cvjetko P, Tkalec M, Pavoković D, Letofsky-Papst I, Balen B (2020) Coating-dependent effects of silver nanoparticles on tobacco seed germination and early growth. Int J Mol Sci 21(10):3441
10. Budhani S, Egboluche NP, Arslan Z, Yu H, Deng H (2019) Phytotoxic effect of silver nanoparticles on seed germination and growth of terrestrial plants. J Environ Sci Health C 37(4):330–355
11. Buriak JM, Liz-Marzán LM, Parak WJ, Chen X. Nano and plants. ACS Publications, pp 1681–1684
12. Carlson C, Hussain SM, Schrand AM, K. Braydich-Stolle L, Hess KL, Jones RL et al (2008) Unique cellular interaction of silver nanoparticles: size-dependent generation of reactive oxygen species. J Phys Chem B 112(43):13608–13619

13. Castro L, Blázquez ML, González FG, Ballester A (2014) Mechanism and applications of metal nanoparticles prepared by bio-mediated process. Rev Adv Sci Eng 3(3):199–216
14. Cui D, Zhang P, Ma Y-h, He X, Li Y-y, Zhao Y-C et al (2014) Phytotoxicity of silver nanoparticles to cucumber (Cucumis sativus) and wheat (Triticum aestivum). J Zhejiang Univ Sci A 15(8):662–670
15. de Paiva SK, de Medeiros CM, Miguel TB, de Freitas Barros FC, Farias CP, Ferreira OP, de Castro ME (2020) Toxic effects of silver nanoparticles on the germination and root development of lettuce (Lactuca sativa). Aust J Bot 68(2):127–136
16. Gruyer N, Dorais M, Bastien C, Dassylva N, Triffault-Bouchet G (eds) Interaction between silver nanoparticles and plant growth. In: International symposium on new technologies for environment control, energy-saving and crop production in greenhouse and plant 1037
17. Gupta SD, Agarwal A, Pradhan S (2018) Phytostimulatory effect of silver nanoparticles (AgNPs) on rice seedling growth: an insight from antioxidative enzyme activities and gene expression patterns. Ecotoxicol Environ Saf 15(161):624–633
18. Hojjat SS, Ganjali A (2016) The effect of silver nanoparticle on lentil seed germination under drought stress. Int J Farm Allied Sci 5(3):208–212
19. Hojjat SS, Hojjat H (2015) Effect of nano silver on seed germination and seedling growth in Fenugreek seed. Int J Food Eng 1(2):106–110
20. Husen A (2020) Introduction and techniques in nanomaterials formulation. In: Husen, A, Jawaid M (eds) Nanomaterials for agriculture and forestry applications. Elsevier Inc. 50 Hampshire St., 5th Floor, Cambridge, MA 02139, USA, pp 1–14. https://doi.org/10.1016/B978-0-12-817852-2.00001-9
21. Husen A (2022) Engineered nanomaterials for sustainable agricultural production, soil improvement and stress management. Elsevier Inc., 50 Hampshire St., 5th Floor, Cambridge, MA 02139, USA. https://doi.org/10.1016/C2021-0-00054-7
22. Husen A (2023) Nanomaterials and nanocomposites exposures to plants (Response, Interaction, Phytotoxicity and Defense Mechanisms). Springer Nature Singapore Pte Ltd., 152 Beach Road, #21-01/04 Gateway East, Singapore 189721, Singapore. https://doi.org/10.1007/978-981-99-2419-6
23. Husen A, Iqbal M (2019) Nanomaterials and plant potential. Springer International Publishing AG, Gewerbestrasse 11, 6330 Cham, Switzerland. https://doi.org/10.1007/978-3-030-05569-1
24. Jung JH, Oh HC, Noh HS, Ji JH, Kim SS (2006) Metal nanoparticle generation using a small ceramic heater with a local heating area. J Aerosol Sci 37(12):1662–1670
25. Kapoor D, Yadav S, Sharma MM, Sharma P (2023) Interaction between metal nanoparticles and PGPR on the plant growth and development. In: Nanomaterials and nanocomposites exposures to plants: response, interaction, phytotoxicity and defense mechanisms. Springer Nature Singapore, Singapore, pp 327–351
26. Karami S, Heidari R, Rahmani F, Najafi S (2016) Effect of chemical synthesis silver nanoparticles on germination indices and seedlings growth in seven varieties of Lycopersicon esculentum Mill (tomato) plants. J Cluster Sci 27:327–340
27. Korbekandi H, Iravani S, Abbasi S (2012) Optimization of biological synthesis of silver nanoparticles using Lactobacillus casei subsp. casei. J Chem Technol Biotechnol 87(7):932–937
28. Koul B, Poonia AK, Yadav D, Jin J-O (2021) Microbe-mediated biosynthesis of nanoparticles: applications and future prospects. Biomolecules 11(6):886
29. Krutyakov YA, Olenin AY, Kudrinskii A, Dzhurik P, Lisichkin G (2008) Aggregative stability and polydispersity of silver nanoparticles prepared using two-phase aqueous organic systems. Nanotechnol Russ 3:303–310
30. Li P, Du Y, Huang L, Mitter N, Xu ZP (2016) Nanotechnology promotes the R&D of new-generation micronutrient foliar fertilizers. RSC Adv 6(73):69465–69478
31. Loyal A, Pahuja SK, Rani N, Pooja, Srivastava RK, Sharma P (2023a) Titanium oxide nanoparticles: plant response, interaction, phytotoxicity, and defence mechanisms. In: Nanomaterials and nanocomposites exposures to plants: response, interaction, phytotoxicity and defense mechanisms. Springer Nature Singapore, Singapore, pp 263–284

32. Loyal A, Pahuja SK, Sharma P, Malik A, Srivastava RK, Mehta S (2023b) Potential environmental and human health implications of nanomaterials used in sustainable agriculture and soil improvement. In: Engineered nanomaterials for sustainable agricultural production, soil improvement and stress management. Academic Press, pp 387–412
33. Mafuné F, Kohno J-Y, Takeda Y, Kondow T, Sawabe H (2000) Structure and stability of silver nanoparticles in aqueous solution produced by laser ablation. J Phys Chem B 104(35):8333–8337
34. Mahakham W, Sarmah AK, Maensiri S, Theerakulpisut P (2017) Nanopriming technology for enhancing germination and starch metabolism of aged rice seeds using phytosynthesized silver nanoparticles. Sci Rep 7(1):8263
35. Mohanpuria P, Rana NK, Yadav SK (2008) Biosynthesis of nanoparticles: technological concepts and future applications. J Nanopart Res 10:507–517
36. Nair B, Pradeep T (2002) Coalescence of nanoclusters and formation of submicron crystallites assisted by Lactobacillus strains. Cryst Growth Des 2(4):293–298
37. Nair PM, Chung IM (2015) Physiological and molecular level studies on the toxicity of silver nanoparticles in germinating seedlings of mung bean (Vigna radiata L.). Acta physiologiae plantarum 37:1
38. Noshad A, Hetherington C, Iqbal M (2019) Impact of AgNPs on seed germination and seedling growth: a focus study on its antibacterial potential against Clavibacter michiganensis subsp. michiganensis infection in Solanum lycopersicum. J Nanomater:1–2
39. Parveen A, Rao S (2015) Effect of nanosilver on seed germination and seedling growth in Pennisetum glaucum. J Cluster Sci 26:693–701
40. Prażak R, Święciło A, Krzepiłko A, Michałek S, Arczewska M (2020) Impact of Ag nanoparticles on seed germination and seedling growth of green beans in normal and chill temperatures. Agriculture 10(8):312
41. Rani K, Sharma P, Kumar S, Wati L, Kumar R, Gurjar DS, Kumar D, Kumar R (2019) Legumes for sustainable soil and crop management. In: Sustainable management of soil and environment, pp 193–215
42. Rani K, Rani A, Sharma P, Dahiya A, Punia H, Kumar S, Sheoran S, Banerjee A (2022) Legumes for agroecosystem services and sustainability. In: Advances in legumes for sustainable intensification. Academic Press, pp 363–380
43. Razak NQA, Yusoff MHM, Aziz WNAA, Kamal ML, Hasan S, Uyup NH et al (2023) Effects of silver nanoparticles on seed germination and seedling growth: a review. J Indian Chem Soc 100(1):100866
44. Salem SS, Fouda A (2021) Green synthesis of metallic nanoparticles and their prospective biotechnological applications: an overview. Biol Trace Elem Res 199:344–370
45. Saravanan M, Barik SK, MubarakAli D, Prakash P, Pugazhendhi A (2018) Synthesis of silver nanoparticles from Bacillus brevis (NCIM 2533) and their antibacterial activity against pathogenic bacteria. Microb Pathog 116:221–226
46. Savithramma N, Ankanna S, Bhumi G (2012) Effect of nanoparticles on seed germination and seedling growth of Boswellia ovalifoliolata an endemic and endangered medicinal tree taxon. Nano Vision 2(1):2
47. Shahverdi AR, Fakhimi A, Shahverdi HR, Minaian S (2007) Synthesis and effect of silver nanoparticles on the antibacterial activity of different antibiotics against Staphylococcus aureus and Escherichia coli. Nanomed: Nanotechnol, Biol Med 3(2):168–171
48. Sharma P, Sharma MM, Patra A, Vashisth M, Mehta S, Singh B, Tiwari M, Pandey V (2020a) The role of key transcription factors for cold tolerance in plants. In: Transcription factors for abiotic stress tolerance in plants. Academic Press, pp 123–152
49. Sharma P, Sharma MM, Anamika, Kapoor D, Rani K, Singh D, Barkodia M (2020b) Role of microbes for attaining enhanced food crop production. In: Microbial biotechnology: basic research and applications, pp 55–78
50. Sharma P, Pandey V, Sharma MMM, Patra A, Singh B, Mehta S, Husen A (2021a) A review on biosensors and nanosensors application in agroecosystems. Nanoscale Res Lett 16(136):1–24. https://doi.org/10.1186/s11671-021-03593-0

51. Sharma P, Sharma MM, Malik A, Vashisth M, Singh D, Kumar R, Singh B, Patra A, Mehta S, Pandey V (2021b) Rhizosphere, rhizosphere biology, and rhizospheric engineering. In: Plant growth-promoting microbes for sustainable biotic and abiotic stress management. Springer International Publishing, Cham, pp 577-624
52. Sharma MM, Sharma P, Kapoor D, Beniwal P, Mehta S (2021c) Phytomicrobiome community: an agrarian perspective towards resilient agriculture. In: Plant performance under environmental stress: hormones, biostimulants and sustainable plant growth management, pp 493–534
53. Sharma P, Meyyazhagan A, Easwaran M, Sharma MM, Mehta S, Pandey V, Liu WC, Kamyab H, Balasubramanian B, Baskaran R, Klemeš JJ (2022) Hydrogen Sulfide: a new warrior in assisting seed germination during adverse environmental conditions. Plant Growth Regul 98(3):401–420
54. Sharma P, Sangwan S, Kumari A, Singh S, Kaur H (2022b) Impact of climate change on soil microorganisms regulating nutrient transformation. In: Plant stress mitigators: action and application. Springer Nature Singapore, Singapore, pp 145–172
55. Sharma MM, Kapoor D, Rohilla R, Sharma P (2023a) Nanomaterials and their toxicity to beneficial soil microbiota and fungi associated plants rhizosphere. In: Nanomaterials and nanocomposites exposures to plants: response, interaction, phytotoxicity and defense mechanisms. Springer Nature Singapore, Singapore, pp 353–380
56. Sharma P, Sangwan S, Mehta S (2023b) Emerging role of phosphate nanoparticles in agriculture practices. In: Engineered nanomaterials for sustainable agricultural production, soil improvement and stress management. Academic Press, pp 71–97
57. Sharma P, Sangwan S, Kaur H, Patra A, Mehta S (2023c) Microbial remediation of agricultural residues. In: Sustainable agriculture reviews 60: microbial processes in agriculture. Springer Nature Switzerland, Cham, pp 325–358
58. Sharma P, Sangwan S, Kaur H, Patra A, Anamika, Mehta S (2023d) Diversity and evolution of nitrogen fixing bacteria. In: Sustainable agriculture reviews 60: microbial processes in agriculture, pp 95–120
59. Sharma P, Patra A, Singh B, Mehta S (2023e) Microbial rejuvenation of soils for sustainable agriculture. In: Sustainable agriculture reviews 60: microbial processes in agriculture. Springer Nature Switzerland, Cham, pp 293–323.
60. Siddiqi KS, Rahman A, Tajuddin, Husen A (2016) Biogenic fabrication of iron/iron oxide nanoparticles and their application. Nanoscale Res Lett 11(498):1–13 https://link.springer.com/article/10.1186/s11671-016-1714-0
61. Siddiqi KS, Husen A (2016a) Green synthesis, characterization and uses of palladium/platinum nanoparticles. Nanoscale Res Lett 11(482):1–13. https://link.springer.com/article/10.1186/s11671-016-1695-z
62. Siddiqi KS, Husen A (2016b) Fabrication of metal and metal oxide nanoparticles by algae and their toxic effects. Nanoscale Res Lett 11(363):1–11 https://link.springer.com/article/10.1186/s11671-016-1580-9
63. Siddiqi KS, Husen A (2016c) Fabrication of metal nanoparticles from fungi and metal salts: scope and application. Nanoscale Res Lett 11(98):1–15
64. Singh A, Sharma P, Kumari A, Kumar R, Pathak DV (2019) Management of root-knot nematode in different crops using microorganisms. In: Plant biotic interactions: state of the art, pp 85–99
65. Singh P, Kim Y-J, Zhang D, Yang D-C (2016) Biological synthesis of nanoparticles from plants and microorganisms. Trends Biotechnol 34(7):588–599
66. Singh S, Sangwan S, Sharma P, Devi P, Moond M (2021) Nanotechnology for sustainable agriculture: an emerging perspective. J Nanosci Nanotechnol 21(6):3453–3465
67. Sriram MI, Kalishwaralal K, Barathmanikanth S, Gurunathani S (2012) Size-based cytotoxicity of silver nanoparticles in bovine retinal endothelial cells. Nanosci Methods 1(1):56–77
68. Suresh AK, Pelletier DA, Wang W, Morrell-Falvey JL, Gu B, Doktycz MJ (2012) Cytotoxicity induced by engineered silver nanocrystallites is dependent on surface coatings and cell types. Langmuir 28(5):2727–2735
69. Tabata Y, Ikada Y (1988) Macrophage phagocytosis of biodegradable microspheres composed of L-lactic acid/glycolic acid homo-and copolymers. J Biomed Mater Res 22(10):837–858

70. Thuesombat P, Hannongbua S, Akasit S, Chadchawan S (2014) Effect of silver nanoparticles on rice (Oryza sativa L. cv. KDML 105) seed germination and seedling growth. Ecotoxicol Environ Saf 104:302–309
71. Tian Z, Wang JW, Li J, Han B (2021) Designing future crops: challenges and strategies for sustainable agriculture. Plant J 105(5):1165–1178
72. Wang X, Han H, Liu X, Gu X, Chen K, Lu D (2012) Multi-walled carbon nanotubes can enhance root elongation of wheat (Triticum aestivum) plants. J Nanopart Res 14:1–10
73. Younis M, Abdel-Aziz H, Heikal Y (2019) Nanopriming technology enhances vigor and mitotic index of aged Vicia faba seeds using chemically synthesized silver nanoparticles. S Afr J Bot 125:393–401
74. Yu Z, Li Q, Wang J, Yu Y, Wang Y, Zhou Q et al (2020) Reactive oxygen species-related nanoparticle toxicity in the biomedical field. Nanoscale Res Lett 15(1):115
75. Zhang Q, Li N, Goebl J, Lu Z, Yin Y (2011) A systematic study of the synthesis of silver nanoplates: is citrate a "magic" reagent? J Am Chem Soc 133(46):18931–18939
76. Zuverza-Mena N, Armendariz R, Peralta-Videa JR, Gardea-Torresdey JL (2016) Effects of silver nanoparticles on radish sprouts: root growth reduction and modifications in the nutritional value. Front Plant Sci 16(7):90
77. Sangwan S, Sharma P, Wati L, Mehta S (2023 Jan 1) Effect of chitosan nanoparticles on growth and physiology of crop plants. InEngineered nanomaterials for sustainable agricultural production, soil improvement and stress management, Academic Press, pp 99–123
78. Kapoor D, Sharma P, Sharma MM, Yadav S, Husen A (2024 Sep 11) Exploring soil microbiota and their role in plant growth, stress tolerance, disease control and nutrient immobilizer. Biocatal Agric Biotechnol 103358.
79. Sharma MM, Kapoor D, Sharma P, Husen A (2024 May 1) Role of carbon nanomaterials in the prevention of plant disease. In: Carbon-based nanomaterials: synthesis, agricultural, biomedical, and environmental interventions. Springer Nature Singapore, Singapore, pp 105–121
80. Sharma MM, Kapoor D, Loyal A, Kumar R, Sharma P, Husen A (2024 May 1) Environmental toxicity of engineered carbon nanoparticles. In: Carbon-Based nanomaterials: Synthesis, agricultural, biomedical, and environmental interventions. Springer Nature Singapore, Singapore, pp 337–353

Mitigation of Metal Toxicity in Plants Using Silver Nanoparticles

Kulsum Hashmi, Tahmeena Khan, Saman Raza, and Seema Joshi

Abstract Certain metals like Ca, Mg, Cu, and Fe are essential for plant growth while several others, like As, Cd, and Pb are not; the presence of these metals in soil above a certain threshold concentration is toxic to plants. They lead to several cellular and structural changes in the plants by increasing oxidative stress and modifications in certain metabolic pathways. The plant in turn responds to this stress by several methods like increasing the production of antioxidants and other enzymes. Metal toxicity hampers water uptake, nutrient assimilation and consequently plant growth. As a result, there is a decrease in plant productivity as well as quality, leading to financial loss for the farmers and health problems for the consumers. Several methods are used for the remediation of this toxicity in the soil and the plants, including leaching, use of chelators, phytoremediation, etc. However, most of these techniques have drawbacks like being expensive, hazardous to the environment, unsuitable for varied use and not being target-specific. In recent years, nanotechnology has emerged as a safe and effective tool with many desirable outcomes. Its use is being explored in agriculture as well, with nanoparticles being employed in soil fertilizers, disease management and removal of toxins. This chapter describes in detail, the effect of metal toxicity in plants and the use of nanoparticles, for its remediation. AgNPs are of particular interest here, owing to their beneficial effects on plants which have been widely investigated.

Keywords Heavy metals · Nanotechnology · AgNPs · Synthetic routes · Toxicity · Mechanistic insight

---

K. Hashmi · S. Raza · S. Joshi
Department of Chemistry, Isabella Thoburn College, Lucknow, UP 226007, India

T. Khan (✉)
Department of Chemistry, Integral University, Lucknow, UP 226026, India
e-mail: tahminakhan30@yahoo.com

A. Husen (ed.), *Plant Response to Silver Nanoparticles*, Smart Nanomaterials Technology, https://doi.org/10.1007/978-981-97-7352-7_9

1 Introduction

Soil and plants are affected by high concentrations of heavy metals. The permitted limits of their concentrations (mg kg^{-1}) in soil and plants (Fig. 1) have been prescribed by the WHO [196]. Of the heavy metals, Pb has the highest permitted levels in soil, followed by Zn and Cu, while Cd has the lowest permissible value. These data indicate that even at low concentrations the accumulation of Cd in the soil is more harmful than that of Cu, Zn, and Pb. Whereas in plants the levels of Cu can be most, followed by Cd, Pb or Zn. Cd accumulation in plants is a very serious issue; whereas Cu has the safest limits for soil, followed by Pb and Zn.

The plant naturally needs certain elements for its growth and development. While these elements play an important role in plant growth, they can also have a detrimental effect above the permissible limits. The various sources through which the heavy metal can enter the environment are depicted in Fig. 2 [127]. The toxic impact of heavy metals on plants begins in the rhizosphere, where chemicals and metalliferous minerals mingle via root exudates and the researchers also examined the traits of the Zn–Pb mining-polluted rhizosphere soils and the detrimental impact on plant roots [31]. The production of free radicals by heavy metals leads to an increase in intracellular reactive oxygen species (ROS), which causes harm to biological molecules such as nucleic acids, proteins, enzymes, and lipids. Numerous physiological issues, including DNA damage, cell death, and the suppression of enzyme activity, are reported which may finally cause the death of the plant [197].

For environmental, ecological, nutritional, and toxicological reasons, metals are significant environmental contaminants, and their toxicity is a serious issue. Metals can affect a wide range of biochemical and physiological processes in plants, and their toxicity depends on the type of plant, the specific metal, concentration, and chemical structure. Several studies have been carried out worldwide to investigate the adverse effects of harmful metals on plants and their cultivation [63]. Metal toxicity in plants is caused by elements like Fe, Mn, Cu, Co, Al, Ni and Zn which are also considered micronutrients, as well as Cd, As, and Pb which have no biological significance to plants but are responsible for harmful consequences. For physiological and biochemical processes, small amounts of heavy metals are required

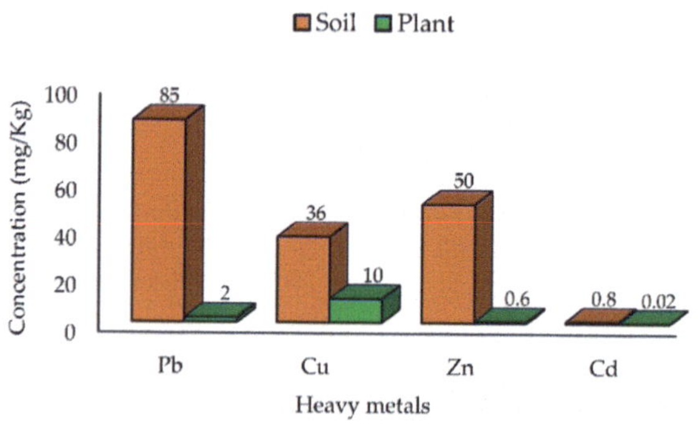

Fig. 1 Permissible limits for concentrations of heavy metals in the soil and plants

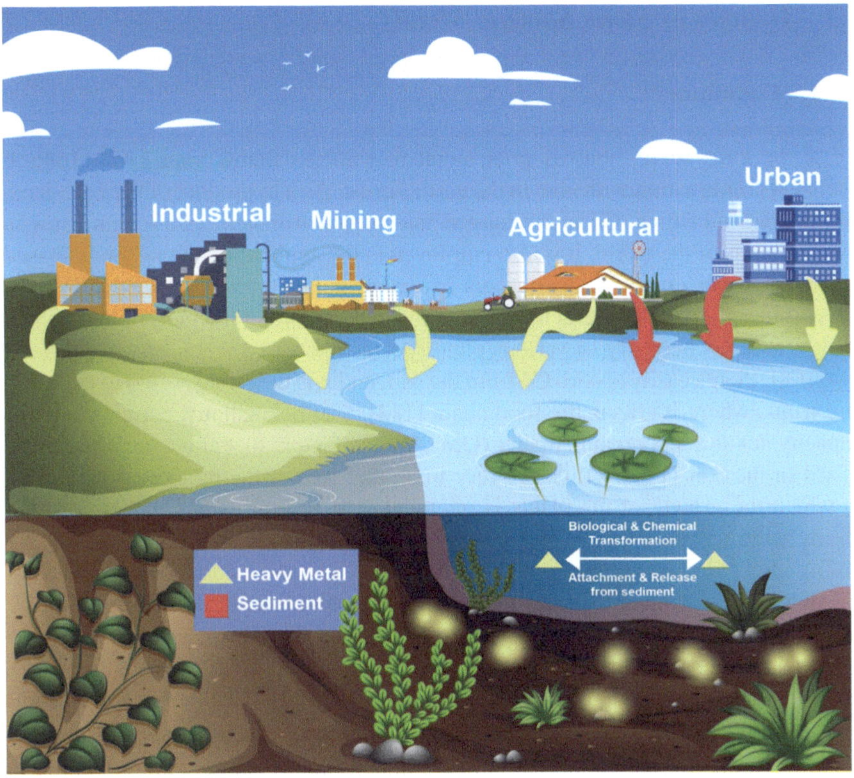

Fig. 2 Various sources of heavy metals in the environment [127]

[141, 200]. All of these substances have different levels of toxicity in plants when present above their upper limits, which primarily depend on their concentrations and bioavailability in different types of soil. These toxic substances have negative effects on the normal activity of metabolic processes by interfering with or transforming pigments or enzymes and modifying their original modes of action [136, 149]. Heavy metals are taken up and accumulated by plant roots via an interconnected chain of molecular and physiological processes, such as toxic metal compartmentalization from the cytoplasm to the vacuoles, metal binding to extracellular exudates, and cell wall components; the generation of metal complexes and organic acids (OAs), MTs, or phytochelatin (PC), amino acids; the formation of heavy metal-induced anti-oxidative enzymes; and modification of the plant metabolism as well as recover and repair the cellular structure. These processes cause modifications of biochemical and physiological processes, like protein variants, metabolic composition, and gene regulation. These processes are responsible for the development of the proper signals that promote protection and tolerance mechanisms in plants [25, 57]. Excessive levels of metal in plants cause chlorosis and inhibit plant growth through the dissociation of photosynthetic enzymes [204].

1.1 *Toxicity of Some Specific Metals*

1.1.1 Cadmium

Cd is the most toxic, non-essential metal that affects many crops [164, 168]. It occurs in combination with zinc in the earth's crust. As a byproduct of the zinc, lead, and copper industries. Cd is discharged into the environment and accumulates in many regions of the world. It is also employed in the application of phosphate-based fertilizers, urban composts, wastewater irrigation and metal-based insecticides [44]. Several investigations have demonstrated that Cd can affect the morphology, physiology, and biochemistry of various plant species like *Brassica* and *Lactuca sativa* [94, 168]. Plants easily absorb Cd from the soil, and it is then transported to the seeds and fruits, where it affects the food chain [144]. Soil pH, salinity, cation exchange capacity, texture and organic matter (OM), are the factors that affect the availability of Cd in the soil [114]. Cd (II) is very toxic that affects the biological processes, in which the growth of the plants is inhibited. The responses of Cd accumulation ability were tested in 28 vegetable species, and it was found that genotypic variability affected plant response to Cd stress [203]. The most common symptoms of Cd stress in aerial plant parts are leaf rolling and chlorosis [115]. Cd affects the Fe (II) uptake and inhibits the root Fe (III) reductase activity. This leads to a deficiency of Fe (II), which has a significant impact on photosynthesis [32]. In another study, it was demonstrated that at lower concentrations Cd stimulated the growth of roots and improved nutrient and water uptake. This is because of the activation of the interchange in the root cell wall to reject Cd (II) [115, 137]. Cd has a detrimental effect on the intake, movement, and utilization of various elements like Mg, K, Ca and P as well as water by plants. Nitrate intake and transport from root to shoot are impeded because Cd inhibits the activity of the enzyme nitrate reductase [169]. Cd is toxic to plants because it causes oxidative stress in their cells, which encourages the production of ROS [134], and the antioxidant enzymes are "inhibited" or "inactivated" [84] so, it causes oxidative damage to the cells and lipoprotein peroxidation [34]. The toxic effects of Cd in plants are illustrated in Fig. 3.

1.1.2 Lead

Pb affects the plant growth and development. According to WHO [196], the prescribed level of Pb for plants is about 2 mg kg^{-1}, whereas agricultural soils have a range of 50–300 mg kg^{-1} [85]. All morphological, physiological, and biochemical systems are severely impacted when the Pb level increases [109]. Pb contamination in soil causes changes in plant water relationship, formation of ROS, and decreased nutrient absorption, etc., which causes cell death and decreased photosynthesis [70, 187]. Additionally, it decreases the activity of enzymes that are involved in the metabolism of carbohydrates, acid invertases and phosphatases [128]. Germination inhibition by Pb has been observed in several crops, such as elsholtzia (*Elsholtzia*

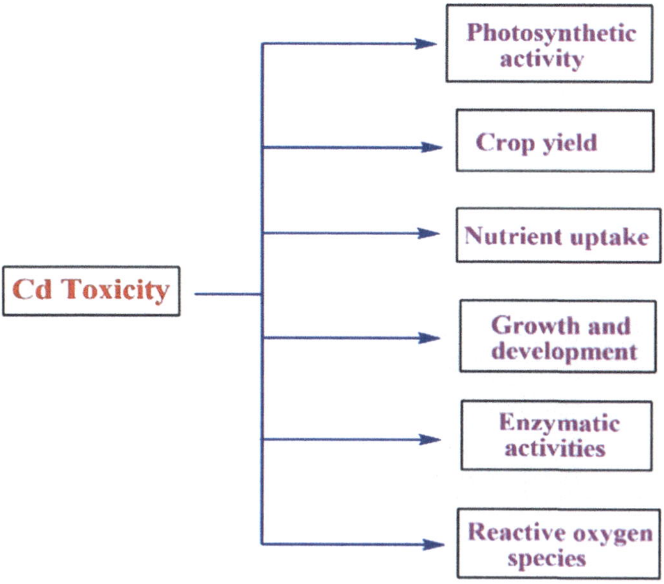

Fig. 3 Effects of Cd toxicity in plants

argyi H. Lév) [86], rice (*Oryza sativa* L.) [183], mungbean (*Vigna radiata* (L.) R. Wilczek) [14], alfalfa (*Medicago sativa* L.) and lentil (*Lens culinarus* Medikus) [162]. Pb toxicity reduces food mobilization, radical formation, proteolytic activity degradation, and cellular osmoregulation, inhibiting germination and seedling development [38]. Researchers also demonstrated the toxic effect of Pb on the early growth of plants, such as radish plants [183]. The Pb toxicity causes decreased chlorophyll production, transpiration and cell division [70]. The growth, sprouting, and seedling development of wheat (*Triticum aestivum* L.) were also affected by Pb exposure [43]. At low concentrations, Pb inhibits the growth of roots and apical sections of the plants and causes short, stubby, and swollen roots [103] such as reduced root elongation in Mesquite (*Prosopis* sp.) roots [11]. At higher concentrations, plants exhibited growth impairment [86]. Pb reduced the turgor of protective cells in leaves, which decreased cell membrane flexibility and stomatal closure [172]. Due to a decrease in the turgor Pb induced reduces the concentration of sugar, amino acid and a variety of chemical compounds. Pb-induced toxicity decreases crop productivity due to reduced intake of nutrients and water which increases oxidative stress, and C-fixation. Pb exposure reduced the economic yield of wheat and sugarcane by approximately 25–30% [125, 151]. The effects of Pb on plants and different mechanisms are illustrated in Fig. 4 [106].

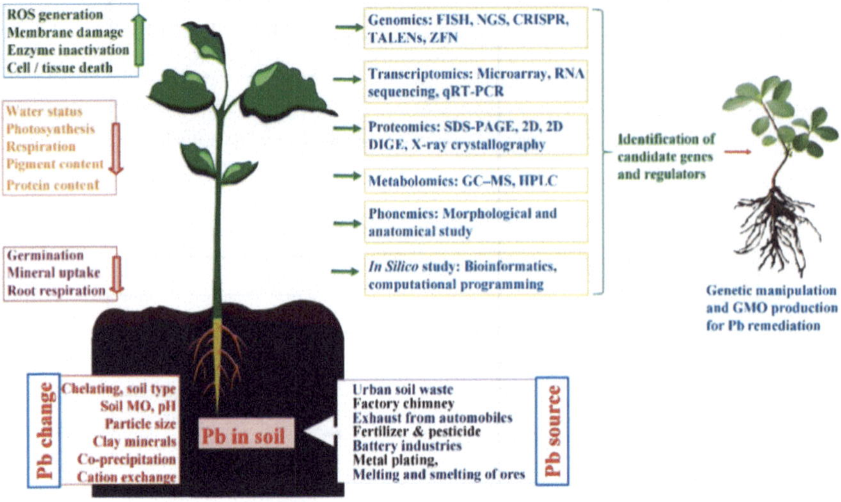

Fig. 4 Various pathways of Pb uptake and approaches for its remediation [106]

1.1.3 Zinc

For all living things, including plants, zinc is an essential micronutrient [180]. Zn is the second most abundant transition metal after Fe, and it is involved in all enzymatic functions. It is absorbed by plants through their roots as Zn (II) from the soil [29] and is involved in the redox processes that occur during photosynthesis [71]. Zn accumulation in the root or shoot of plants can lead to serious damage. High levels of Zn in the plant cells disturb the physiological processes of plants, resulting in plant death [112]. It has been demonstrated that Zn poisoning first caused chlorosis in leaves at the beginning of the exposure period, which subsequently spread to the older leaves [49]. Zn also reduces the length of the roots and the rate of photosynthesis [180]. Zn toxicity reduces the relative water content (RWC) of plant tissues [59] by decreasing the transpiration rates, water transporters and dehydration [61] which is caused by Zn deposition in aerial tissues (Fig. 6). Reduction in RWC as a result of excessive Zn exposure may also be caused by a decrease in the vascular system and absorbing surface of roots which is caused by a blockage in the xylem elements and affects the ascent of sap [102]. When water is reduced, there is a decrease in turgor or turgidity, which reduces the rate of cell proliferation, resulting in reduced growth and biomass production. The reduction in water capacity of leaves caused by excessive Zn exposure is due to increased osmotic pressure in the surroundings and decreased water uptake [123]. Physical drought in plants is caused by excess Zn, which disrupts the water balance by either preventing water from moving through the body of the plant or limiting the tonoplast's ability to move water. The molecular causes of the Zn-mediated loss in the water transport capacity of roots. Cu/Zn stress-induced *Mesembryanthemum crystallinum* severely suppressed the aquaporin (AQP) genes [101] and it was also discovered that under Zn toxicity plasma membrane intrinsic proteins (PIP2) AQP was decreased in lettuce [116]. The reduction in root and shoot hydraulic conductivity, which was facilitated by down-regulation of AQPs or gene expression and also enhanced suberization in endodermis, was attributed to the decrease in barley root-intrinsic transport of water due to excess Zn. It was also proposed that Zn excess decreased the flow rate of water and affected both the hydraulic properties of the root (AQPs) and shoot, in contrast to losing shoot surface area and reduced water-absorbing root [62]. Recently, observed that PIP1 isoforms were involved in understanding a prolonged response of Zn treatments in *Brassica rapa* plants, as well as a decreased root hydraulic conductivity. According to these observations, excess Zn results in physiological and morphological changes that affect the plant's capacity to transport and absorb water [54].

1.1.4 Copper

After steel and aluminum, copper is the third most consumed metal all over the world. For the growth of plants and the germination of seeds, a small amount of copper is required but at higher concentrations it is very toxic [35, 198]. Cu intake by plants depends upon physical and chemical composition of the soil and other metabolic features of the plant [26]. Studies have reported higher concentrations of Cu in the roots than in the shoot. Higher Cu concentration has been reported in the root epidermis [104]. Cu toxicity altered the process of photosynthesis, which reduced the production of the crop, the synthesis of chlorophyll, and the productivity of the plant. Concentrations of Cu (5 mg kg^{-1}) are enough to damage plants, like decreasing the growth of plants and its efficacy [3]. It was investigated that excess levels of Cu can influence the morphology and productivity of maize plants and it also reduced the height of the plant [21]. The toxic effect of the high concentration of Cu on maize plants was also reported and the result showed a reduction in shoot length, which demonstrated that it affects the entire plant [9].

1.1.5 Arsenic

The most hazardous metalloid in the world is arsenic (As), widely distributed in numerous ecosystems and is a global threat due to its toxicity to all forms of life. It affects the physiological, morphological and biochemical functions of the plants (Fig. 5) [2]. While a concentration as low as 5 ppm has been shown to be harmful to sensitive crops, the acceptable limit of As in agricultural soils is 20 mg kg^{-1} [60]. Arsenite As (III) and arsenate As (V) are the two naturally occurring inorganic forms of arsenic in soil [122]. As (III) has the ability to deactivate crucial enzymes in plants [170]. It has been reported that As (V) reduced the rate of germination of *Brassica* [181]. A correlation was also found between the decrease in plant height and the increase in As concentrations, as well as the same pattern in root length or leaf number and plant biomass of lentils [4]. Other investigations on the toxicity of As

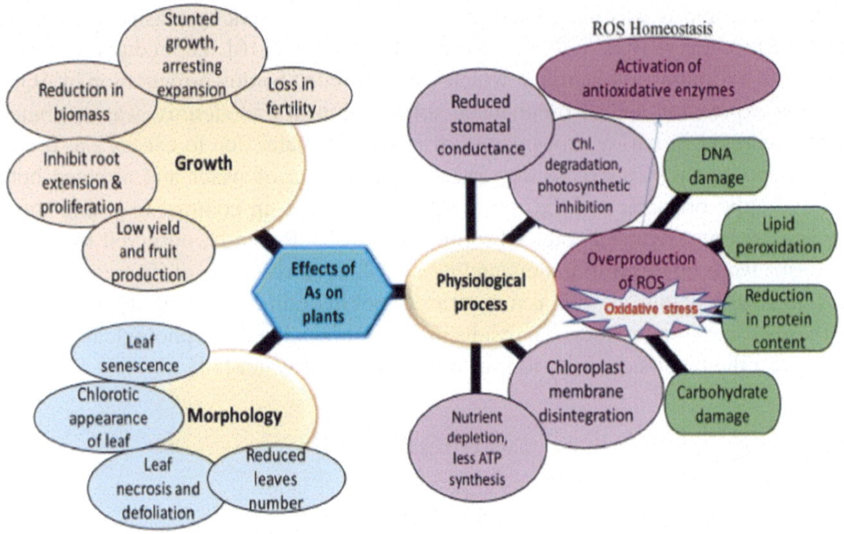

Fig. 5 Arsenic toxicity in plants [2]

noted loss of fresh and dry biomass in the roots and shoots, stop fruit development, and reduced yield [165, 181]. Potato cultivars growing in soil that had been contaminated with As had a 6.29–23.69% decrease in their tuber production compared to controls [72]. Higher concentrations of As have been shown to reduce vegetative growth in tomato and sorghum [126, 165]. Contamination greatly enhanced the physiological issue known as "straight head disease" in rice, which causes the spikelets and florets to be sterile and diminish the production of grains [179].

1.1.6 Chromium

Cr toxicity in plants has a detrimental effect on molecular, physiological, and biochemical functions, which inhibit growth and lower the production of overall yield. Excessive Cr accumulation in plants, affects seed germination and decreases root and shoot growth rates [47]. Numerous studies have revealed that Cr excess in plants has an impact on their chlorophyll level (Chl a and b), which inhibits photosynthesis. Prior research has demonstrated that high Cr accumulation in plant tissues affects the mineral and water balance, cell cycle, nitrogen absorption, antioxidant system, enzyme function, and metabolic activities [186]. The toxic effect of Cr on a variety of plants, including *Nicotiana tabacum*, *Vigna radiate*, *Cicer arietinum*, *B. juncea*, *Brassica oleracea*, *Brassica napus*, *Zea mays*, and *Sorghum bicolour* has been investigated [69]. The impact of Cr on plants is represented in Fig. 6 [7].

The toxic effects of heavy metals in various plant species are illustrated in Table 1.

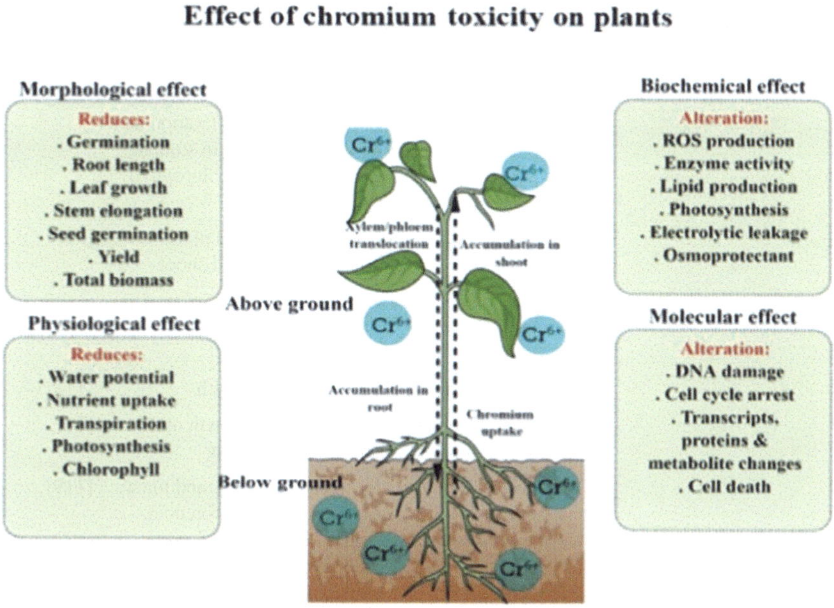

Fig. 6 Impact of Cr toxicity on physiological, morphological and biochemical features in plants [7]

2 Molecular Aspects and Mechanistic Insight of Metal Toxicity in Plants

Certain metals and metalloids such as Cu, Zn, Fe, Mn, Co, Al, and Ni may have some nutritive value for the plants while others like As, Cd, and Pb which have no biological importance to plants are associated with toxic effects [200]. All metals and metalloids, whether essential or non-essential, have a limiting or threshold concentration beyond which they cause toxicity leading to structural changes and impairment of growth in plants [65]. The level of toxicity depends on the concentration of the ions and their bioavailability in the soil. The harmful effects are mainly because these metal ions compete with normally found ions in the plants that are necessary for the normal physiology of the plants, i.e. as cofactors or ligands of pigments or vital enzymes involved in metabolism [146]. Heavy metals may cause inactivation or denaturation of proteins and enzymes. Metals like Al, Cu, and Zn can displace ions like Ca (II) and Mg (II) that are present in membranes and cell walls as they bind more easily to the pectin present in the cell wall [56]. They may also interact with reactive groups or cause conformational changes in the various molecules required in metabolism, such as proteins, and cause an imbalance in their functions, as well as interfere with the uptake and transport of water and nutrients to various plant parts [191]. Metal toxicity has also been attributed to oxidative stress in plant cells by the production of more reactive oxygen species ROS, as well as inhibition of ATP production and

Table 1 Toxic effect of metals on different plants

Sl. No.	Metals	Plant species	Effect on plants	Refs.
1	Ni	Pigeon pea (*Cajanus cajan*)	Reduced enzyme activity that impacted the CO_2 fixation and calvin cycle; drop in stomatal conductance and chlorophyll content	[175]
		Rye grass (*Lolium perenne*)	Chlorosis; reduced plant nutrient uptake and shoot production	[99]
		Wheat (*Triticum* sp.)	Decreased plant nutrient uptake	[138]
		Rice (*Oryza sativa*)	Reduced root growth	[113]
		Bean (*Vicia faba*)	Reduced plant growth and chlorophyll content	[74]
2	Zn	Cluster bean (*Cyamopsis tetragonoloba*)	Decreased biomass and plant height; reduced carotenoid, starch, chlorophyll, amino acid, and sugar content	[119]
		Pea (*Pisum sativum*)	Reduced photosystem II; altered chloroplast structure; decreased chlorophyll content and plant development	[46]
		Rye grass (*Lolium perenne*)	Decreased plant growth and photosynthetic energy conversion; drop in the amount of nutrients in the plant	[27]
3	Pb	Maize (*Zea mays*)	Decreased plant growth and biomass; reduced plant protein content and germination percentage	[83]
		Portia tree (*Thespesia populnea*)	Decreased plant height, biomass, leaf area, and the amount of leaves	[95]
		Oat (*Avena sativa*)	Inhibited enzyme activity which influences CO_2 fixation	[130]
4	As	Tomato (*Lycopersicon esculentum*)	Decreased leaf fresh weight and fruit yield	[22]
		Canola (*Brassica napus*)	Chlorosis; wilting; reduced growth	[39]
		Lentil (*Lens culinaris*)	Root length, leaf number, and plant biomass	[4]
		Rice (*Oryza sativa*)	Sterility of the florets or spikelets, reduced yield	[179]

(continued)

Table 1 (continued)

Sl. No.	Metals	Plant species	Effect on plants	Refs.
		Oilseed (*Brassica napus*)	Reduced thiol transferase activity and cell death	[53]
5	Co	Radish (*Raphanus sativus*)	Decreased the amount of nutrients, chlorophyll, protein, sugar, amino acid content in plants; reduced root, shoot, and total leaf area	[90]
		Mung bean (*Vigna radiata*)	Reduction in plant sugar, amino acids, protein, and starch content; decline in antioxidant enzyme actions	[89]
		Tomato (*Lycopersicon esculentum*)	Reduced nutrient uptake by plants	[91]
6	Cd	Maize (*Zea mays*)	Inhibition of root growth; reduced shoot development	[193]
		Wheat (*Triticum* sp.)	Reduction in nutrient content of plant; decline in seed germination	[207]
		Garlic (*Allium sativum*)	Decreased growth of the shoot	[93]
		Rice (*Oryza sativa*)	Necrotic dark spots on older leaves; chlorosis in young leaves	[156]
7	Mn	Spearmint (*Mentha spicata*)	Decreased carotenoid as well as chlorophyll content	[15]
		Tomato (*Lycopersicon esculentum*)	Reduced chlorophyll concentration; slower plant growth	[174]
		Broad bean (*Vicia faba*)	Chlorosis; Mn deposition in the root or shoot	[12]
		Pea (*Pisum sativum*)	Reduced photosynthetic activity as well as growth rate	[45]
		Rice (*Oryza sativa*)	Inhibited the process of photosynthesis	[148]
		Wheat (*Triticum aestivum*)	Increased antioxidative enzyme activity; ROS overproduction	[52]
8	Hg	Rice (*Oryza sativa*)	Reduced tiller and panicle formation; reduction in plant height; reduced yield	[48]
		Tomato (*Lycopersicon esculentum*)	Reduced plant height; chlorosis; drop in germination percentage	[173]

(continued)

Table 1 (continued)

Sl. No.	Metals	Plant species	Effect on plants	Refs.
		Barley (*Hordeum vulgare*)	Reduced yield	[199]
9	Fe	Rice (*Oryza sativa*)	Short, flimsy, blunted roots with dark brown colour and bronzing of leaves	[51]
10	Cr	Tomato (*Lycopersicon esculentum*)	Decreased plant nutrients	[167]
		Indian mustard (*Brassica juncea*)	Reduced stomatal transpiration and CO_2 assimilation	[159]
11	Cu	Rice (*Oryza sativa*)	Chlorosis and rolling of leaves	[129]

DNA damage [37]. All these effects of metals alter the normal functions of the plant like photosynthesis, respiration, growth, and development. The response of the plants to heavy metal toxicity can involve different molecular mechanisms, like confinement of toxic metals inside vacuoles, formation of complexes of metals and amino acids and organic acids or phytochelatin (chemicals produced in plants for heavy metal detoxification) production of antioxidants and stress enzymes, etc. which cause modifications in gene expression, protein synthesis, as well as metabolite composition. These changes help in the restoration and repair of plant cells and protection against metal toxicity [25]. The common mechanisms of metal toxicity are summarized in Fig. 7.

2.1 *Oxidative Stress*

Heavy metals disturb the redox state of the plant cells by increasing the production of ROS like hydrogen peroxide, singlet oxygen, and free radicals like hydroxyl radical and superoxide radical [68]. This oxidative stress produced by heavy metals leads to several disruptions at the molecular level, like the destruction of biologically significant molecules, lipid peroxidation, DNA cleavage, and breakdown of membranes which leads to ion leakage, and cell death [28, 155]. The redox-active heavy metals like Fe, Cu, Cr, and Co cause redox reactions directly resulting in the generation of superoxide radicals and H_2O_2 and then hydroxide radicals in the cells by the Haber–Weiss reaction followed by the Fenton reaction [143] (Eqs. (1) and (2)).

$$Fe^{3+} + \cdot O_2^- \rightarrow Fe^{2+} + O_2 \tag{1}$$

$$Fe^{2+} + H_2O_2 \rightarrow Fe^{3+} + OH^- + \cdot OH \tag{2}$$

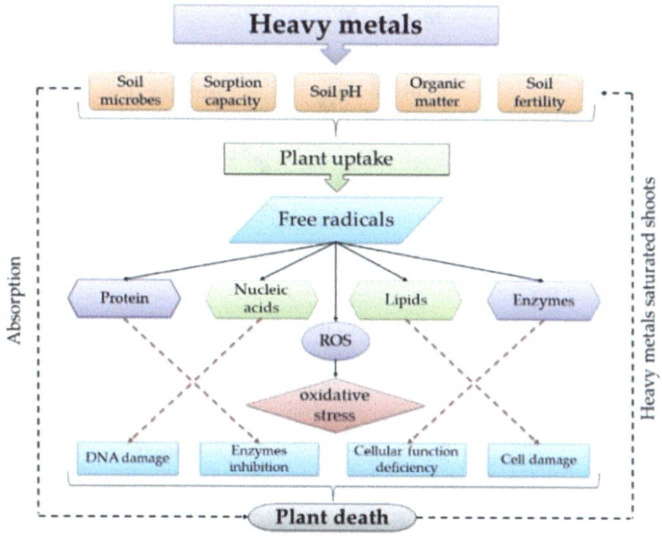

Fig. 7 Pathway and mechanism of heavy metals toxicity in soil and plant

On the other hand, redox inactive heavy metals like Cd, Zn, Al, etc., cause oxidative stress in an indirect manner. This may include interaction with antioxidants produced by the plants, which may alter them or destroy them, leading to an increase in ROS. Beside this, there may be disruption of electron transport within the cell or increased production of enzymes leading to lipid peroxidation such as lipoxygenase (LOX). For example, in a study on *Arabidopsis thaliana*, the expression of LOX in the plant tissues was found to be induced by Cd, leading to the peroxidation of polyunsaturated fatty acids present in lipids [178]. In response to ROS, plants produce radical scavenger enzymes as well as non-enzymatic scavengers. The elevated levels of these free radical scavengers have been linked to heavy metal toxicity in plants; for example, after intense Pb exposure to pea plants, the activity of antioxidants was found to be increased [120]. In another study on Pb toxicity in plants, it was found that Pb causes excessive production of ROS in plants, which increases lipid peroxidation and oxidative stress. It oxidizes proteins and nucleic acids of the cell and as a result, it affects the function of cell organelles like mitochondria, chloroplasts and peroxisomes.

Plants also produce signalling molecules under stress, which help the plant cells to identify the problem and generate a remedial response accordingly. Methylglyoxal (MG) is a signalling molecule whose concentration has been found to increase in response to abiotic stress in different studies [201]. In a study by Hossain et al. on pumpkin, its seedlings were subjected to various stresses like drought, salinity, high temperature, white light, and heavy metal, and the levels of MG were monitored. It was found that heavy metal led to a 1.55-fold induction of MG level [76]. An increased level of MG has been attributed to abiotic stress, however, studies have shown that

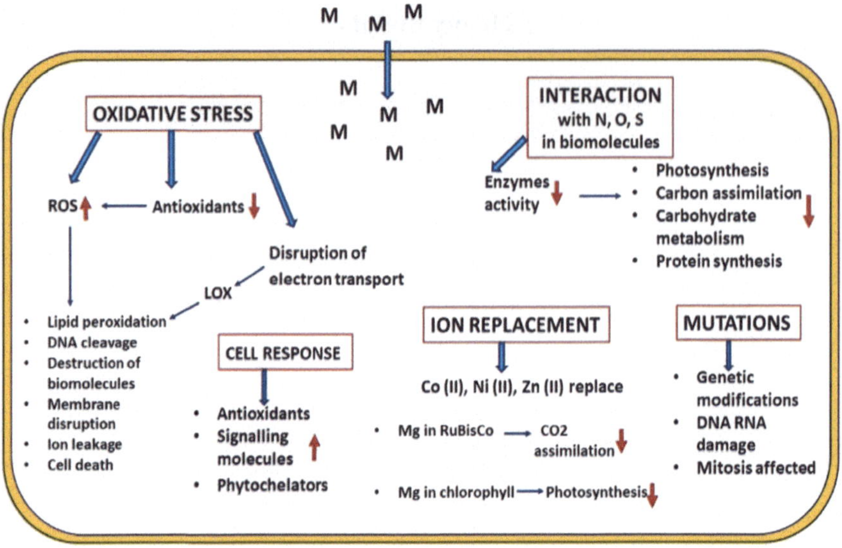

Fig. 8 ROS generation due to heavy metal toxicity

an increase in MG may affect certain pathways in the plant which interfere with the antioxidant mechanism, further increasing the production of ROS [72] (Fig. 8).

2.2 *Binding with Biologically Significant Molecules*

Apart from causing oxidative stress, heavy metals cause toxicity due to their strong binding ability with the ligand atoms like oxygen, nitrogen, and sulphur, present in various macromolecules like enzymes, pigments, and proteins. These metals bind to the sulfhydryl-, histidyl-, and carboxyl-groups present in proteins and enzymes and deactivate them. For example, Cd binds to sulfhydryl groups of structural proteins and enzymes that cause their misfolding and inhibition of activity [41]. Such interactions affect enzymatic reactions that are vital for the various cellular pathways responsible for the growth, development, and vitality of the plant, such as respiration, photosynthesis, and transport. The detrimental impacts of As in plants are caused by its interaction with the sulfhydryl groups of tissue proteins and enzymes, which inhibits cellular activity. Heavy metals have been found to be involved in the disruption of plant growth by inhibiting enzymes responsible for carbon dioxide fixation which decreases carbon assimilation. A study on the effect of Zn and Cd toxicity in durum wheat showed up to 30% inhibition of CO_2 assimilation leading to decrease in growth [140]. The disruption of enzyme activity also affects carbohydrate metabolism. For example, the activity of enzymes like α-amylase, β-amylase, and sucrose phosphate synthase, which are involved in carbohydrate metabolism have been found to be decreased in a study on As toxicity in rice seedlings. In the same study, inhibition of nitrogen-fixing enzymes like nitrate reductase and nitrite reductase, resulted in decreased nitrogen assimilation [92]. Disruption of enzyme activity affects DNA and RNA present in the cell nucleus as well. The inhibition of the enzymes ribonuclease and protease, in rice seedlings, caused by Ni toxicity was found to suppress the hydrolysis of RNA and proteins [118].

Heavy metals also damage cell membranes by interaction with protein thiols and lipids present in the membrane, as well as by the inhibition of membrane proteins like H^+-ATPase, which causes structural changes [121].

2.3 Replacement of Metal Ions

Heavy metals also replace essential metals present in molecules like cofactors, needed for proper functioning of enzymes, vitamins, and pigments, such as FAD, NAD, coenzyme A, chlorophyll, etc., resulting in their loss of activity. For example, divalent heavy metal cations like Co (II), Ni (II), and Zn (II) can substitute Mg (II) in ribulose-1,5-bisphosphate-carboxylase/oxygenase (RuBisCO), an enzyme responsible for CO_2 assimilation in Calvin cycle, which results in loss of activity and decreases photosynthesis [188]. Another essential metal ion Ca (II) is present in calmodulin, an important protein in abiotic stress-induced cellular signalling, its replacement by Cd (II), causes the inhibition of the protein, as studied in radish [153]. Similarly, some divalent metals can replace Mg in the pigment chlorophyll, which affects the rate of photosynthesis. Chlorophyll synthesis is also reduced by the inhibition of enzymes involved, together, both mechanisms cause the browning of green leaves and reduce photosynthesis and affect the growth of the plant [166].

Apart from the above-mentioned effects on plant cells, heavy metals also cause mutations leading to genetic modifications, DNA, RNA, and protein modification, changes in cell division, and cell cycle. For example, in a study on Cd toxicity in *Pisum sativum*, an alteration was observed in the mitosis of meristematic cells and nuclear ploidy [58].

3 Role of Nanotechnology in Agricultural Practices

Nanoparticles (NPs) range between 1 and 100 nm in dimension and upon nanotization, their properties are known to change owing to the quantum confinement. Nano-science has led to a revolution in the field of material sciences for the development of novel processes and products which cannot be achieved through conventional methods. The use of nanotechnology in agriculture has also evolved with time leading to better yield and protection of plants. For instance, through nano-formulation, new pesticides and herbicides have been developed which led to improved productivity. Encapsulated NPs have been developed for sustained release of water and nutrients,

Fig. 9 Role of nanotechnology in agriculture [1]

gene delivery, and for biosensors development. The United States Department of Agricultural roadmap first addressed the applications of nanotechnology in agricultural sciences and food industries [161]. Food safety and bio-security instruments have also been developed with the help of nanotechnology. Crop productivity has been found to increase by incorporating nanosciences in agricultural practices. Silver nanoparticles (AgNPs) have been used as antifungal and antibacterial and to regulate nutrition in plants. Carbon-nanotubes-based nanosensors are used to catch and measure proteins and small molecules. ZnO NPs have been shown to promote seed germination and plant growth and increase stem and root growth. MgO and ZnO NPs have been found to be effective against microorganisms and used in food packaging [97]. Nanofertilizers of different elements like Fe, Ag, Mn, and carbon nanotubes have been found to improve plant growth and plant protection [139]. Therefore, it can be safely said that appropriate implementation of nanomaterials in agricultural sector may have the potential to revolutionize agricultural practices and food production Fig. 9 [1].

3.1 AgNPs in Agriculture

The AgNPs have been employed in pest control, improved crop production, and identification of specific diseases. Although they can be prepared through physical, chemical methods, yet biological methods are the preferred ones. This eco-friendly approach has been used to synthesize AgNPs from different plant extracts and microorganisms. AgNPs are very frequently employed in bio-systems and have strong bactericidal effects and antimicrobial activities. They are also known for their antioxidant, antibacterial, and anti-inflammatory properties.

3.2 AgNPs: Synthetic Routes

3.2.1 Chemical and Biological Silver Nanoparticles

AgNPs can be prepared through physical and chemical methods, (Fig. 10) [1] however the use of toxic chemicals makes these approaches unfavourable. The commonly employed methodologies are top-down, bottom-up, and reduction, etc. The absorption of toxic materials on the surface can hinder the medicinal utility of AgNPs [142]. Biologically, AgNPs are synthesized by bottom-up approach involving reduction or oxidation. The microbial enzymes or the phytochemicals lead to the formation of the desired NPs. Solvent medium, eco-friendly reducing agent, and a nontoxic stabilizing agent are required for the biological synthesis. The protein capping over such NPs makes the entry feasible inside pathogenic microorganisms. The presence of protein cap can be detected through FT-IR analysis which gives a characteristic peak at 1654 cm^{-1} [19].

3.3 AgNPs in Agricultural Practices

3.3.1 Positive Effect on Plant Growth

NPs have been found to have beneficial effect on seed germination, crop production and disease management [163]. The effect of AgNPs on the growth metabolism of *Bacopa monnieri* has been studied [105] showing a positive effect with induced protein and carbohydrate content and decreased phenol content and catalase activity. Another study reported favourable seedling growth of *Boswellia ovalifoliolata* [158]. AgNPs also increased the plant morphological and biochemical attributes like chlorophyll, carbohydrate, and protein content in *Brassica juncea* [171]. A study reported both the positive and negative impact of AgNPs on root elongation [66]. Effect of AgNPs on germination and early growth of wetland plants has also been reported [206]. AgNPs have been found to induce root growth by blocking ethylene signalling in *Cricus sativus* [152]. The effect of AgNPs may be positive or negative as dependent on size, concentration, chemical composition, and stability of NPs [40]. The impact may also be negative as found in a few cases in the form of reduced growth and formation of pigments [184]. Treated AgNPs with different compounds may have different impacts on plants. For instance, AgNPs in the presence of a magnetic field improved the quantitative yield in *Zea mays* [24]. Lipid peroxidation was observed in the wheat pathogen photosystem as compared to NPs or pathogen alone [23]. The biogenic AgNPs have been found to have improved protein and carbohydrate synthesis and reduced phenol content which may be due to the alteration in size (2–50 nm) [87]. Small-sized (<30 nm) AgNPs in high concentration have been found to restrict root and shoot growth [192].

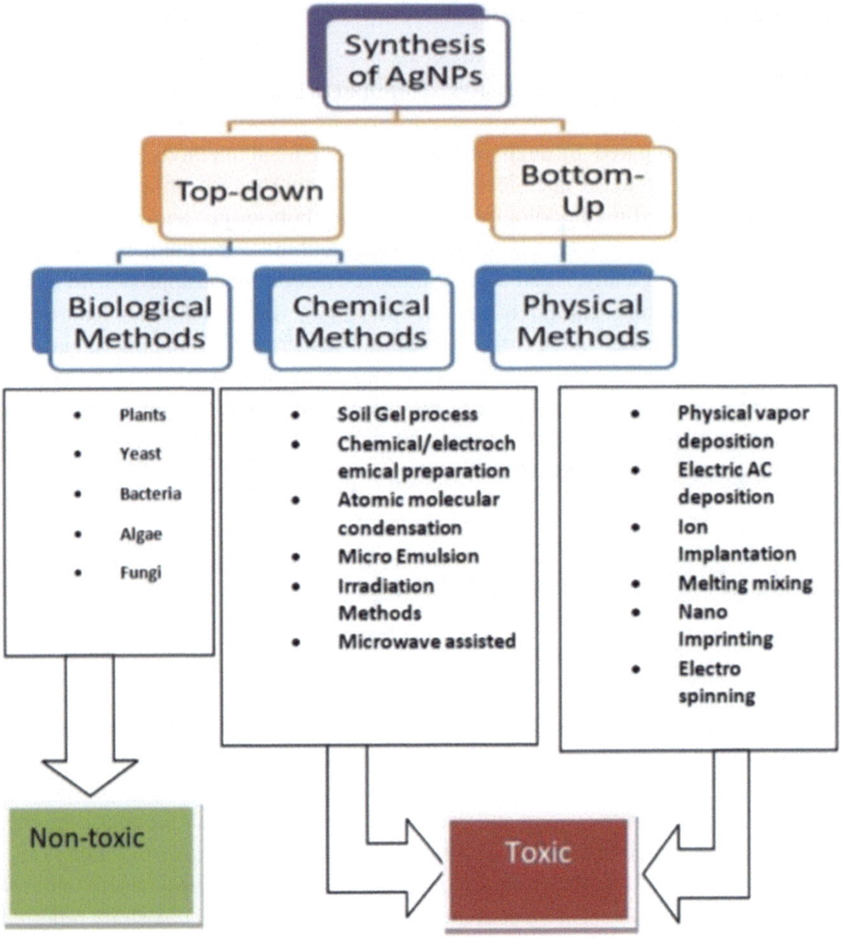

Fig. 10 Preparatory methods of AgNPs [1]

3.3.2 Plant Pathogen Detection

Nanomaterials may be useful in the detection of plant pathogens and disease diagnostic [100]. Through striping volumetry metallic NPs like AgNPs and AuNPs can be used in it including inorganic nanocrystals for analytical identification [160]. A helicase-dependent isothermal amplification in conjugation with on-chip hybridization for the detection of *Phytophthora species* was reported which allowed efficient amplification of the yeast GTP-binding protein (Ypt1) target gene region. The assay's specificity was determined by on-chip DNA hybridization and subsequent AgNP deposition which served as stable endpoint signals enabling the readout.

3.3.3 Controlled Released Nanofertilizers

AgNPs may be used for the sustained release of agrochemicals in the form of nanocapsulation. Optimized delivery of agrochemicals, pesticides, and fertilizers is also achieved through this approach [36]. Nanocapsulated agrochemicals possess desired properties like effective optimum concentration, controlled release, and targeted drug delivery with less toxicity [185]. The AgNPs cause the slow release of agrochemicals to the host by dissolution, biodegradation or diffusion, etc. The encapsulated agrochemicals cause reduced risk to the other plant tissues, chemical contamination, and hazards to the environment [147]. A combination of inorganic fertilizers mainly supplies N, K or P to different crops, hence boosting nutrient management.

4 Remedial Strategies for Toxicity Removal Using Nanotechnology

4.1 AgNPs and Their Role in Metal Toxicity Mitigation

Rapid urbanization has led to the release of toxic pollutants including pesticides, heavy metals, etc., and their deposition in plant tissues, water bodies, and soil [108]. Heavy metals have higher atomic densities exceeding 4000 kg/m^3. Even in trace amounts many of them can cause toxicity. Several heavy metals like Cu, Fe, B, Ni, and Mo are also required for plant growth. However, above the permissible level they can cause negative impact on plants [73]. Heavy metal exposure in human beings can also cause damage to brain, spinal cord, and other organs and also be a cause of cancer, sclerosis and Alzheimer's disease [107]. The use of NPs for the remediation of heavy metals has been found to be cost-efficient and an environment-friendly [17, 194]. Due to the potential morphological characteristics and presence of adsorptive sites metallic NPs have been found to adsorb heavy metals like Cd, Pb, As, Cr, etc., efficiently. They also show nutrient support and pesticidal actions against phytopathogens and can promote morphological and biochemical indices in plants after stimulation of desirable enzymes [208]. Metallic NPs can be applied through soil or foliage for remediation [133]. Although a number of techniques have been employed for heavy metal remediation, yet there are gaps such as environmental sustainability, cost-effectiveness, target delivery, and ecological suitability that need to be tackled. Nanoparticles owing to their properties like size and their application in small quantities as herbicides, pesticides, fertilizers, etc., are used in agriculture and biotechnology [10, 20, 77–81, 84, 181]. AgNPs have been used as stimulators of plant-growth [117, 177], fruit ripening agents, fungicides, etc. Metallic NPs have found widespread therapeutic and environmental applications and it is very important to select an eco-friendly and simple methodology of synthesis. NPs have been found to be effective and promising adsorbents in the form of carbon nanotubes, multi-walled carbon nanotubes, graphene oxide, fullerene, TiO_2, Ag, magnetite, iron phosphate, CuO, Si, Zn and Al_2O_3 NPs which act as efficient adsorbents and immobilizing agents [17]. The properties of NPs largely depend on their synthetic route and of all the methodologies the biosynthesis using plant extract which acts as reductants, stabilizing, and capping agents has marked advantages over the conventional methods due to their eco-friendliness [110]. AgNPs obtained from the extract of *Catharanthus roseus* were used for cadmium and chromium remediation [5]. Another study reported the preparation of AgNPs and CuO NPs from the *Catharanthus roseus* leaf extract. It was used for the removal of Cd and Cr. The phytochemicals in plant extract acted as stabilizing agents. The DLS data suggested an average distribution of 602.9 nm for silver. 47.84 and 5.68% removal were obtained for Cr and Cd in case of biogenic AgNPs [190]. There is an increased demand for healthy and functional food in phytochemicals to curb degenerative diseases. *Moringa oleifera* Linn is one such plant having antiinflamamtory, antidiabetic and antioxidant properties. The plant extract is rich in polyohenols, flavonoids, β-carotene, and linoleic acid [55, 150]. However, these properties can be greatly affected by biotic and abiotic factors like presence of heavy metals which affect the fruit quality and soil nutrients [135, 209]. AgNPs extracted from the extract of cocoa pods were used to immobilize and adsorb Cd and Pb in addition to their phytostimulatory influence on the plant. Heavy metals are accumulated in soil and they are being generated by a number of anthropogenic activities like mining, vehicular exhaust, and sewage disposal. These heavy metals cause detrimental toxicity to plants and result in reduced soil quality and fertility. It also leads to adverse human health effects [13]. Cd and Pb are two such heavy metals that cause toxic oxidative stress in plants causing morphological and physiological changes affecting their germination, growth, and translocation of nutrients. These heavy metals generate free radicals inhibiting photosynthetic pigment formation and causing cell damage. They are also not easily degraded in soil [98, 154, 195]. However, plants are equipped to tolerate heavy metals-induced toxicity through the induction of secondary metabolites like polyphenols, carotenoids, and vitamins, etc. The secretion of antioxidant enzymes like superoxide dismutase and catalase also scavenges free radicals produced by heavy metals, yet excessive heavy metal contamination results in plant deformation, discolouration of leaves and reduced plant growth [42, 145, 205]. The germination potential of AgNPs has been associated with their ability to modify the physiological processes like gene regulation which is also responsible for cell proliferation. They also enhance photosynthesis via translocation to the leaves [67, 75, 202]. AgNPs have also been used to remove heavy metals owing to their high adsorptive capacity and affinity as compared to other NPs. They are also associated with good antimicrobial properties adding to their benefit [18]. Biologically synthesized AgNPs have been associated with a number of agricultural applications and also act as adsorbents for pollutants in soils [111]. Phtoremediation of Cd (II) ions from polluted water by the use of peels and wastes of different plants has been reported. Zero valent AgNPs synthesized by eco-friendly leaf extract of Ficus tree (*Ficus benjamina*) have been used for Cd (II) ions removal. The removal increased when the dose of the bio sorbent increased (Fig. 11) [8]. The Freundlich model was applied to study the isotherm adsorption having a constant correlation (R2) being higher than 0.973 (Fig. 12a, b [8]).

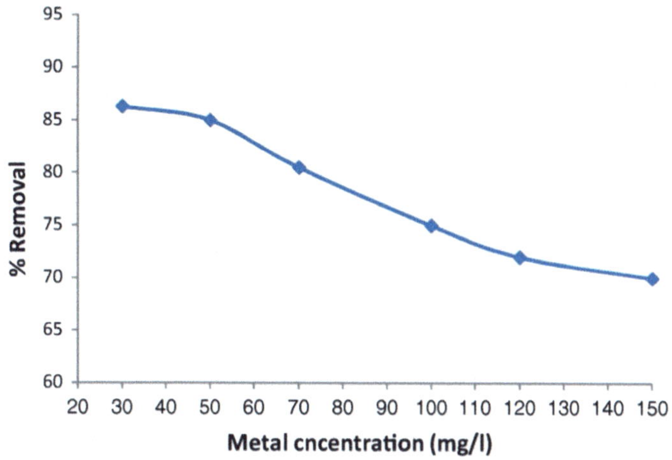

Fig. 11 The effect of metal concentration on Cd (II) remediation by AgNPs [8]

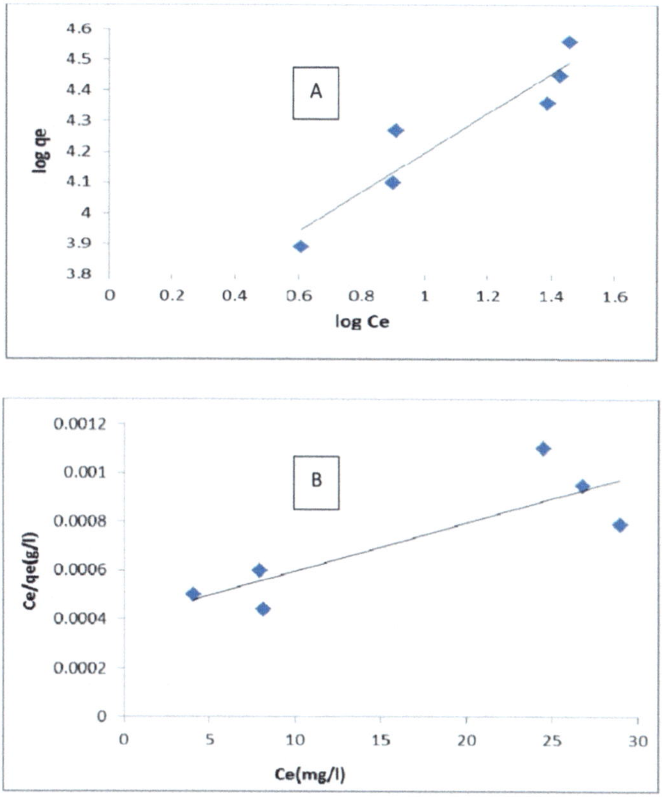

Fig. 12 Freundlich (**a**) and Langmuir (**b**) curve for Cd (II) removal from polluted water [8]

Phytoremediation of *Lemna minor* L., *Azolla filiculoides* Lam. and *Pistia stratiotes* L. was done against Cr, Cd, Pb, and Zn after modifying with interfacial layer of AgNPs. The samples were tested for the absorbance of heavy metals and *L. minor* (Fig. 13) and *P. stratiotes* (Fig. 14) showed selective absorption against Pb, *Azolla filiculoides* (Fig. 15) showed higher absorption against Cr. The absorption of the heavy metals increased in the presence of green synthesized AgNPs [50].

Araucaria polysaccharide gum either dissolves in water or can be water-insoluble and is reported to possess antiulcerogenic properties and anticytotoxic activity [7]. The gum is also used for the green synthesis of AgNPs and for the removal of heavy metals including chromium [6]. In a study, AgNPs were synthesized from the gums of *Araucaria heterophylla, Azadirachta indica*, and *Prosopis chilensis*. The synthesized AgNPs were tested for the removal of Cr whose adsorption was tested with chromium solution as the adsorbate at 50 ppm. The absorbance was determined by a UV spectrophotometer. The data was used to form a Langmuir isotherm model

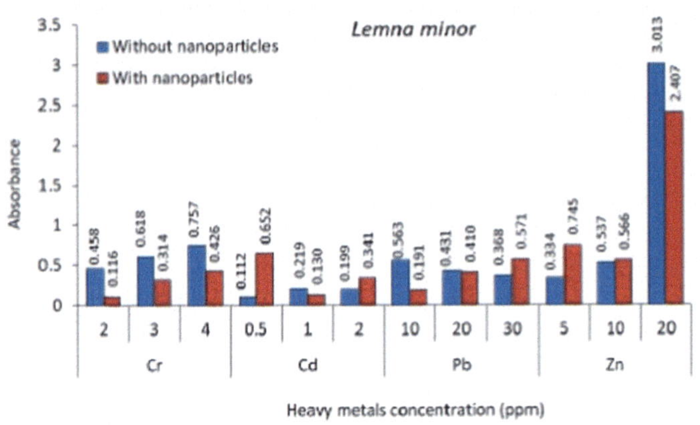

Fig. 13 Changes in absorbance upon treatment of *Lemna minor* with AgNPs [50]

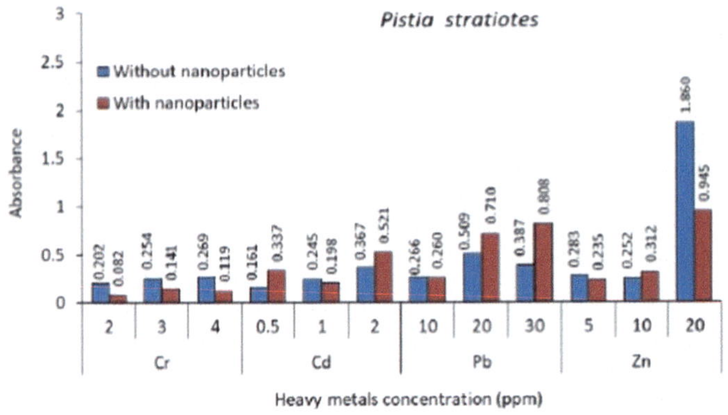

Fig. 14 Changes in absorbance upon treatment of *Pistia stratiotes* with AgNPs [50]

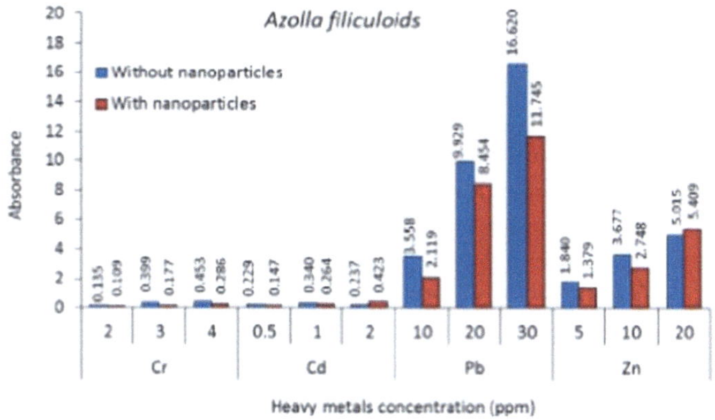

Fig. 15 Changes in absorbance upon treatment of *Azolla filiculoides* powders with AgNPs [50]

to understand the adsorption efficiency of Cr onto the synthesized AgNPs [96]. In another study, nanofibrillated cellulose (NFC) was extracted by a chemical method using alkali and acid hydrolysis. The crystallinity index of treated NFC increased from 39 to 75%. AgNPs were also synthesized by using the extract of *Citrus sinensis* skins as a reducing agent. The synthesized NPs were used to remediate Cd and Cr with nanofibrillated cellulose. 83% removal was found for Cd, i.e. by using Ag and NFC together as adsorbents, it was followed by Cr (47%) for NFC only. The isolation of NFC and synthesis of AgNPs proved efficient for heavy metal sorption by the reuse of waste skins [182]. Heavy metal contamination in soil also affects the living system and the food chain. The application of AgNPs in two genotypes (G1 and G2) of Mung bean (*Vigna radiata*) was studied to mitigate Pb toxicity. Different doses of Pb viz. 0, 25, and 50 μM were treated with AgNPs. Both genotypes exhibited significant modulations for Pb tolerance. In G2 a significant increase in plant height (43.79%), fresh biomass (49.6%), and total chlorophyll (20%) was observed at L2 (AgNPs 10 mg/L) as compared to the control. AgNPs also caused enhanced uptake of minerals and improved growth of *Vigna* genotype. The maximum reduction in growth of the plant was obtained for 50 μM Pb along with decrease in photosynthetic ability. The findings suggested that AgNPs contributed towards Pb tolerance in plants [33]. The phytoremediation of *Abelmoschus esculentus* grown on gold-mined soil was also done through foliar applications of green synthesized AgNPs which had an average particle size of 55 ± 2.3 nm [16]. Different concentrations of AgNPs viz. 0.75, 0.50, and 0.25 mg/mL were used. The metal contamination was reduced by 60% and 44% respectively in *A. esculentus*. The foliar application also enhanced enzymatic and non-enzymatic antioxidant activities. Zero-valent AgNPs mediated from cocoa pods were checked for their heavy metal (Cd and Pb) removal potential and immobilization and their influence on phytochemical contents in *Moringa oleifera* plant. The seeds of *M. oleifera* were planted in spiked soil with different concentrations of $CdCl_2$ and $PbCl_2$. Significant ($p < 0.05$) repression in shoot and root lengths, germination percentage, vigour and growth tolerance indices, relative water content, total carotenoid, and flavonoid content. AgNPs enhanced the physiological and biochemical parameters in *M. oleifera* which were grown on Cd and Pb-treated soil and in which increased lipid peroxidation (MDA) levels were obtained. The results showed that AgNPs exhibited excellent immobilizing ability and acted as outstanding modulators of heavy metals-induced toxicities [17]. It was found that AgNPs substantially alleviated Cd and Pb contamination by promoting germination and tolerance indices. The results were related to other studies which showed the ability of AgNPs to strengthen *M. oleifera* delirious stresses of the heavy metals by decreasing absorption and mobility [64]. In another study, yellow-lupin (*Lupinus luteus* L.) was grown on Cd, Pb, Zn, and Ni and AgNPs-infused soils. The study assessed a plant's response to contamination including its antioxidative response and molecular mechanisms involved in the remediation of metals through the expression of the levels of metallothioneins (MTs). Upon treatment with AgNPs no over-expression of MTs was obtained, whereas upon treatment with heavy metals a significant increase in the level of MTs was obtained [88].

4.2 Mechanistic Details of AgNPs in Heavy Metal Toxicity Removal

The transformation of bulk materials into smaller ones imparts unique physical and chemical properties. NPs may enter into terrestrial and aquatic ecosystems which need to be thoroughly explored as they may also have toxicological effects on the biota. NPs of oxides like TiO_2, ZnO, and CeO_2 and metallic NPs of Ag can induce oxidative stress signalling pathways [30]. NPs can also form pores in cell walls for the internal compartmentalization of large-size NPs [132]. Growth enhamncement has been obtained in *B. juncea* [171], *Eruca sativa* [189], wetland plants [206], *Phaseolus vulgaris* and *Zea mays* [157]. There is a growing evidence showing that AgNPs can interact with membrane proteins and activate signalling pathways which result in cell proliferation. In *E. sativa*, the AgNPs altered the abundance of protein related to redox regulation and metabolism of sulphur especially those related to the endoplasmic reticulum and vacuole [189]. In rice plants upon the treatment of AgNPs the proteins that were altered were mainly related to the oxidative stress response pathway, cell division, and apoptosis [124]. In another study, exposure to 15 nm AgNPs induced plant growth under flooding stress conditions [131]. The proteomics revealed that treatment with AgNPs led to a metabolic shift from fermentative pathways towards normal cellular processes. The result suggested that the exposure of NPs with different perturbations may have positive effects on plant growth. The metals like Al, Cd, As and Zn have been reported to produce oxidative stress. Several defense mechanisms including peroxidase, L-ascorbate peroxidase, glutathione S-transferase, and quinone oxidoreductase are regulated in plants under heavy metal toxicity. Glyoxalase level were enhanced in flooding stress in comparison to those in plants treated with AgNPs [131] which exhibits that the the glutathione cycle acts as an important detoxification pathway under the heavy metal stress.

5 Future Prospects

Nanotechnology can make a vast difference in agricultural sector is becoming important for agribusiness. AgNPs because of their remarkable characteristics have potential uses in agriculture and have been employed to enhance crop yield, in plant protection from pests and microbial infections. AgNPs can also be employed for nano-packaging to increase the shelf life of food. Although AgNPs have a number of benefits, yet there is a need to assess the impact of AgNPs on the environment and human health before proceeding for their mass production in agricultural practices. Although nanotechnology has been explored in different sectors including electronics, energy, and environment, yet it remains unexplored in agricultural sciences. Although AgNPs-mediated responses of plants have been explored against heavy metal toxicity and its phytoremediation in metal-contaminated soils, but long-term studies are required to correlate the AgNPs with the genetic potential and metal stress tolerance.

6 Conclusion

Metal-based NPs have emerged as a feasible way to reduce environmental pollution. Due to their robust structure, compact size, and high surface area, they have shown ecotoxicological effects and making them ideal for heavy metal removal from plants. These heavy metals can enter plant tissues through stomata and cause a reduction in ROS in cells. Ag owing to its related antimicrobial properties has been employed for ages which are also correlated to its unique physicochemical properties. The AgNPs have also been found to be infused with the potential properties and have been synthesized using different approaches. They have been chiefly explored for their medicinal properties. This chapter has summarized the applications of AgNPs for the remediation of heavy metal toxicity in plants. Other aspects like pathogen detection, the role of AgNPs as nanofertilizers, and other agricultural practices have also been discussed. AgNPs have also been found to possess excellent adsorbents to remove contaminants from environmental components and they also are eco-friendly and associated with facile synthesis having catalytic and potential antioxidant properties as well.

Acknowledgements The authors are thankful to Dr. (Mrs.) Vinita Prakash, principal, Isabella Thoburn College, and Prof. Abdul Rahman Khan, Head, Department of Chemistry, Integral University, Lucknow for the support.

References

1. Abasi F, Raja NI, Mashwani ZUR, Amjad MS, Ehsan M, Mustafa N, Proćków J (2022) Biogenic silver nanoparticles as a stress alleviator in plants: a mechanistic overview. Molecules 27:3378
2. Abbas G, Murtaza B, Bibi I, Shahid M, Niazi NK, Khan MI, Natasha (2018) Arsenic uptake, toxicity, detoxification, and speciation in plants: physiological, biochemical, and molecular aspects. Int J Environ Res Public Health 15:59
3. Adrees M, Ali S, Rizwan M, Ibrahim M, Abbas F, Farid M, Zia-ur-Rehman M, Irshad MK, Bharwana SA (2015) The effect of excess copper on growth and physiology of important food crops: a review. Environ Sci Pollut Res 22:8148–8162
4. Ahmed SFR, Killham K, Alexander I (2006) Influences of arbuscular fungus *Glomus mosseae* on growth and nutrition of lentil irrigated with arsenic contaminated water. Plant Soil 258:33–41
5. Ajaib M, Khan Z, Khan NASRULLAH, Wahab M (2010) Ethnobotanical studies on useful shrubs of district Kotli, Azad Jammu & Kashmir, Pakistan. Pak J Bot 42:1407–1415
6. Ali A, Mannan A, Hussain I, Hussain I, Zia M (2018) Effective removal of metal ions from aquous solution by silver and zinc nanoparticles functionalized cellulose: isotherm, kinetics and statistical supposition of process. Environ Nanotechnol Monit Manag 9:1–11
7. Ali S, Mir RA, Tyagi A, Manzar N, Kashyap AS, Mushtaq M, Bae H (2023) Chromium toxicity in plants: signaling, mitigation, and future perspectives. Plants 12:1502
8. Al-Qahtani KM (2017) Cadmium removal from aqueous solution by green synthesis zero valent silver nanoparticles with benjamina leaves extract. Egypt J Aquat Res 43:269–274
9. Aly AA, Mohamed AA (2012) The impact of copper ion on growth, thiol compounds and lipid peroxidation in two maize cultivars (*Zea mays* L.) grown in vitro. Aust J Crop Sci 6:541–549
10. Amjad S, Tiwari S, Serajuddin M (2021) Applications of nanobiotechnology in overcoming drought stress in crops. In: Nanobiotechnology. Springer, pp 399–415
11. Arias JA, Peralta-Videa JR, Ellzey JT, Ren M, Viveros MN, Gardea-Torresdey JL (2010) Effects of Glomus deserticola inoculation on Prosopis: enhancing chromium and lead uptake and translocation as confirmed by X-ray mapping, ICP-OES and TEM techniques. Environ Exp Bot 68:139–148
12. Arya SK, Roy BK (2011) Manganese induced changes in growth, chlorophyll content and antioxidants activity in seedlings of broad bean (*Vicia faba* L.). J Environ Biol 32:707–711
13. Ashfaque F, Inam A, Sahay S, Iqbal S (2016) Influence of heavy metal toxicity on plant growth, metabolism and its alleviation by phytoremediation—a promising technology. J Agric Ecol Res Int 6:1–19
14. Ashraf R, Ali TA (2007) Effect of heavy metals on soil microbial community and mung beans seed germination. Pak J Bot 39:629–636
15. Asrar Z, Khavari-Nejad RA, Heidari H (2005) Excess manganese effects on pigments of *Mentha spicata* at flowering stage. Arch Agron Soil Sci 51:101–107
16. Azeez L, Adebisi SA, Adetoro RO, Oyedeji AO, Agbaje WB, Olabode OA (2022) Foliar application of silver nanoparticles differentially intervenes remediation statuses and oxidative stress indicators in *Abelmoschus esculentus* planted on gold-mined soil. Int J Phytorem 24:384–393
17. Azeez L, Adejumo AL, Lateef A, Adebisi SA, Adetoro RO, Adewuyi SO, Tijani KO, Olaoye S (2019) Zero-valent silver nanoparticles attenuate Cd and Pb toxicities on *Moringa oleifera* via immobilization and induction of phytochemicals. Plant Physiol Biochem 139:283–292
18. Azeez L, Lateef A, Adebisi SA (2017) Silver nanoparticles (AgNPs) biosynthesized using pod extract of Cola nitida enhances antioxidant activity and phytochemical composition of *Amaranthus caudatus* Linn. Appl Nanosci 7:59–66
19. Aziz N, Faraz M, Pandey R, Sakir M, Fatma T, Varma A, Barman I, Prasad R (2015) Facile algae-derived route to biogenic silver nanoparticles: synthesis, antibacterial and photocatalytic properties. Langmuir 31:11605–11612

20. Bachheti RK, Bachheti A, Husen A (2024) Metal and metal-oxide based nanomaterials. In: Synthesis, agricultural, biomedical and environmental interventions. Springer Nature Singapore Pte Ltd., Singapore
21. Barbosa RH, Tabaldi LA, Miyazaki FR, Pilecco M, Kassab SO, Bigaton D (2013) Absorção foliar de cobre por plantas de milho: efeitos no crescimento e rendimento. Cienc Rural 43:1561–1568
22. Barrachina AC, Carbonell FB, Beneyto JM (1995) Arsenic uptake, distribution, and accumulation in tomato plants: effect of arsenite on plant growth and yield. J Plant Nutr 18:1237–1250
23. Belava VN, Panyuta OO, Yakovleva GM, Pysmenna YM, Volkogon MV (2017) The effect of silver and copper nanoparticles on the wheat—*Pseudocercosporella herpotrichoides* pathosystem. Nanoscale Res Lett 12:250
24. Berahmand AA, Ghafariyan Panahi A, Sahabi H, Feizi H, Rezvani Moghaddam P, Shahtahmassebi N, Gallehgir O (2012) Effects silver nanoparticles and magnetic field on growth of fodder maize (*Zea mays* L.). Biol Trace Elem Res 149:419–424
25. Berni R, Luyckx M, Xu X, Legay S, Sergeant K, Hausman J-F, Lutts S, Cai G, Guerriero G (2019) Reactive oxygen species and heavy metal stress in plants: impact on the cell wall and secondary metabolism. Environ Exp Bot 161:98–106
26. Bjuhr J (2007) Trace metals in soils irrigated with waste water in a periurban area downstream Hanoi City, Vietnam. Semin Pap 1–50
27. Bonnet M, Camares O, Veisseire P (2000) Effects of zinc and influence of *Acremonium lolii* on growth parameters, chlorophyll a fluorescence and antioxidant enzyme activities of ryegrass (*Lolium perenne* L. cv Apollo). J Exp Bot 51:945–953
28. Braconi D, Bernardini G, Santucci A (2011) Linking protein oxidation to environmental pollutants: redox proteomic approaches. J Proteomics 74:2324–2337
29. Broadley MR, White PJ, Hammond JP, Zelko I, Lux A (2007) Zinc in plants: Tansley review. New Phytol 173:677–702
30. Buzea C, Pacheco II, Robbie K (2007) Nanomaterials and nanoparticles: sources and toxicity. Biointerphases 2:MR17–MR71
31. Cabala J, Teper L (2007) Metalliferous constituents of rhizosphere soils contaminated by Zn–Pb mining in southern Poland. Water Air Soil Pollut 178:351–362
32. Chang YC, Zouari M, Gogorcena Y, Lucena JJ, Abadía J (2003) Effects of cadmium and lead on ferric chelate reductase activities in sugar beet roots. Plant Physiol Biochem 41:999–1005
33. Chen F, Aqeel M, Maqsood MF, Khalid N, Irshad MK, Ibrahim M, Lam SS (2022) Mitigation of lead toxicity in Vigna radiata genotypes by silver nanoparticles. Environ Pollut 308:119606
34. Chien HF, Lin CC, Wang JW, Chen CT, Kao CH (2002) Changes in ammonium ion content and glutamine synthetase activity in rice leaves caused by excess cadmium are a consequence of oxidative damage. Plant Growth Regul 36:41–47
35. Chiou WY, Hsu FC (2019) Copper toxicity and prediction models of copper content in leafy vegetables. Sustainability 11:6215
36. Chowdappa P, Shivakumar G (2013) Nanotechnology in crop protection: status and cope. Pest Manag Hortic Ecosys 19:131–151
37. Clemens S, Ma JF (2016) Toxic heavy metal and metalloid accumulation in crop plants and foods. Annu Rev Plant Biol 67:489–512
38. Cokkizgin A, Cokkizgin H (2015) Effects of lead ($PbCl_2$) stress on germination of lentil (*Lens culinaris Medic.*) lines. Afr J Biotechnol 9:8608–8612
39. Cox MS, Bell PF, Kovar JL (1996) Differential tolerance of canola to arsenic when grown hydroponically or in soil. J Plant Nutr 19:1599–1610
40. Da Costa MVJ, Sharma PK (2016) Effect of copper oxide nanoparticles on growth, morphology, photosynthesis, and antioxidant response in Oryza sativa. Photosynthetica 54:110–119
41. DalCorso G, Farinati S, Maistri S, Furini A (2008) How plants cope with cadmium: staking all on metabolism and gene expression. J Integr Plant Biol 50:1268–1280

42. Desoky EM, Elrys AS, Rady MM (2019) Integrative moringa and licorice extracts application improves *Capsicum annuum* fruit yield and declines its contaminant contents on a heavy metals contaminated saline soil. Ecotoxicol Environ Saf 169:50–60
43. Dey SK, Dey J, Patra S, Pothal D (2007) Changes in the antioxidative enzyme activities and lipid peroxidation in wheat seedlings exposed to cadmium and lead stress. Braz J Plant Physiol 19:53–60
44. Di Toppi LS, Gabbrielli R (1999) Response to cadmium in higher plants. Environ Exp Bot 41:105–130
45. Doncheva S, Georgieva K, Vassileva V, Stoyanova Z, Popov N, Ignatov G (2005) Effects of succinate on manganese toxicity in pea plants. J Plant Nutr 28:47–62
46. Doncheva S, Stoynova Z, Violeta V (2001) Influence of Succinate on zinc toxicity of pea plants. J Plant Nutr 24:789–804
47. Dotaniya ML, Thakur JK, Meena VD, Jajoria DK, Rathor G (2014) Chromium pollution: a threat to environment—a review. Agric Rev 35:153
48. Du X, Zhu YG, Liu WJ, Zhao XS (2005) Uptake of mercury (Hg) by seedlings of rice (*Oryza sativa* L.) grown in solution culture and interactions with arsenate uptake. Environ Exp Bot 54:1–7
49. Ebbs SD, Kochian LV (1997) Toxicity of zinc and copper to brassica species: implications for phytoremediation. J Environ Qual 26:776–781
50. El Awady FR, Abbas MA, Abdelghany AM, El-Amier YA (2021) Silver modified hydrophytes for heavy metal removal from different water resources. Biointerface Res Appl Chem 11:14555–14563
51. Fageria NK, Santos AB, Filho MP, Guimaraes CM (2008) Iron toxicity in lowland rice. J Plant Nutr 31:1676–1697
52. Faria JM, Pinto AP, Teixeira D, Brito I, Carvalho M (2022) Diversity of native arbuscular mycorrhiza extraradical mycelium influences antioxidant enzyme activity in wheat grown under Mn toxicity. Bull Environ Contam Toxicol 108:451–456
53. Farooq MA, Hong Z, Islam F, Noor Y, Hannan F, Zhang Y, Song W (2021) Comprehensive proteomic analysis of arsenic induced toxicity reveals the mechanism of multilevel coordination of efficient defense and energy metabolism in two Brassica napus cultivars. Ecotoxicol Environ Saf 208:111744
54. Fatemi H, Zaghdoud C, Nortes PA, Carvajal M, Martínez-Ballesta MDC (2020) Diferential aquaporin response to distinct efects of two Zn concentrations after foliar application in Pak Choi (*Brassica rapa* L.) plants. Agronomy 10:450
55. Fitriana WD, Ersam T, Shimizu K, Fatmawati S (2016) Antioxidant activity of *Moringa oleifera* extracts. Indones J Chem 16:297–301
56. Fryzova R, Pohanka M, Martinkova P, Cihlarova H, Brtnicky M, Hladky J, Kynicky J (2017) Oxidative stress and heavy metals in plants. Residue Rev 245:129–156
57. Fu SF, Chen PY, Nguyen QTT, Huang LY, Zeng GR, Huang TL, Lin CY, Huang HJ (2014) Tran-scriptome profiling of genes and pathways associated with arsenic toxicity and tolerance in *Arabidopsis*. BMC Plant Biol 14:94
58. Fusconi A, Repetto O, Bona E, Massa N, Gallo C, Dumas-Gaudot E, Berta G (2006) Effects of cadmium on meristem activity and nucleus ploidy in roots of *Pisum sativum* L. cv frisson seedlings. Environ Exp Bot 58:253–260
59. Garg N, Singh S (2018) Arbuscular mycorrhiza Rhizophagus irregularis and silicon modulate growth, proline biosynthesis and yield in *Cajanus cajan* L. Millsp. (*pigeonpea*) genotypes under cadmium and zinc stress. J Plant Growth Regul 37:46–63
60. Garg N, Singla P (2011) Arsenic toxicity in crop plants: physiological effects and tolerance mechanisms. Environ Chem Lett 9:303–321
61. Ghnaya AB, Hourmant A, Cerantola S, Karvarec N, Cabon JY, Branchard M, Charles G (2010) Infuence of zinc on soluble carbohydrate and free amino acid levels in rapeseed plants regenerated in vitro in the presence of zinc. Plant Cell Tiss Organ Cult 102:191–197
62. Gitto A, Fricke W (2018) Zinc treatment of hydroponically grown barley plants causes a reduction in root and cell hydraulic conductivity and isoform-dependent decrease in aquaporin gene expression. Physiol Plant 164:176–190

63. Gjorgieva Ackova D (2018) Heavy metals and their general toxicity on plants. Plant Sci Today 5:15–19
64. Gong X, Huang D, Liu Y, Peng Z, Zeng G, Xu P, Cheng M, Wang R, Wan J (2018) Remediation of contaminated soils by biotechnology with nanomaterials: biobehavior, applications, and perspectives. Crit Rev Biotechnol 38:455–468
65. Gong Z, Xiong L, Shi H, Yang S, Herrera-Estrella LR, Xu G, Zhu JK (2020) Plant abiotic stress response and nutrient use efficiency. Sci China Life Sci 63:635–674
66. Gruyer N, Dorais M, Bastien C, Dassylva N, Triffault-Bouchet G (2013) Interaction between sliver nanoparticles and plant growth. In: International symposium on new technologies for environment control, energy-saving and crop production in greenhouse and plant, vol 1037, pp 795–800
67. Gupta SD, Agarwal A, Pradhan S (2018) Phytostimulatory effect of silver nanoparticles (AgNPs) on rice seedling growth: an insight from antioxidative enzyme activities and gene expression patterns. Ecotoxicol Environ Saf 161:624–633
68. Gupta SD (2010) Metal toxicity, oxidative stress and antioxidative defense system in plants. In: Reactive oxygen species and antioxidants in higher plants. CRC Press, pp 193–220
69. Habiba U, Ali S, Rizwan M, Ibrahim M, Hussain A, Shahid MR, Alamri SA, Alyemeni MN, Ahmad P (2019) Alleviative role of exogenously applied mannitol in maize cultivars differing in chromium stress tolerance. Environ Sci Pollut Res 26:5111–5121
70. Hadi F (2015) A mini review on lead toxicity in plants. J Biol Life Sci 6:91–101
71. Hafeez B (2013) Role of zinc in plant nutrition—a review. Am J Exp Agric 3:374–391
72. Haque MdN, Ali MdH, Roy TS (2015) Yield reduction and arsenic accumulation in potatoes (*Solanum tuberosum* L.) in an arsenic contaminated soil. Agron Colomb 33:315–321
73. Hawkes SJ (1997) What is a "heavy metal"? J Chem Educ 74:1374
74. Helaoui S, Hattab S, Mkhinini M, Boughattas I, Majdoub A, Banni M (2022) The effect of nickel exposure on oxidative stress of Vicia faba plants. Bull Environ Contam Toxicol 108:1074–1080
75. Hernández HH, Benavides-Mendoza A, Ortega-Ortiz H, Hernández-Fuentes AD, Juárez-Maldonado A (2017) Cu Nanoparticles in chitosan-PVA hydrogels as promoters of growth, productivity and fruit quality in tomato. Emir J Food Agric 29:573–580
76. Hossain MA, Hossain MZ, Fujita M (2009) Stress-induced changes of methylglyoxal level and glyoxalase I activity in pumpkin seedlings and cDNA cloning of glyoxalase I gene. Aust J Crop Sci 3:53–64
77. Husen A (2020) Interactions of metal and metal-oxide nanomaterials with agricultural crops: an overview. In: Husen A, Jawaid M (eds) Nanomaterials for agriculture and forestry applications. Elsevier Inc., Cambridge, MA, USA, pp 167–197
78. Husen A (2023) Nanomaterials and nanocomposites exposures to plants. In: Response, interaction, phytotoxicity and defense mechanisms. Springer Nature Singapore Pte Ltd., Singapore. https://doi.org/10.1007/978-981-99-2419-6
79. Husen A, Bachheti RK, Bachheti A (2023) Current trends in green nano-emulsions. In: Food, agriculture and biomedical sectors. Springer Nature Singapore Pte Ltd., Singapore
80. Husen A, Iqbal M (2019) Nanomaterials and plant potential. Springer International Publishing AG, Switzerland. https://doi.org/10.1007/978-3-030-05569-1
81. Husen A, Jawaid M (2020) Nanomaterials for agriculture and forestry applications. Elsevier Inc., Cambridge, MA, USA. https://doi.org/10.1016/C2018-0-02349-X
82. Husen A, Siddiqi KS (2023) Advances in smart nanomaterials and their applications. Elsevier Inc., Cambridge, MA, USA. https://doi.org/10.1016/C2021-0-02202-1
83. Hussain A, Abbas N, Arshad F, Akram M, Khan ZI, Ahmad K, Mirzaei F (2013) Effects of diverse doses of lead (Pb) on different growth attributes of *Zea mays* L. Agric Sci 4:262–265
84. Iannelli MA, Pietrini F, Fiore L, Petrilli L, Massacci A (2002) Antioxidant response to cadmium in Phragmites australis plants. Plant Physiol Biochem 40:977–982
85. Inglezakis VJ, Zorpas AA, Karagianides A, Samaras P, Voukalli I (2011) European Union legislation on sewage sludge management. Fresenius Environ Bull 23:635–639

86. Islam E, Yang X, Li T, Liu D, Jin X, Meng F (2007) Effect of Pb toxicity on root morphology, physiology and ultrastructure in the two ecotypes of Elsholtzia argyi. J Hazard Mater 147:806–816
87. Jasim B, Thomas R, Mathew J, Radhakrishnan EK (2016) Plant growth and diosgenin enhancement effect of silver nanoparticles in Fenugreek (*Trigonella foenumgraecum* L.). Saudi Pharm J 25:443–447
88. Jaskulak M, Rorat A, Grobelak A, Chaabene Z, Kacprzak M, Vandenbulcke F (2019) Bioaccumulation, antioxidative response, and metallothionein expression in *Lupinus luteus* L. exposed to heavy metals and silver nanoparticles. Environ Sci Pollut Res 26:16040–16052
89. Jayakumar K, Jaleel CA, Azooz MM (2008) Phytochemical changes in green gram (*Vigna radiata*) under cobalt stress. Glob J Mol Sci 3:46–49
90. Jayakumar K, Jaleel CA, Vijayarengan P (2007) Changes in growth, biochemical constituents, and antioxidant potentials in radish (*Raphanus sativus* L.) under cobalt stress. Turk J Biol 31:127–136
91. Jayakumar K, Rajesh M, Baskaran L, Vijayarengan P (2013) Changes in nutritional metabolism of tomato (*Lycopersicon esculantum* Mill.) plants exposed to increasing concentration of cobalt chloride. Int J Food Nutr Saf 4:62–69
92. Jha AB, Dubey RS (2004) Arsenic exposure alters activity behaviour of key nitrogen assimilatory enzymes in growing rice plants. Plant Growth Regul 43:259–268
93. Jiang W, Liu D, Hou W (2001) Hyperaccumulation of cadmium by roots, bulbs and shoots of garlic. Bioresour Technol 76:9–13
94. Jibril SA, Hassan SA, Ishak CF, Megat Wahab PE (2017) Cadmium toxicity affects phytochemicals and nutrient elements composition of Lettuce (*Lactuca sativa* L.). Adv Agric 2017:1–3
95. Kabir M, Iqbal MZ, Shafiq M (2009) Effects of lead on seedling growth of *Thespesia populnea* L. Adv Environ Biol 3:184–190
96. Kakavandi B, Jonidi A, Rezaei R, Nasseri S, Ameri A, Esrafily A (2013) Synthesis and properties of Fe_3O_4-activated carbon magnetic nanoparticles for removal of aniline from aqueous solution: equilibrium, kinetic and thermodynamic studies. Iran J Environ Health Sci Eng 10:1–9
97. Kale SK, Parishwad GV, Husainy ASN, Patil AS (2021) Emerging agriculture applications of silver nanoparticles. ES Food Agrofor 3:17–22
98. Khalid S, Shahid M, Niazi NK, Murtaza B, Bibi I, Dumat C (2017) A comparison of technologies for remediation of heavy metal contaminated soils. J Geochem Explor 182:247–268
99. Khalid BY, Tinsley J (1980) Some effects of nickel toxicity on rye grass. Plant Soil 55:139–144
100. Khiyami MA, Almoammar H, Awad YM, Alghuthaym MA, Abd-Elsalam KA (2014) Plant pathogen nanodiagnostic techniques: forthcoming changes? Biotechnol Biotechnol Equip 28:775–785
101. Kholodova V, Volkov K, Abdeyeva A, Kuznetsov V (2011) Water status in Mesembryanthemum crystallinum under heavy metal stress. Environ Exp Bot 71:382–389
102. Koleva L, Semerdjieva I, Nikolova A, Vassilev A (2010) Comparative morphological and histological study of zinc- and cadmium-treated durum wheat plants with similar growth inhibition. Gen Appl Plant Physiol 36:8–11
103. Kopittke PM, Asher CJ, Kopittke RA, Menzies NW (2007) Toxic effects of Pb^{2+} on growth of cowpea (*Vigna unguiculata*). Environ Pollut 150:280–287
104. Kopittke PM, Menzies NW, de Jonge MD, Mckenna BA, Donner E, Webb RI, Paterson DJ, Howard DL, Ryan CG, Glover CJ (2011) In situ distribution and speciation of toxic copper, nickel, and zinc in hydrated roots of cowpea. Plant Physiol 156:663–673
105. Krishnaraj C, Jagan EG, Ramachandran R, Abirami SM, Mohan N, Kalaichelvan PT (2012) Effect of biologically synthesized silver nanoparticles on *Bacopa monnieri* (Linn.) Wettst. Plant growth metabolism. Process Biochem 47:51–658
106. Kumar A, Kumar A, MMS CP, Chaturvedi AK, Shabnam AA, Subrahmanyam G, Yadav KK (2020) Lead toxicity: health hazards, influence on food chain, and sustainable remediation approaches. Int J Environ Res Public Health 17:2179

107. Kumar L, Bharadvaja N (2020) Microbial remediation of heavy metals. In: Microbial bioremediation and biodegradation. Springer, Singapore, pp 49–72
108. Kumar L, Bharadvaja N (2019) Enzymatic bioremediation: a smart tool to fight environmental pollutants. In: Smart bioremediation technologies. Academic Press, pp 99–118
109. Kushwaha A, Hans N, Kumar S, Rani R (2018) A critical review on speciation, mobilization and toxicity of lead in soil-microbe-plant system and bioremediation strategies. Ecotoxicol Environ Saf 147:1035–1045
110. Lateef A, Akande MA, Azeez MA, Ojo SA, Folarin BI, Gueguim-Kana EB, Beukes LS (2016) Phytosynthesis of silver nanoparticles (AgNPs) using miracle fruit plant (*Synsepalum dulcificum*) for antimicrobial, catalytic, anticoagulant, and thrombolytic applications. Nanotechnol Rev 5:507–520
111. Lawal AT, Azeez L, Sulaiman WK (2021) Silver nanoparticles (AgNPs) alleviate naphthalene-triggered oxidative stress and physiological deficiencies in Moringa oleifera. Chem Ecol 37:15–31
112. Liang J, Yang W (2019) Effects of zinc and copper stress on antioxidant system of olive leaves. IOP Conf Ser Earth Environ Sci 300:52058
113. Lin YC, Kao CH (2005) Nickel toxicity of rice seedlings: cell wall peroxidase, lignin, and $NiSO_4$-inhibited root growth. Crop Environ Bioinform 2:131–136
114. Lin YW, Liu TS, Guo HY, Chiang CM, Tang HJ, Chen HT, Chen JH (2015) Relationships between Cd concentrations in different vegetables and those in arable soils, and food safety evaluation of vegetables in Taiwan. Soil Sci Plant Nutr 61:983–998
115. Liu S, Shi X, Pan Y, Ding J, He Y, Wang L (2013) Effects of cadmium stress on growth, accumulation and distribution of biomass and nutrient in *Catharanthus roseus*. Acta Pratacul Sin 22:154–161
116. Lucini L, Bernardo L (2015) Comparison of proteome response to saline and zinc stress in lettuce. Front Plant Sci 6:240
117. Mahajan S, Kadam J, Dhawal P, Barve S, Kakodkar S (2022) Application of silver nanoparticles in in-vitro plant growth and metabolite production: revisiting its scope and feasibility. Plant Cell Tissue Organ Cult (PCTOC) 150:15–39
118. Maheshwari R, Dubey RS (2007) Nickel toxicity inhibits ribonuclease and protease activities in rice seedlings: protective effects of proline. Plant Growth Regul 51:231–243
119. Manivasagaperumal R, Balamurugan S, Thiyagarajan G, Sekar J (2011) Effect of zinc on germination, seedling growth and biochemical content of cluster bean (*Cyamopsis tetragonoloba* (L.) Taub). Curr Bot 2:11–15
120. Małecka A, Derba-Maceluch M, Kaczorowska K (2009) Reactive oxygen species production and antioxidative defense system in pea root tissues treated with lead ions: mitochondrial and peroxisomal level. Acta Physiol Plant 31:1065–1075
121. Meharg AA (1993) The role of the plasmalemma in metal tolerance in angiosperms. Physiol Plant 88:191–198
122. Meharg AA, Lombi E, Williams PN, Scheckel KG, Feldmann J, Raab A, Islam R (2008) Speciation and localization of arsenic in white and brown rice grains. Environ Sci Technol 42:1051–1057
123. Mir BA, Khan TA, Fariduddin Q (2015) 24-Epibrassinolide and spermidine modulate photosynthesis and antioxidant systems in *Vigna radiata* under salt and zinc stress. Int J Adv Res 3:592–608
124. Mirzajani F, Askari H, Hamzelou S, Schober Y, Römpp A, Ghassempour A, Spengler B (2014) Proteomics study of silver nanoparticles toxicity on *Oryza sativa* L. Ecotoxicol Environ Saf 108:335–339
125. Misra P, Nath K, Tandon PK (2010) Effect of heavy metals (Ni and Pb) stress on sugarcane (*Saccharum officinarum* L.). Res Environ Life Sci 3:183–188
126. Miteva E (2002) Accumulation and effect of arsenic on tomatoes. Commun Soil Sci Plant Anal 33:1917–1926
127. Mitra S, Chakraborty AJ, Tareq AM, Emran TB, Nainu F, Khusro A, Idris AM, Khandaker MU, Osman H, Alhumaydhi FA, Gandara SM (2022) Impact of heavy metals on the environment

and human health: novel therapeutic insights to counter the toxicity. J King Saud Univ - Sci 34:101865
128. Mohamed HI (2011) Molecular and biochemical studies on the effect of gamma rays on lead toxicity in cowpea (*Vigna sinensis*) plants. Biol Trace Elem Res 144:1205–1218
129. Mostofa MG, Seraj ZI, Fujita M (2014) Exogenous sodium nitroprusside and glutathione alleviate copper tox-icity by reducing copper uptake and oxidative damage in rice (*Oryza sativa* L.) seedlings. Protoplasma 251:1373–1386
130. Moustakas M, Lanaras T, Symeonidis L, Karataglis S (1994) Growth and some photosynthetic characteristics of field grown Avena sativa under copper and lead stress. Photosynthetica 30:389–396
131. Mustafa G, Sakata K, Hossain Z, Komatsu S (2015) Proteomic study on the effects of silver nanoparticles on soybean under flooding stress. J Proteomics 122:100–118
132. Navarro E, Baun A, Behra R, Hartmann NB, Filser J, Miao AJ, Sigg L (2008) Environmental behavior and ecotoxicity of engineered nanoparticles to algae, plants, and fungi. Ecotoxicology 17:372–386
133. Ogunkunle CO, Varun M, Ogundele IG, Olorunmaiye KS, Paul MS (2018) Citrus epicarp-derived biochar reduced Cd uptake and ameliorates oxidative stress in young *Abelmoschus esculentus* (L.) Moench (okra) under low Cd stress. Bull Environ Contam Toxicol 100:827–833
134. Olmos E, Martínez-Solano JR, Piqueras A, Hellin E (2003) Early steps in the oxidative burst induced by cadmium in cultured tobacco cell (BY-2 line). J Exp Bot 54:291–301
135. Olowolaju ED, Okunlola GO, Adejumo AM, Adelusi AA (2018) Effects of prior heat stress on the growth and phytochemical contents accumulation of *Amaranthus hybridus* (Linn.). Not Sci Biol 10:240–244
136. Palansooriya KN, Shaheen SM, Chen SS, Tsang DC, Hashimoto Y, Hou D, Bolan NS, Rinklebe J, Ok YS (2020) Soil amendments for immobilization of potentially toxic elements in contaminated soils: a critical review. Environ Int 134:105046
137. Pan X, Zhao Q, Zhao T (2021) Plant growth, antioxidative enzyme, and cadmium tolerance responses to cadmium stress in Canna orchioides. Hortic Plant J 7:256–266
138. Pandolfini T, Gabbrielli R, Comparini C (1992) Nickel toxicity and peroxidase activity in seedlings of *Triticum aestivum* L. Plant Cell Environ 15:719–725
139. Parikh RY, Singh S, Prasad BLV, Patole MS, Sastry M, Shouche YS (2008) Extracellular synthesis of crystalline silver nanoparticles and molecular evidence of silver resistance from *Morganella* sp.: towards understanding biochemical synthesis mechanism. ChemBioChem 9:1415–1422
140. Paunov M, Koleva L, Vassilev A, Vangronsveld J, Goltsev V (2018) Effects of different metals on photosynthesis: cadmium and zinc affect chlorophyll fluorescence in durum wheat. Int J Mol Sci 19:787
141. Pauwels M, Saumitou-Laprade P, Holl AC, Petit D, Bonnin I (2005) Multiple origin of metallicolous populations of the pseudometallophyte arabidopsis halleri (*Brassicaceae*) in Central Europe: the cpDNA testimony. Mol Ecol 14:4403–4414
142. Prasad R, Pandey R, Barman I (2016) Engineering tailored nanoparticles with microbes: quo vadis. WIREs Nanomed Nanobiotechnol 8:316–330
143. Prasad MNV, Hagemeyer J, Dietz KJ, Baier M, Krämer U (1999) Free radicals and reactive oxygen species as mediators of heavy metal toxicity in plants. In: Heavy metal stress in plants: from molecules to ecosystems, pp 73–97
144. Rabêlo FHS, Azevedo RA, Monteiro FA (2017) The proper supply of S increases amino acid synthesis and antioxidant enzyme activity in Tanzania Guinea grass used for Cd phytoextraction. Water Air Soil Pollut 228:394
145. Rady MM, Taha SS, Kusvuran S (2018) Integrative application of cyanobacteria and antioxidants improve common bean performance under saline conditions. Sci Hortic 233:61–69
146. Rahman Z, Singh VP (2019) The relative impact of toxic heavy metals (THMs) (arsenic (As), cadmium (Cd), chromium (Cr) (VI), mercury (Hg), and lead (Pb)) on the total environment: an overview. Environ Monit Assess 191:1–21

147. Rai M, Ingle A (2012) Role of nanotechnology in agriculture with special reference to management of insect pests. Appl Microbiol Biotechnol 94:287–293
148. Rajpoot R, Srivastava RK, Rani A, Pandey P, Dubey RS (2021) Manganese-induced oxidative stress, ultrastructural changes, and proteomics studies in rice plants. Protoplasma 258:319–335
149. Rascio N, Navari-Izzo F (2011) Heavy metal hyperaccumulating plants: how and why do they do it? And what makes them so interesting? Plant Sci 180:169–181
150. Rehman H, Alharby HF, Alzahrani Y, Rady MM (2018) Magnesium and organic biostimulant integrative application induces physiological and biochemical changes in sunflower plants and its harvested progeny on sandy soil. Plant Physiol Biochem 126:97–105
151. Rehman MZU, Rizwan M, Ali S, Sabir M, Sohail MI (2017) Contrasting effects of organic and inorganic amendments on reducing lead toxicity in wheat. Bull Environ Contam Toxicol 99:642–647
152. Rezvani N, Sorooshzadeh A, Farhadi N (2012) Effect of nano-silver on growth of saffron in flooding stress. World Acad Sci Eng Technol 1:517–522
153. Rivetta A, Negrini N, Cocucci M (1997) Involvement of Ca^{2+}-calmodulin in Cd^{2+} toxicity during the early phases of radish (*Raphanus sativus* L.) seed germination. Plant, Cell Environ 20:600–608
154. Rizwan M, Ali S, Fatima N, Yousaf B, Naeem A, Sabir M, Ahmad HR, Ok YS (2016) Contrasting effects of biochar, compost and farm manure on alleviation of nickel toxicity in maize (*Zea mays* L.) in relation to plant growth, photosynthesis and metal uptake. Ecotoxicol Environ Saf 133:218–225
155. Romero-Puertas MC, Palma JM, Gómez M, Del Rio LA, Sandalio LM (2002) Cadmium causes the oxidative modification of proteins in pea plants. Plant, Cell Environ 25:677–686
156. Sagonda T, Adil MF, Sehar S, Rasheed A, Joan HI, Ouyang Y, Shamsi IH (2021) Physio-ultrastructural footprints and iTRAQ-based proteomic approach unravel the role of Piriformospora indica-colonization in counteracting cadmium toxicity in rice. Ecotoxicol Environ Saf 220:112390
157. Salama HM (2012) Effects of silver nanoparticles in some crop plants, common bean (*Phaseolus vulgaris* L.) and corn (*Zea mays* L.). Int Res J Biotechnol 3:190–197
158. Savithramma N, Ankanna S, Bhumi G (2012) Effect of nanoparticles on seed germination and seedling growth of *Boswellia ovalifoliolata* an endemic and endangered medicinal tree taxon. Nano Vision 2:61–68
159. Schiavon M, Agostini G, Pittarello M, Dalla Vecchia F, Pastore P, Malagoli M (2009) Interactions between chromate and sulfate affect growth, photosynthesis and ultrastructure in Brassica juncea (L.) Czern. In: Sulfur metabolism in plants. Backhuys Publishers, Leiden
160. Schwenkbier L, Pollok S, König S, Urban M, Werres S, Dana Cialla-May D, Karina Weber K, Popp J (2015) Towards on-site testing of *Phytophthora* species. Anal Methods 7:211–217
161. Scott N, Chen H (2013) Nanoscale science and engineering for agriculture and food systems. Ind Biotechnol 9:17–18
162. Sedzik M, Smolik B, Krupa-Małkiewicz M (2015) Effect of lead on germination and some morphological and physiological parameters of 10 day-old seedlings of various plant species. Environ Prot Nat Resour 26:22–27
163. Servin A, Elmer W, Mukherjee A, De La Torre-Roche R, Hamdi H, White JC, Bindraban P, Dimkpa C (2015) A review of the use of engineered nanomaterials to suppress plant disease and enhance crop yield. J Nanopart Res 17:1–21
164. Sethy SK, Ghosh S (2013) Effect of heavy metals on germination of seeds. J Nat Sci Biol Med 4:272–275
165. Shaibur MR, Kitajima N, Sugewara R, Kondo T, Alam S, Imamul Huq SM, Kawai S (2008) Critical toxicity of arsenic and elemental composition of arsenic-induced chlorosis in hydroponic Sorghum. Water Air Soil Pollut 191:279–292
166. Shakya K, Chettri MK, Sawidis T (2008) Impact of heavy metals (copper, zinc, and lead) on the chlorophyll content of some mosses. Arch Environ Contam Toxicol 54:412–421
167. Shanker AK, Cervantes C, Loza Tavera H, Avudainayagam S (2005) Chromium toxicity in plants. Environ Int 31:739–753

168. Shanmugaraj BM, Harish MC, Balamurugan S, Sathishkumar R (2013) Cadmium induced physio-biochemical and molecular response in *Brassica juncea*. Int J Phytoremediation 15:206–221
169. Shanying HE, Xiaoe YANG, Zhenli HE, Baligar VC (2017) Morphological and physiological responses of plants to cadmium toxicity: a review. Pedosphere 27:421–438
170. Sharma I (2012) Arsenic induced oxidative stress in plants. Biologia 67:447–453
171. Sharma P, Bhatt D, Zaidi MGH, Saradhi PP, Khanna PK, Arora S (2012) Silver nanoparticle-mediated enhancement in growth and antioxidant status of *Brassica juncea*. Appl Biochem Biotechnol 167:2225–2233
172. Sharma P, Dubey RS (2005) Lead toxicity in plants. Braz J Plant Physiol 17:35–52
173. Shekar CHC, Sammaiah D, Shasthree T, Reddy KJ (2011) Effect of mercury on tomato growth and yield attributes. Int J Pharma Bio Sci 2:358–364
174. Shenker M, Plessner OE, Tel-Or E (2004) Manganese nutrition effects on tomato growth, chlorophyll concentration, and superoxide dismutase activity. J Plant Physiol 161:197–202
175. Sheoran IS, Singal HR, Singh R (1990) Effect of cadmium and nickel on photosynthesis and the enzymes of the photosynthetic carbon reduction cycle in pigeonpea (*Cajanus cajan* L.). Photosynth Res 23:345–351
176. Siddiqi KS, Husen A (2016) Engineered gold nanoparticles and plant adaptation potential. Nanoscale Res Lett 11(400):1–10
177. Siddiqi KS, Husen A (2022) Plant response to silver nanoparticles: a critical review. Crit Rev Biotechnol 42(7):773–990. https://doi.org/10.1080/07388551.2021.1975091
178. Skórzyńska-Polit E, Pawlikowska-Pawlęga B, Szczuka E, Drążkiewicz M, Krupa Z (2006) The activity and localization of lipoxygenases in Arabidopsis thaliana under cadmium and copper stresses. Plant Growth Regul 48:29–39
179. Smith SE, Christophersen HM, Pope S, Smith FA (2010) Arsenic uptake and toxicity in plants: integrating mycorrhizal influences. Plant Soil 327:1–21
180. Song C, Yan Y, Rosado A, Zhang Z, Castellarin SD (2019) ABA alleviates uptake and accumulation of zinc in grapevine (*Vitis vinifera* L.) by inducing expression of ZIP and detoxification-related genes. Front Plant Sci 10:872
181. Srivastava S, Srivastava AK, Suprasanna P et al (2009) Comparitive biochemical and transcriptional profiling of two contrasting varieties of Brassica juncea L. in response to arsenic exposure reveals mechanisms of stress perception and tolerance. J Exp Bot 60:3419–3431
182. Tavker N, Yadav VK, Yadav KK, Cabral-Pinto MM, Alam J, Shukla AK, Ali FAA, Alhoshan M (2021) Removal of cadmium and chromium by mixture of silver nanoparticles and nano-fibrillated cellulose isolated fromwaste peels of Citrus sinensis. Polymers 13:234
183. Tomulescu IM, Radoviciu EM, Merca VV, Tuduce AD (2004) Effect of copper, zinc and lead and their combinations on the germination capacity of two cereals. J Agric Sci 15:39–42
184. Tripathi DK, Singh S, Singh S, Srivastava PK, Singh VP, Singh S, Chauhan DK (2017) Nitric oxide alleviates silver nanoparticles (AgNps)-induced phytotoxicity in *Pisum sativum* seedlings. Plant Physiol Biochem 110:167–177
185. Tsuji K (2001) Microencapsulation of pesticides and their improved handling safety. J Microencapsul 18:137–147
186. Ugwu EI, Agunwamba JC (2020) A review on the applicability of activated carbon derived from plant biomass in adsorption of chromium, copper, and zinc from industrial wastewater. Environ Monit Assess 192:240
187. Uzu G, Sobanska S, Aliouane Y, Pradere P, Dumat C (2009) Study of lead phytoavailability for atmospheric industrial micronic and sub-micronic particles in relation with lead speciation. Environ Pollut 157:1178–1185
188. Van Assche F, Clijsters H (1986) Inhibition of photosynthesis in Phaseolus vulgaris by treatment with toxic concentration of zinc: effect on ribulose-1, 5-bisphosphate carboxylase/oxygenase. J Plant Physiol 125:355–360
189. Vannini C, Domingo G, Onelli E, Prinsi B, Marsoni M, Espen L, Bracale M (2013) Morphological and proteomic responses of Eruca sativa exposed to silver nanoparticles or silver nitrate. PLoS ONE 8:e68752

190. Verma A, Bharadvaja N (2022) Plant-mediated synthesis and characterization of silver and copper oxide nanoparticles: antibacterial and heavy metal removal activity. J Cluster Sci 33:1697–1712
191. Villiers F, Ducruix C, Hugouvieux V, Jarno N, Ezan E, Garin J, Bourguignon J (2011) Investigating the plant response to cadmium exposure by proteomic and metabolomic approaches. Proteomics 11:1650–1663
192. Vinković T, Novák O, Strnad M, Goessler W, Jurašin DD, Paradiković N, Vrček IV (2017) Cytokinin response in pepper plants (*Capsicum annuum* L.) exposed to silver nanoparticles. Environ Res 156:10–18
193. Wang W, Zou J, Duan X, Jiang W, Liu D (2007) Cadmium accumulation and its effects on metal uptake in maize (*Zea mays* L.). Bioresour Technol 98:82–88
194. White JC, Gardea-Torresdey J (2018) Achieving food security through the very small. Nat Nanotechnol 13:627–629
195. Winska-Krysiak M, Koropacka K, Gawronski S (2015) Determination of the tolerance of sunflower to lead-induced stress. J Elementol 20
196. World Health Organization (WHO) (1996) Permissible limits of heavy metals in soil and plants. World Health Organization, Geneva, Switzerland
197. Wu X, Cobbina SJ, Mao G, Xu H, Zhang Z, Yang L (2016) A review of toxicity and mechanisms of individual and mixtures of heavy metals in the environment. Environ Sci Pollut Res 23:8244–8259
198. Wuana RA, Okieimen FE (2011) Heavy metals in contaminated soils: a review of sources, chemistry, risks and best available strategies for remediation. ISRN Ecol 2011:1–20
199. Wyszkowski M, Radziemska M (2010) Effects of chromium (III and VI) on spring barley and maize biomass yield and content of nitrogenous compounds. J Toxicol Environ Health Part A 73:1274–1282
200. Yadav SK (2010) Heavy metals toxicity in plants: an overview on the role of glutathione and phytochelatins in heavy metal stress tolerance of plants. South Afr J Bot 76:167–179
201. Yadav SK, Singla-Pareek SL, Ray M, Reddy MK, Sopory SK (2005) Methylglyoxal levels in plants under salinity stress are dependent on glyoxalase I and glutathione. Biochem Biophys Res Commun 337:61–67
202. Yang J, Cao W, Rui Y (2017) Interactions between nanoparticles and plants: phytotoxicity and defence mechanisms. J Plant Interact 12:158–169
203. Yang Y, Wei X, Lu J, You J, Wang W, Shi R (2010) Lead-induced phytotoxicity mechanism involved in seed germination and seedling growth of wheat (*Triticum aestivum* L.). Ecotoxicol Environ Saf 73:1982–1987
204. Yang Y, Zhang L, Huang X, Zhou Y, Quan Q, Li Y, Zhu X (2020) Response of photosynthesis to different concentrations of heavy metals in *Davidia involucrata*. PLoS ONE 15:e0228563
205. Yin D, Wang X, Chen C, Peng B, Tan C, Li H (2016) Varying effect of biochar on Cd, Pb and as mobility in a multi-metal contaminated paddy soil. Chemosphere 152:196–206
206. Yin L, Colman BP, McGill BM, Wright JP, Bernhardt ES (2012) Effects of silver nanoparticle exposure on germination and early growth of eleven wetland plants *Lolium multiflorum, Panicum virgatum, Carex lurida, C. scoparia, C. vulpinoidea, C. crinita, Eupatorium fistulosum, Phytolacca americana, Scirpus cyperinus, Lobelia cardinalis, Juncus effusus*). PLoS One 7:e47674
207. Yourtchi MS, Bayat HR (2013) Effect of cadmium toxicity on growth, cadmium accumulation and macronutrient content of durum wheat (Dena CV.). Int J Agric Crop Sci 6:1099–1103
208. Zare Z, Pishkar L, Iranbakhsh A, Talei D (2020) Physiological and molecular effects of silver nanoparticles exposure on purslane (*Portulaca oleracea* L.). Russ J Plant Physiol 67:521–528
209. Zhu JK (2016) Abiotic stress signaling and responses in plants. Cell 167:313–324

Mitigation of Drought Stress in Plants Using Silver Nanoparticles

Aisha Kamal, Nida Sultan, Sazia Siddiqui, Ayeesha Khatoon, and Bushra Ahmad

Abstract Nanotechnology, characterized by its dynamic nature, has emerged as a focal point in material science research with significant applications in agriculture. The global surge in nanoparticle production reflects the swift evolution of this field, marked by innovations such as nanopesticides, nanofertilizers, and nanosensors aimed at enhancing agronomic practices. Silver nanoparticles (AgNPs), a widely recognized nanomaterial, find applications in agriculture, showcasing potential benefits in plant growth, seed germination, stress tolerance, pest control, and mitigating microbial infections. Additionally, AgNPs play a crucial role in alleviating abiotic stresses, particularly drought, by regulating water balance, enhancing photosynthetic efficiency, and activating antioxidant defense systems. Understanding the molecular mechanisms underlying the interaction between AgNPs and drought-responsive genes remains a research challenge. Current studies indicate that AgNPs modulate gene expression, especially those associated with stress response and phytohormone signaling pathways. While AgNPs show promise in enhancing crop resilience to drought, concerns about environmental impact and nanoparticle accumulation in edible plant parts underscore the need for rigorous evaluation methodologies. Ongoing research is crucial to establish safety limits and address environmental considerations, ensuring the responsible and sustainable use of nanotechnology in agriculture.

Keywords Silver nanoparticle · Abiotic stress · Drought · Stress-tolerance · Stress-responsive genes

A. Kamal (✉) · B. Ahmad
Department of Bioengineering, Integral University, Lucknow, India
e-mail: aishakamal04@gmail.com

N. Sultan · S. Siddiqui · A. Khatoon
Department of Bioscience, Integral University, Lucknow, India

A. Husen (ed.), *Plant Response to Silver Nanoparticles*, Smart Nanomaterials Technology, https://doi.org/10.1007/978-981-97-7352-7_10

1 Introduction

Nanotechnology is a multi-disciplinary science. It stands as the focal point of material science research, characterized by its dynamic nature. The global surge in the production of nanoparticles (NPs) signifies the swift evolution of this field. NPs exhibit novel or enhanced properties attributed to their distinct attributes, including their size, which typically ranges between 1 and 100 nm, as well as their shape and structure [1–3]. Over the past decade, numerous innovations and products have successfully integrated engineered nanoparticles (NPs) into various agricultural practices (Fig. 1). These include nanopesticides, nanofertilizers, and nanosensors, all designed with the primary objective of enhancing the agricultural systems' sustainability and quality. These advancements aim to reduce the resource-intensive nature of traditional products and approaches while minimizing waste generation [4, 5]. The utilization of nanomaterials in agriculture offers farmers a myriad of advantages, including increased crop yield, reduced water usage, and a host of other benefits that contribute to enhanced agricultural practices. Dwivedi et al. [6] have highlighted that the primary concerns for crop production lie in the realm of pesticides and fertilizers. Consequently, the development of nanopesticides and nanofertilizers has gained prominence [7]. It is anticipated that nanostructured catalysts will soon emerge to enhance the efficacy of pesticides and herbicides, facilitating their usage at lower doses. This represents a promising avenue for more sustainable and effective agricultural practices. Nanofertilizers have the potential to enhance the efficiency of nutrient uptake by plants and mitigate the adverse impacts associated with the application of conventional fertilizers [8, 4, 9]. Moreover, Nanopesticides serve as crucial agents for the efficient and sustainable control of various pests, demonstrating the potential to reduce the reliance on synthetic chemicals and mitigate the correlated environmental risks [10]. Due to their smaller particle size, increased surface area, and well-documented reactivity, nanoparticles (NPs) have gained significant attention for their potential to remediate contaminated soils. This potential stems from their capacity for sorption, reduction, and chemical oxidation [11].

Abiotic stresses, including salinity, drought, and extreme temperatures, represent significant concerns that detrimentally impact the development, growth, and yield of plants [13, 14, 15–17, 18, 19, 20]. A variety of metal-based nanoparticles, including Ag, Au, Fe, Cu, TiO_2, Zn, and ZnO, have been increasingly employed for promoting germination, growth, and stress tolerance in cereal crops [2]. This chapter delves into the specific application of silver nanoparticles (AgNPs) in agriculture, shedding light on their potential to address abiotic stresses, particularly drought. As plants face an increasing array of environmental stresses, AgNPs emerge as versatile tools in bolstering stress tolerance and mitigating the adverse effects of factors like salinity, extreme temperatures, and water scarcity. The intricate interplay between AgNPs and plant physiology, including their impact on water balance, photosynthetic efficiency, antioxidant defense systems, and stress-responsive genes, forms the core of this exploration. However, it is crucial to acknowledge the subtle nature of nanoparticle interactions, as higher concentrations may lead to inhibitory effects.

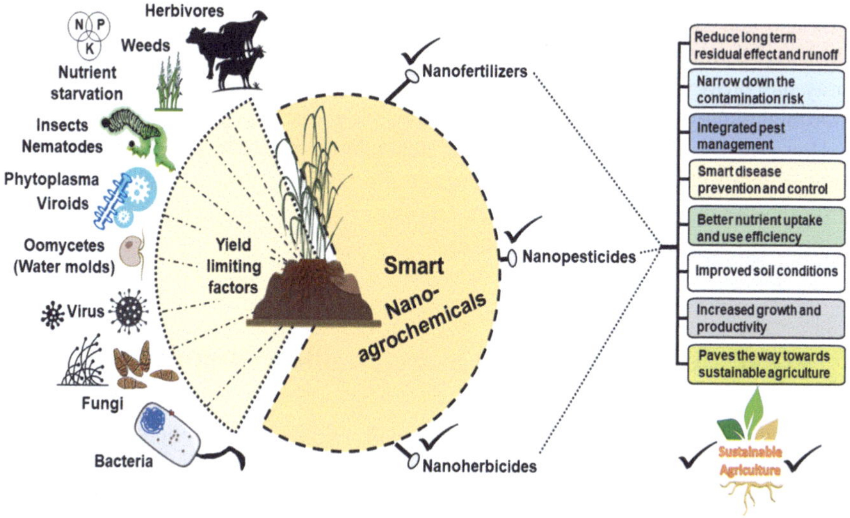

Fig. 1 Various applications of nanotechnology in the agriculture sector [12]

The intricate balance between the positive and negative impacts of AgNPs in agriculture underscores the need for a nuanced understanding and responsible application of nanotechnology.

Furthermore, the chapter acknowledges existing knowledge gaps and underscores the significance of continuous research to address environmental concerns and establish safe thresholds for nanoparticle use in agriculture. This investigation lays the groundwork for a thorough comprehension of the complex interplay between nanotechnology and agriculture, facilitating informed decision-making and the adoption of sustainable agricultural practices.

2 Silver Nanoparticles and Its Application in Agriculture

Silver nanoparticles (AgNPs) are most widely recognized as a commercialized nanomaterial, finding extensive application in various domains such as antimicrobial agents in medical and Healthcare gear, building supplies, water filtration, and personal hygiene items. Many metallic nanoparticles (NPs), including AgNPs, CuNPs, ZnNPs, and others, have drawn a lot of interest recently because of their environmentally beneficial uses in the agriculture industry [21, 22].

Silver nanoparticles (AgNPs) have demonstrated the potential to promote plant growth, promote seed germination, improve photosynthesis, and inhibit microbial infections in various plant species. It has emerged as promising alternatives to traditional pesticides thereby potentially reducing the reliance on chemical fertilizers and promoting sustainable farming practices. Additionally, these nanoparticles have demonstrated efficacy in eradicating detrimental microorganisms present in both soil-based and hydroponic cultivation systems. Notably, their application as a foliar spray has proven effective in curbing the proliferation of fungi, molds, and other microbial pathogens that contribute to various plant diseases [23].

However, the impact of AgNPs is dose-dependent; lower concentrations stimulate plant growth and yield, while higher concentrations inhibit growth [24]. Vannini et al. [25] reported that the effect of AgNPs on plants may be influenced by factors such as their size and the duration of exposure.

3 AGNP: A Potential Aid in Stress Resilience Improvement

In the natural environment, plants contend with a range of environmental stressors such as low temperatures, salinity, heat, and drought. The frequency of these stressors is increasing significantly in the global environment [26]. Nanoparticles, characterized by their unique properties stemming from their small size, offer enhanced solubility, surface area, and reactivity compared to bulk materials. This distinctiveness positions nanoparticles as promising contributors to the pursuit of sustainable agriculture, aiming to alleviate the negative impacts of abiotic stress.

Silver nanoparticles (AgNPs) are increasingly recognized in the agricultural sector for their significant role in enhancing stress tolerance in plants. Extensive research has been conducted on various forms of AgNPs, highlighting their potential to fortify crops against abiotic stresses [26]. These nanoparticles have shown promising effects in addressing nutrient deficiencies, enhancing enzymatic activities, and making it easier for microorganisms that support plant development to cling to plant roots, especially under challenging conditions such as salt stress [13], cold stress [27], and heat stress [28]. Moreover, AgNPs priming has been linked to the promotion of seed germination in various crops like wheat, tomato, maize [29, 30, 31] by reducing oxidative damage through the regulation of hydrogen peroxide and other reactive substances [23]. This is attributed to the activation of antioxidant systems, including enzymes like SOD, CAT, APX, DHAR, and MDHAR, along with compounds such as proline, glycine, betaine, and anthocyanin, showcasing the potential of AgNPs in bolstering plant resilience to abiotic stresses [13, 32].

In addition, AgNPs exhibit potent larvicidal and pupicidal properties, presenting an eco-friendly solution for pest control [33, 34] (Vimala et al. 2015). Field trials have demonstrated the effective inhibition of fungi by AgNPs, particularly when applied preventively before the onset of plant diseases [35]. This emphasizes the potential of AgNPs as a cost-effective and environmentally friendly approach to pest management.

4 Drought Stress and Its Impact on Plants

One of the most significant environmental stressors is drought, which can occur for a variety of causes salinity, hot and low temperatures, little precipitation, and intense light. Water deficient stress is tricky to define, however several perspectives may be used to characterize drought, including meteorological, agricultural, hydrological, and macroeconomic ones [36, 37, 38] (Jaleel et al. 2009). According to agricultural and physiological perspectives, drought stress often happens when the amount of water that plants can access in the soil at a given moment is reduced owing to low soil moisture [37, 39]. However, when the rate at which water is absorbed by roots is less than the rate at which leaves transpire a plant is said to be under water stress (deficiency) [40].

Drought is a recurring and ongoing problem in many parts of the planet, including India. Drought is a multifaceted abiotic stress that impacts plants at many levels of their organization. Indeed, many plants dry and eventually die during extended droughts [41]. Plants under water stress have a decreased water potential and strength, which increases the concentrations of solutes in the cell wall and extracellular matrix. Cell enlargement consequently declines, which inhibits growth and impairs reproduction. Enhance the production of reactive oxygen species (ROS) leads to the development of chemicals that scavenge free radicals, including ascorbate and glutathione, worsening the negative impact [42, 43–47, 48] (Ahmad et al. 2019). In addition to decreasing water content, turgor, and total water, drought also affects stomatal closure, restricts gaseous exchange, lowers transpiration, and stops the rate of carbon absorption thereby adversely affecting photosynthesis [49, 50, 51, 52]. The general morphology of plants, including plant growth, biomass and yield, roots, and grain formation, is adversely affected by drought stress. Due to poor or insufficient circulation of water onto the xylem or neighbouring cell, plants under water stress shows reduced growth, leaf size, leaf area, leaf number, plant matter, cell growth, and elongation. Leaf rolling, stomata shutting, leaf tip drying, and basal shrinking are other evident morphological changes [53].

5 Adaptive Strategies of Plants Under Drought Stress

Because they are sessile creatures and are subject to environmental stressors, plants have evolved a variety of defence systems. Furthermore, because they are multicellular animals, plants have complicated reactions when faced with environmental challenges like drought [54, 40, 55, 56]. Stress prevention and stress tolerance are the two basic tactics that make up a plant's resistance to environmental stress. Physiological, appearance, phenological, biochemicals, and molecular responses enable plants to adapt to a water scarcity. The reactions may occur at the molecular level or throughout the entire plant. Preserving plants' high water potential in situations of water scarcity is the primary goal of the strategy. Reducing plant water loss through stomatal regulation and preserving soil water absorption are the major features of the technique [57, 58]. To effectively absorb water from a significant depth of soil and at a great distance from the plant, a deep and thick root system is beneficial [59, 58, 60]. High regional water potential for drought resistance is maintained by the cuticle and hairy leaves [58, 54, 40]. Other mechanisms that contribute to plant tolerance to drought include the formation of appropriate solute and pressure modification, the induction of an anti-oxidative system, modifications to the Krebs cycle, a boost in the root/shoot ratio, and stomata closure [58, 40]. Early blooming brought on by a short life cycle can help plants survive dryness; this is known as stress-avoidance adaptation to drought [57, 58, 54]. Thus, even if the yield is often lower, early maturity and subsequently early blooming assist plants escape drought stress [59, 57, 54]. Film farming [61], the creation of crops that are drought-resistant [62], the use of super-absorbent hydrogels [63, 64], the use of rhizobacteria that promote plant growth [65] and the use of small particles [66] are some recent strategies to mitigate the effects of drought stress.

6 Silver Nanoparticles as an Effective Tool in Alleviating Drought Stress

6.1 Regulation of Water Balance and Stomatal Behaviour

AgNPs modify the morphophysiological, biochemical, and molecular condition of plants. AgNPs increase grain production in wheat and other cereal crops under drought stress because of increases in gas exchange components, membrane stability, and water relations [67, 68]. According to Mattila et al. [69], Reduced transpiration, which is gaseous exchange, and loss of water are the results of alterations in stomatal closure and gene expression caused by ROS buildup following AgNP treatment. A notable decrease in transpiration rate was seen in *Cucurbita pepo,* as a result of the AgNP, which induces stomatal closure [70]. Wheat's leaf water potential is decreased by moderate drought stress and is further restricted by severe drought stress [71, 72]. Exogenously administered AgNPs improved the gradient for water flow from the xylem to the leaves during moderate to severe drought conditions despite decreasing osmotic potential. The advantages of AgNPs could be exploited to increase drought tolerance in plants grown in areas with limited water resources.

6.2 Modulation of Photosynthetic Efficiency

According to Ashkavand et al. [73], nanoparticles of silver, and zinc improve photosynthetic rates, lower malondialdehyde content, which is a sign of lipid impairment, raise relative water content, improve the shoot and root system, and contribute to the general modification and growth of plants. According to Kalaji et al. [74], photosynthesis is regarded as a reliable indicator of a plant's general performance. It affects all facets of plant metabolism and physiology because it is the only energy source for plants. Top leaves' senescence, gas exchange, and chlorophyll all suffered noticeably from drought levels, and genotypes, and when AgNPs treatment was done a remarkable improvement was observed in plants. According to several studies [75, 76, 77] AgNPs may contribute to the stabilization of cell and chloroplast membranes, enhancing photosynthetic efficiency. In drought-stressed plants, AgNPs enhance nutrient absorption and mobility, increasing photosynthetic pigments, stomatal opening, photosynthetic rate, and yield of photosynthetic machinery Ahmed et al. [68].

6.3 Enhancement of Antioxidant Defence System

Antioxidant activity has the power to lower the proportion of oxidative processes by lowering initiating events or interacting with radicals that carry chains of reactions [78]. Several researchers have reported elevation of CAT and SOD enzyme levels following the application of silver nanoparticles (AgNPs) [79, 80]. The antioxidant activity of the enzymes catalase (CAT) and superoxide dismutase (SOD) was found to be increased in boysenberry leaves subjected to silver nanoparticles (AgNPs) during a drought. Additionally, a decrease in malondialdehyde (MDA) content was reported, indicating a potential mitigation in response to oxidative stress to AgNPs treatment [81]. There are a variety of pathways that contribute to nanomaterials' antioxidant potential, but catalase mimic (CAT-mimic) activity is primarily responsible for the antioxidant properties of biogenic silver nanoparticles (AgNPs) and other metallic nanoparticles (Amorati and Valgimigli 78).

Additionally, it has been documented that AgNPs raise the levels of phenolic and flavonoid compounds in rice [79] and potatoes [82] as well as increase oxidative stress. It is well known that phenols and flavonoids can penetrate the lipid bilayer to prevent ROS-mediated peroxidation of the lipids and to preserve the fluidity and functionality of the membrane [83, 84]. Increased synthesis of flavonoids and phenols in plant tissues can strengthen the antioxidant system during droughts [85, 86].

6.4 Activation of Stress Responsive Genes and Signalling Pathways

Drought stress initiates a complex network of signalling pathways that are absolutely essential in modulating the expression of genes responsive to both drought and phytohormones. Silver and silver oxide nanoparticles (NPs) have been found to affect the ABA-dependent and ABA-independent pathways, which independently affect the expression of genes linked to drought. These pathways collectively contribute to the intricate regulatory mechanisms involved in the way the plant reacts to dryness conditions influenced by metal and metal oxide nanoparticles [87]. In the study conducted by Garcia-Sanchez et al. [88], they observed that out of 351 genes that were differentially regulated in Arabidopsis thaliana, 16 of them were responsive to drought when treated with TiO_2 NPs, AgNPs, and multi-walled carbon nanotubes. Nair and Chung [89] found that when rice seedlings were exposed to different concentrations of citrate-coated AgNPs, their roots and shoots expressed more SOD genes (FSD, MSD1, and CSD1). SOD gene expression was found to be upregulated in *A. thaliana* leaves [90] and seedlings [91] after exposure to AgNP. Enzymes encoded by SOD genes are considered to be the primary line of defense for cells against oxidative stress. Additionally, cells exposed to silver nanoparticles (AgNPs) showed an upregulation of the auxin-responsive gene IAA8. It's interesting to note that lateral root development suppression has been connected to IAA8 overexpression [92]. Importantly, IAA8 is implicated in the abscisic acid (ABA) signaling pathway, suggesting a potential interplay between AgNPs treatment, auxin response, and ABA signaling in cellular processes.

However, there is a scarcity of molecular studies investigating this phenomenon, and only a limited number of studies involve AgNPs treatment. This highlights the limited understanding of the molecular mechanisms underlying the interaction between nanoparticles and drought-responsive genes, emphasizing the need for further research in this area.

Certainly, beyond the limited set of coding genes that exhibit responsiveness, it's reasonable to assume that non-coding elements, such as microRNAs (miRNAs), may play a crucial role in regulating essential physiological genes. MiRNAs are known for their involvement in post-transcriptional gene regulation through mechanisms like transcriptional repression. This regulatory activity is a common phenomenon in plants, particularly under stress conditions. Therefore, considering the complexity of plant stress responses, the involvement of non-coding elements like miRNAs adds another layer of regulatory complexity in understanding the molecular mechanisms associated with nanoparticle treatments and stress responses in plants [93, 94].

6.5 Improvement in Water Uptake and Retention

The intricate and poorly elucidated mechanism by which nanoparticles confer drought tolerance remains a subject of ongoing research. Nonetheless, a number of investigations have suggested that silver nanoparticles are involved in the control of aquaporins, which are channel proteins that facilitate seed germination and water permeability. This regulation appears to enhance the supply of nutrients and water for seeds to germinate, improving germination rates even under drought-stressed circumstances [95, 96]. In fact, membrane channel opening or activation—more especially, water-permeability aquaporins (AQPs), appears to be a potential mechanism by which plants can partially regulate their water flow. This process is especially relevant in mitigating the negative impacts of drought, as it facilitates the movement of water across cellular membranes, aiding in the maintenance of proper hydration levels within plant cells. The regulation of AQPs is a crucial aspect of a plant's adaptive response to water scarcity, allowing for a more controlled and efficient water management system [32].

Silver nanoparticles (AgNPs) exert a remarkable influence on the permeability and fluidity of membranes, consequently affecting the absorption of nutrients and water. In the context of *Raphanus sativus* seedlings, it has been observed that treatment with AgNPs led to a dose-dependent decrease in water content. This suggests that AgNPs have the potential to modulate the water balance within plant tissues, impacting their overall hydration status and nutrient absorption capabilities. Such findings underscore the multifaceted ways in which AgNPs can influence physiological processes in plants, including water regulation [97].

7 Conclusion

Nanotechnology has emerged as a revolutionary science with innovative applications spanning various scientific fields, including medicine, energy, electronics, and biology. The unique physical and chemical properties of silver nanoparticles (AgNPs) have led to their extensive exploration and application in the field of agriculture. Some notable contributions and applications include plant growth and productivity, crop protection, nutrient delivery, soil remediation, water management, tolerating both abiotic and biotic stresses, etc. The application of nanoparticles, particularly silver nanoparticles (AgNPs), in agriculture has shown promise in enhancing plant drought tolerance through various mechanisms. AgNPs have demonstrated the ability to improve the antioxidant system, nutrient uptake, elevated chlorophyll content, higher sugar levels, and accumulation of osmolytes, while also reducing reactive oxygen species. Additionally, they modulate proteins and signaling pathways associated with drought tolerance. AgNPs treatments have been linked to elevated articulation of drought genes related to stress, leading to elevated chlorophyll content and protein levels, as well as the promotion of nitrogen metabolism. These beneficial effects highlight the potential of AgNPs in drought stress management for the improvement of crop performance.

While the efficacy of nanoparticles in enhancing resistance to abiotic stress is well-established, the majority of the laboratory phase of these investigations is still ongoing. Concerns about the potential negative environmental effects of nanoparticles and the potential for the accumulation of nanoparticles in edible plant parts have been raised by the widespread use of these particles. Therefore, there is a pressing need for focused research to develop suitable techniques for evaluating and assessing the impact of nanoparticles on the environment and human health and establishing acceptable limits. Ongoing research is crucial to address environmental and safety considerations associated with the use of nanoparticles in agriculture. However, most documented studies stated the favourable impacts of nanoparticles on the growth of plants, with only a limited number of studies highlighting potential adverse impacts. To comprehend the potential advantages of integrating nanotechnology into agriculture, it is imperative to first examine the infiltration and movement of nanoparticles within plants. Understanding these dynamics can provide crucial insights into the broader implications of employing nanotechnology in agricultural practices.

References

1. Nejatzadeh F (2021) Effect of silver nanoparticles on salt tolerance of *Satureja hortensis* L. during in vitro and in vivo germination tests. *Heliyon* 7(2):e05981. https://doi.org/10.1016/j.heliyon.2021.e05981
2. Taran N, Storozhenko V, Svietlova N, Batsmanova L, Shvartau V, Kovalenko M (2017) Effect of zinc and copper nanoparticles on drought resistance of wheat seedlings. Nanoscale Res Lett 12(1):60. https://doi.org/10.1186/s11671-017-1839-9
3. Nadiyah M, Alabdallah, Md. Mahadi Hasan (2021) Plant-based green synthesis of silver nanoparticles and its effective role in abiotic stress tolerance in crop plants. Saudi J Biol Sci 28(10):5631–5639
4. Liu R, Lal R (2015) Potentials of engineered nanoparticles as fertilizers for increasing agronomic productions. Sci Total Environ 514:131–139. https://doi.org/10.1016/j.scitotenv.2015.01.104
5. Servin A, Elmer W, Mukherjee A, De la Torre-Roche R, Hamdi H, White JC, Dimkpa C (2015) A review of the use of engineered nanomaterials to suppress plant disease and enhance crop yield. J Nanopart Res 17:1–21
6. Dwivedi S, Saquib Q, Al-Khedhairy AA, Musarrat J (2016) Understanding the role of nanomaterials in agriculture. In: Microbial inoculants in sustainable agricultural productivity, vol 2 (Functional applications), pp 271–288
7. Husen A, Jawaid M (eds) (2020) Nanomaterials for agriculture and forestry applications. Elsevier
8. Kah M, Kookana RS, Gogos A, Bucheli TD (2018) A critical evaluation of nanopesticides and nanofertilizers against their conventional analogues. Nat Nanotechnol 13(8):677–684. https://doi.org/10.1038/s41565-018-0131-1
9. Rico CM, Majumdar S, Duarte-Gardea M, Peralta-Videa JR, Gardea-Torresdey JL (2011) Interaction of nanoparticles with edible plants and their possible implications in the food chain. J Agric Food Chem 59(8):3485–3498. https://doi.org/10.1021/jf104517j
10. Kah M, Tufenkji N, White JC (2019) Nano-enabled strategies to enhance crop nutrition and protection. Nat Nanotechnol 14(6):532–540. https://doi.org/10.1038/s41565-019-0439-5
11. Guerra FD, Attia MF, Whitehead DC, Alexis F (2018) Nanotechnology for environmental remediation: materials and applications. Molecules 23(7):1760. https://doi.org/10.3390/molecules23071760
12. Sharma P, Pandey V, Sharma MMM, Patra A, Singh B, Mehta S, Husen A (2021) A review on biosensors and nanosensors application in agroecosystems. Nanoscale Res Lett 16(136):1–24. https://doi.org/10.1186/s11671-021-03593-0
13. Alharbi BM, Elhakem AH, Alnusairi GSH, Soliman MH, Hakeem KR, Hasan MM, Abdelhamid MT (2021) Exogenous application of melatonin alleviates salt stress-induced decline in growth and photosynthesis in *Glycine max* (L.) seedlings by improving mineral uptake, antioxidant and glyoxalase system. Plant, Soil Environ 67(4):208–220. https://doi.org/10.17221/659/2020-PSE
14. Choudhary A, Sharma S, Kaur H, Sharma N, Gadewar MM, Mehta S, Husen A (2024) Plant system, abiotic stress resilience, reactive oxygen species, and coordination of engineered nanomaterials: a review. S Afr J Bot 171:45–59. https://doi.org/10.1016/j.sajb.2024.05.053

15. Hafeez MB, Ghaffar A, Zahra N, Ahmad N, Al-Qahtani WH, Sharma P, Husen A, Li J (2024) Exploring the efficacy of plant growth promoters in nutrient acquisition and antioxidant defense responses of late sown wheat under saline conditions. S Afr J Bot 171:812–824
16. Hasan MM, Alharby HF, Uddin MN, Ali MA, Anwar Y, Fang XW, Hakeem KR, Alzahrani Y, Hajar AS (2020) Magnetized water confers drought stress tolerance in *Moringa biotype* via modulation of growth, gas exchange, lipid peroxidation and antioxidant activity. Pol J Environ Stud 1:29
17. Hasan MM, Ali MA, Soliman MH, Alqarawi AA, Abd, Fang X (2020) Insights into 28-homobrassinolide (HBR)-mediated redox homeostasis, AsA–GSH cycle, and methylglyoxal detoxification in soybean under drought-induced oxidative stress. J Plant Interact 15(1):371–385. https://doi.org/10.1080/17429145.2020.1832267
18. Jahan MS, Shu S, Wang Y, Hasan MM, El-Yazied AA, Alabdallah NM, Hajjar D, Altaf MA, Sun J, Guo S (2021) Melatonin pretreatment confers heat tolerance and repression of heat-induced senescence in tomato through the modulation of ABA- and GA-mediated pathways. Front Plant Sci 12:650955. https://doi.org/10.3389/fpls.2021.650955
19. Zahra N, Kausar A, Abdelghani HTM, Singh S, Vashishth DS, Bachheti A, Bachheti RK, Husen A (2024) Serotonin improves plant growth, foliar functions and antioxidant defence system in Ethiopian mustard (*Brassica carinata* A. Br.). S Afr J Bot 170:1–9. https://doi.org/10.1016/j.sajb.2024.05.002
20. Zahra N, Kausar A, Sharma P, Wabaidur SM, Husen A (2024) Application of serotonin to *Guizotia abyssinica* improves growth, physiological functions, antioxidant system and alleviates the deleterious effects induced by salt stress. Plant Biosyst - Int J Deal All Asp Plant Biol. https://doi.org/10.1080/11263504.2024.2384972
21. Chouhan N, Meena RK (2015) Biosynthesis of silver nanoparticles using *Trachyspermum ammi* and evaluation of their antibacterial activities. Int J Pharm Biol Sci 62:1077–1086
22. Mahakham W, Sarmah AK, Maensiri S, Theerakulpisut P (2017) Nanopriming technology for enhancing germination and starch metabolism of aged rice seeds using phytosynthesized silver nanoparticles. Sci Rep 7(1):8263. https://doi.org/10.1038/s41598-017-08669-5
23. Wahid I, Kumari S, Ahmad R, Hussain SJ, Alamri S, Siddiqui MH, Khan MIR (2020) Silver nanoparticle regulates salt tolerance in wheat through changes in ABA concentration, ion homeostasis, and defense systems. Biomolecules 10(11):1506. https://doi.org/10.3390/biom10111506
24. Alabdallah NM, Hasan MM (2021) Plant-based green synthesis of silver nanoparticles and its effective role in abiotic stress tolerance in crop plants. Saudi J Biol Sci 28(10):5631–5639. https://doi.org/10.1016/j.sjbs.2021.05.081
25. Vannini C, Domingo G, Onelli E, De Mattia F, Bruni I, Marsoni M, Bracale M (2014) Phytotoxic and genotoxic effects of silver nanoparticles exposure on germinating wheat seedlings. J Plant Physiol 171(13):1142–1148. https://doi.org/10.1016/j.jplph.2014.05.002
26. Khan A, Anwar Y, Hasan MM, Iqbal A, Ali M, Alharby HF, Hakeem K, Hasanuzzaman M (2017) Attenuation of drought stress in Brassica seedlings with exogenous application of Ca^{2+} and H_2O_2. Plants 6(4):20. https://doi.org/10.3390/plants6020020
27. Prażak R, Święciło A, Krzepiłko A, Michałek S, Arczewska M (2020) Impact of Ag nanoparticles on seed germination and seedling growth of green beans in normal and chill temperatures. Agriculture 10(8):312. https://doi.org/10.3390/agriculture10080312
28. Iqbal M, Umar S, Mahmooduzzafar (2018) Nano-fertilization to enhance nutrient use efficiency and productivity of crop plants. In: Nanomaterials and plant potential, pp 473–505
29. Abou-Zeid H, Ismail G (2018) The role of priming with biosynthesized silver nanoparticles in the response of *Triticum aestivum* L. to salt stress. Egypt J Bot 58:73–85
30. Almutairi ZM (2016) Influence of silver nano-particles on the salt resistance of tomato (*Solanum lycopersicum*) during germination. Int J Agric Biol 18(2):449–457. https://doi.org/10.17957/IJAB/15.0114
31. Salami R, Safrinejad A, Hamidi H (2007) Effect of salinity on morphological characteristics of cumin. J Res Dev 72:77–82

32. Hasan MM, Gong L, Nie ZF, Li FP, Ahammed GJ, Fang XW (2021) ABA-induced stomatal movements in vascular plants during dehydration and rehydration. Environ Exp Bot 186:104436. https://doi.org/10.1016/j.envexpbot.2021.104436
33. Gnanadesigan M, Anand M, Ravikumar S, Maruthupandy M, Vijayakumar V, Selvam S, Dhineshkumar M, Kumaraguru AK (2011) Biosynthesis of silver nanoparticles by using mangrove plant extract and their potential mosquito larvicidal property. Asian Pac J Trop Med 4(10):799–803. https://doi.org/10.1016/S1995-7645(11)60197-1
34. Sundaravadivelan C, Padmanabhan MN (2014) Effect of mycosynthesized silver nanoparticles from filtrate of *Trichoderma harzianum* against larvae and pupa of dengue vector *Aedes aegypti* L. Environ Sci Pollut Res Int 21(6):4624–4633. https://doi.org/10.1007/s11356-013-2358-6
35. Lamsal K, Kim SW, Jung JH, Kim YS, Kim KS, Lee YS (2011) Application of silver nanoparticles for the control of Colletotrichum species in vitro and pepper anthracnose disease in field. Mycobiology 39(3):194–199. https://doi.org/10.5941/MYCO.2011.39.3.194
36. Barriopedro D, Gouveia CM, Trigo RM, Wang L (2012) The 2009/10 drought in China: possible causes and impacts on vegetation. J Hydrometeorol 13(4):1251–1267. https://doi.org/10.1175/JHM-D-11-074.1
37. Dai A (2011) Drought under global warming: a review. WIREs Clim Chang 2(1):45–65. https://doi.org/10.1002/wcc.81
38. Rahdari P, Hoseini SM (2012) Drought stress: a review. Int J Agron Plant Prod 3(10):443–446
39. Keyvan S (2010) The effects of drought stress on yield, relative water content, proline, soluble carbohydrates and chlorophyll of bread wheat cultivars. J Anim Plant Sci 8(3):1051–1060
40. Lisar SY, Motafakkerazad R, Hossain MM, Rahman IM (2012) Causes, effects and responses. Water Stress 25(1):33
41. Takahashi F, Suzuki T, Osakabe Y, Betsuyaku S, Kondo Y, Dohmae N, Fukuda H, Yamaguchi-Shinozaki K, Shinozaki K (2018) A small peptide modulates stomatal control via abscisic acid in long-distance signalling. Nature 556(7700):235–238. https://doi.org/10.1038/s41586-018-0009-2
42. Bachheti A, Deepti, Bachheti RK, Husen A (2021) Medicinal plants and their pharmaceutical properties under adverse environmental conditions. In: Husen A (ed) Harsh environment and plant resilience. Springer International Publishing AG, Cham, pp 457–502. https://doi.org/10.1007/978-3-030-65912-7_19
43. Husen A (2021) The harsh environment and resilient plants: an overview. In: Husen A (ed) Harsh environment and plant resilience. Springer International Publishing AG, Cham, pp 1–23. https://doi.org/10.1007/978-3-030-65912-7_1
44. Husen A (2021) Cross talk between autophagy and hormones for abiotic stress tolerance in plants. In: Husen A (ed) Plant performance under environmental stress. Springer International Publishing AG, Cham, pp 1–15. https://doi.org/10.1007/978-3-030-78521-5_1
45. Husen A (2021) Harsh environment and plant resilience. In: Molecular and functional aspects. Springer Nature Switzerland AG, Cham, Switzerland. https://doi.org/10.1007/978-3-030-65912-7
46. Husen A (2021) Plant performance under environmental stress. In: Hormones, biostimulants and sustainable plant growth management. Springer Nature Switzerland AG, Cham, Switzerland. https://doi.org/10.1007/978-3-030-78521-5
47. Husen A, Iqbal M (2023) Medicinal plants—their response to abiotic stress. Springer Nature Singapore Pte Ltd., Singapore. https://doi.org/10.1007/978-981-19-5611-9
48. Hussain HA, Hussain S, Khaliq A, Ashraf U, Anjum SA, Men S, Wang L (2018) Chilling and drought stresses in crop plants: implications, cross talk, and potential management opportunities. Front Plant Sci 9:393. https://doi.org/10.3389/fpls.2018.00393
49. Getnet Z, Husen A, Fetene M, Yemata G (2015) Growth, water status, physiological, biochemical and yield response of stay green sorghum Sorghum bicolor (L.) Moench varieties-a field trial under drought-prone area in Amhara regional state, Ethiopia. J Agron 14:188–202
50. Husen A (2010) Growth characteristics, physiological and metabolic responses of teak (*Tectona grandis* Linn. f.) clones differing in rejuvenation capacity subjected to drought stress. Silvae Genetica 59:124–136

51. Husen A, Iqbal M, Aref MI (2014) Growth, water status and leaf characteristics of *Brassica carinata* under drought stress and rehydration conditions. Braz J Bot 37:217–227
52. Schulze S, Zahn D, Montes R, Rodil R, Quintana JB, Knepper TP, Reemtsma T, Berger U (2019) Occurrence of emerging persistent and mobile organic contaminants in European water samples. Water Res 153:80–90. https://doi.org/10.1016/j.watres.2019.01.008
53. Dash PK, Rai R, Rai V, Pasupalak S (2018) Drought induced signaling in rice: delineating canonical and non-canonical pathways. Front Chem 6:264. https://doi.org/10.3389/fchem.2018.00264
54. Khan MA, Iqbal M, Jameel M, Nazeer W, Shakir S, Aslam MT, Iqbal B (2011) Potentials of molecular based breeding to enhance drought tolerance in wheat (*Triticum aestivum* L.). Afr J Biotechnol 10(55):11340–11344
55. Nezhadahmadi A, Prodhan ZH, Faruq G (2013) Drought tolerance in wheat. Sci World J 2013:610721. https://doi.org/10.1155/2013/610721
56. Rana RM, Rehman SU, Ahmed J, Bilal M (2013) A comprehensive overview of recent advances in drought stress tolerance research in wheat (*Triticum aestivum* L.). Asian J Agric Biol 1(1):29–37
57. Bray EA (2001) Regulation of gene expression during abiotic stresses and the role of the plant hormone abscisic acid. In: Handbook of plant and crop physiology. CRC Press, pp 757–772
58. Farooq M, Wahid A, Kobayashi NSMA, Fujita DBSMA, Basra SMA (2009) Plant drought stress: effects, mechanisms and management. J Sustain Agric 153–188
59. Akhtar I, Nazir N (2013) Effect of waterlogging and drought stress in plants. Int J Water Resour Environ Sci 2(2):34–40
60. Franco JA (2011) Root development under drought stress. Technol Knowl Transf e-Bull 2:1–3
61. Mori Y (2013) New agro-technology (Imec) by hydrogel membrane. React Funct Polym 73(7):936–938. https://doi.org/10.1016/j.reactfunctpolym.2012.11.015
62. Nuccio ML, Paul M, Bate NJ, Cohn J, Cutler SR (2018) Where are the drought tolerant crops? An assessment of more than two decades of plant biotechnology effort in crop improvement. Plant Sci 273:110–119. https://doi.org/10.1016/j.plantsci.2018.01.020
63. Saha A, Sekharan S, Manna U (2020) Superabsorbent hydrogel (SAH) as a soil amendment for drought management: a review. Soil Tillage Res 204:104736. https://doi.org/10.1016/j.still.2020.104736
64. Zhang Y, Ding J, Wang H, Su L, Zhao C (2020) Biochar addition alleviate the negative effects of drought and salinity stress on soybean productivity and water use efficiency. BMC Plant Biol 20(1):288. https://doi.org/10.1186/s12870-020-02493-2
65. Chiappero J, Cappellari L, Sosa Alderete LGS, Palermo TB, Banchio E (2019) Plant growth promoting rhizobacteria improve the antioxidant status in *Mentha piperita* grown under drought stress leading to an enhancement of plant growth and total phenolic content. Ind Crops Prod 139:111553. https://doi.org/10.1016/j.indcrop.2019.111553
66. Saxena R, Tomar RS, Kumar M (2016) Exploring nanobiotechnology to mitigate abiotic stress in crop plants. J Pharm Sci Res 8(9):974
67. Jaskulska I, Jaskulski D (2020) Effects of using nanoparticles of silver (AgNPs) and copper (CuNPs) in foliar fertilizers. Przem Chem 99(2):250–253
68. Ahmed F, Javed B, Razzaq A, Mashwani ZU (2021) Applications of copper and silver nanoparticles on wheat plants to induce drought tolerance and increase yield. IET nanobiotechnology, 15(1):68–78. https://doi.org/10.1049/nbt2.12002
69. Mattila H, Khorobrykh S, Havurinne V, Tyystjärvi E (2015) Reactive oxygen species: reactions and detection from photosynthetic tissues. J Photochem Photobiol: B, Biol 152(B):176–214. https://doi.org/10.1016/j.jphotobiol.2015.10.001
70. Hawthorne J, Musante C, Sinha SK, White JC (2012) Accumulation and phytotoxicity of engineered nanoparticles to *Cucurbita pepo*. Int J Phytorem 14(4):429–442. https://doi.org/10.1080/15226514.2011.620903
71. Alghory A, Yazar A (2019) Evaluation of crop water stress index and leaf water potential for deficit irrigation management of sprinkler-irrigated wheat. Irrig Sci 37(1):61–77. https://doi.org/10.1007/s00271-018-0603-y

72. Arzanesh MH, Alikhani HA, Khavazi K (2011) Wheat (Triticum aestivum L.) growth enhancement by Azospirillum sp. under drought stress. World J Microbiol Biotechnol 27:197–205
73. Ashkavand P, Zarafshar M, Tabari M, Mirzaie J, Nikpour A, Bordbar SK, Struve D, Striker GG (2018) Application of SIO2 nanoparticles AS pretreatment alleviates the impact of drought on the physiological performance of Prunus Mahaleb (Rosaceae). Boletin de la Sociedad Argentina de Botanica 53(2):207–219. https://doi.org/10.31055/1851.2372.v53.n2.20578
74. Kalaji HM, Schansker G, Ladle RJ, Goltsev V, Bosa K, Allakhverdiev SI, Brestic M, Bussotti F, Calatayud A, Dąbrowski P, Elsheery NI, Ferroni L, Guidi L, Hogewoning SW, Jajoo A, Misra AN, Nebauer SG, Pancaldi S, Penella C, Zivcak M (2014) Frequently asked questions about in vivo chlorophyll fluorescence: practical issues. Photosynth Res 122(2):121–158. https://doi.org/10.1007/s11120-014-0024-6
75. Jhanzab HM, Razzaq A, Bibi Y, Yasmeen F, Yamaguchi H, Hitachi K, Tsuchida K, Komatsu S (2019) Proteomic analysis of the effect of inorganic and organic chemicals on silver nanoparticles in wheat. Int J Mol Sci 20(4):825. https://doi.org/10.3390/ijms20040825
76. Kannaujia R, Singh P, Prasad V, Pandey V (2022) Evaluating impacts of biogenic silver nanoparticles and ethylenediurea on wheat (*Triticum aestivum* L.) against ozone-induced damages. Environ Res 203:111857. https://doi.org/10.1016/j.envres.2021.111857
77. Razzaq A, Ammara R, Jhanzab HM, Mahmood T, Hafeez A, Hussain S (2016) A novel nanomaterial to enhance growth and yield of wheat. J Nanosci Technol 2(1):55–58
78. Amorati R, Valgimigli L (2018) Methods to measure the antioxidant activity of phytochemicals and plant extracts. J agric food chem, 66(13):3324–3329. https://doi.org/10.1021/acs.jafc.8b01079
79. Gupta SD, Agarwal A, Pradhan S (2018) Phytostimulatory effect of silver nanoparticles (AgNPs) on rice seedling growth: an insight from antioxidative enzyme activities and gene expression patterns. Ecotoxicol Environ Saf 161:624–633. https://doi.org/10.1016/j.ecoenv.2018.06.023
80. Karami Mehrian S, Heidari R, Rahmani F (2015) Effect of silver nanoparticles on free amino acids content and antioxidant defense system of tomato plants. Indian J Plant Physiol 20(3):257–263. https://doi.org/10.1007/s40502-015-0171-6
81. Şener S, Sayğı H (2023) The role of silver nanoparticles in response of in vitro boysenberry plants to drought stress. Horticulturae 9(11):1177. https://doi.org/10.3390/horticulturae9111177
82. Bagherzadeh Homaee M, Ehsanpour AA (2016) Silver nanoparticles and silver ions: oxidative stress responses and toxicity in potato (*Solanum tuberosum* L.) grown in vitro. Hortic Environ Biotechnol 57(6):544–553. https://doi.org/10.1007/s13580-016-0083-z
83. Ahanger MA, Qi M, Huang Z, Xu X, Begum N, Qin C, Zhang C, Ahmad N, Mustafa NS, Ashraf M, Zhang L (2021) Improving growth and photosynthetic performance of drought stressed tomato by application of nano-organic fertilizer involves up-regulation of nitrogen, antioxidant and osmolyte metabolism. Ecotoxicol Environ Saf 216:112195. https://doi.org/10.1016/j.ecoenv.2021.112195
84. Oteiza PI, Erlejman AG, Verstraeten SV, Keen CL, Fraga CG (2005) Flavonoid-membrane interactions: a protective role of flavonoids at the membrane surface? Clin Dev Immunol 12(1):19–25. https://doi.org/10.1080/10446670410001722168
85. Liu M, Li X, Liu Y, Cao B (2013) Regulation of flavanone 3-hydroxylase gene involved in the flavonoid biosynthesis pathway in response to UV-B radiation and drought stress in the desert plant, *Reaumuria soongorica*. Plant Physiol Biochem 73:161–167. https://doi.org/10.1016/j.plaphy.2013.09.016
86. Song X, Diao J, Ji J, Wang G, Guan C, Jin C, Wang Y (2016) Molecular cloning and identification of a flavanone 3-hydroxylase gene from *Lycium chinense*, and its overexpression enhances drought stress in tobacco. Plant Physiol Biochem 98:89–100. https://doi.org/10.1016/j.plaphy.2015.11.011
87. Linh TM, Mai NC, Hoe PT, Lien LQ, Ban NK, Hien LTT, Chau NH, Van NT (2020) Metal-based nanoparticles enhance drought tolerance in soybean. J Nanomater 2020:1–13. https://doi.org/10.1155/2020/4056563

88. Garcia-Sanchez S, Bernales I, Cristobal S (2015) Early response to nanoparticles in the Arabidopsis transcriptome compromises plant defence and root-hair development through salicylic acid signalling. BMC Genomics 16:341. https://doi.org/10.1186/s12864-015-1530-4
89. Nair PMG, Chung IM (2014) Physiological and molecular level effects of silver nanoparticles exposure in rice (*Oryza sativa* L.) seedlings. Chemosphere 112:105–113. https://doi.org/10.1016/j.chemosphere.2014.03.056
90. Zhang CL, Jiang HS, Gu SP, Zhou XH, Lu ZW, Kang XH, Yin L, Huang J (2019) Combination analysis of the physiology and transcriptome provides insights into the mechanism of silver nanoparticles phytotoxicity. Environ Pollut 252(B):1539–1549. https://doi.org/10.1016/j.envpol.2019.06.032
91. Kaveh R, Li YS, Ranjbar S, Tehrani R, Brueck CL, Van Aken B (2013) Changes in Arabidopsis thaliana gene expression in response to silver nanoparticles and silver ions. Environ Sci Technol 47(18):10637–10644. https://doi.org/10.1021/es402209w
92. Arase F, Nishitani H, Egusa M, Nishimoto N, Sakurai S, Sakamoto N, Kaminaka H (2012) IAA8 involved in lateral root formation interacts with the TIR1 auxin receptor and ARF transcription factors in Arabidopsis. PLoS One. 7(8):e43414. PMID: 22912871; PMCID: PMC3422273. https://doi.org/10.1371/journal.pone.0043414
93. Frazier TP, Burklew CE, Zhang B (2014) Titanium dioxide nanoparticles affect the growth and microRNA expression of tobacco (*Nicotiana tabacum*). Funct Integr Genomics 14(1):75–83. https://doi.org/10.1007/s10142-013-0341-4
94. Sunkar R (2010, October) MicroRNAs with macro-effects on plant stress responses. In: Seminars in cell and developmental biology. vol 21, no 8. Academic Press, pp 805–811. https://doi.org/10.1016/j.semcdb.2010.04.001
95. Khodakovskaya MV, de Silva K, Nedosekin DA, Dervishi E, Biris AS, Shashkov EV, Galanzha EI, Zharov VP (2011) Complex genetic, photothermal, and photoacoustic analysis of nanoparticle-plant interactions. Proc Natl Acad Sci USA 108(3):1028–1033. https://doi.org/10.1073/pnas.1008856108
96. Li H, Guo Q, Jing Y, Liu Z, Zheng Z, Sun Y, Xue Q, Lai H (2020) Application of *Streptomyces pactum* Act12 enhances drought resistance in wheat. J Plant Growth Regul 39(1):122–132. https://doi.org/10.1007/s00344-019-09968-z
97. Zuverza-Mena N, Armendariz R, Peralta-Videa JR, Gardea-Torresdey JL (2016) Effects of silver nanoparticles on radish sprouts: root growth reduction and modifications in the nutritional value. Front Plant Sci 7:90. https://doi.org/10.3389/fpls.2016.00090
98. Bhargava S, Sawant K (2013) Drought stress adaptation: metabolic adjustment and regulation of gene expression. Plant Breed 132(1):21–32. https://doi.org/10.1111/pbr.12004
99. Geisler-Lee J, Wang Q, Yao Y, Zhang W, Geisler M, Li K, Huang Y, Chen Y, Kolmakov A, Ma X (2013) Phytotoxicity, accumulation and transport of silver nanoparticles by *Arabidopsis thaliana*. Nanotoxicology 7(3):323–337. https://doi.org/10.3109/17435390.2012.658094

Mitigation of Salt Stress in Plants Using Silver Nanoparticles

Mustafa Guven Kaysim, Canan Gulmez Samsa, and Muhittin Kulak

Abstract In the era of unpredictable climate variability, abiotic stresses in general and salt stress, in particular, have become critical issues on the agenda of plant scientists and policy makers. Salt stress, with its detrimental effects on plant health and crop productivity, ranks first. However, nanotechnological approaches could be candidates for the mitigation of the effects of salt stress, as is the case in a large number of documents. In this regard, by supplying the plants with silver nanoparticles, the harmful effects caused by salt stress can be buffered to a considerable extent. Significantly, it was highlighted that continuous exposure to silver nanoparticles could be a promising option in this context, as the silver nanoparticles are more stable in low saline water. As with other nanoparticles, silver nanoparticles have received a great deal of attention for their potential applications in plant science. The relevant nanomaterials can improve plant health and productivity when applied at low concentration. Following exposure to silver nanoparticles in salt-stressed plants, significant changes in physiological, biochemical and molecular endpoints were reported. Of the responses, the maintained ratio of Na/K, decreased level of stress indices, enhanced activities of antioxidant enzymes, improved redox status, enhanced photosynthetic attributes, triggered stress memory and finally improved plant health are the reported endpoints. In this chapter, we have a review and discussion of the latest reports on the interaction between silver nanoparticles and salt stress. We then concluded the chapter with questions that need to be addressed in future plant science studies.

Mustafa Guven Kaysim, Canan Gulmez Samsa and Muhittin Kulak are contributed equally to the chapter.

M. G. Kaysim
Department of Field Crops, Faculty of Agriculture, Igdir University, Igdir 76000, Türkiye

C. G. Samsa
Department of Biochemistry, Faculty of Dentistry, Kafkas University, Kars, Türkiye

M. Kulak (✉)
Department of Herbal and Animal Production, Vocational School of Technical Sciences, Igdir University, Igdir 76000, Türkiye
e-mail: muhyttynx@gmail.com

A. Husen (ed.), *Plant Response to Silver Nanoparticles*, Smart Nanomaterials Technology, https://doi.org/10.1007/978-981-97-7352-7_11

Keywords Soil salinity · Abiotic stress · Secondary metabolites · Nano-engineered particles · Nanotechnology

1 Introduction

1.1 Salt Stress and Its Potential Effects on Plants

Of the abiotic stress factors, soil salinization causes critical shifts/deviations in the redox stability of plants, which in turn is manifested as disturbances in physiological, biochemical and molecular status. In general, the changes—either at the macro or micro level—are reflected in a reduction in the yield of the crop. More than 20% of agricultural land already suffers from high salinity, as highlighted by Cebeci et al. [22]. According to recent documents and projections, salinization is likely to increase as the global climate continues to change. Subsequently, secondary effects followed by salinization of soil and water are critical menaces for productivity of crop and non-crop plants. Such assumptions are not far-fetched and are realistic scenarios in agroecosystems. The fact of the expansion rate of 2 million ha per year [51] can support the assumptions or hypotheses considered. De Clercq and Van Meirvenne [26] further hypothesized that insufficient water/low rainfall coupled with high temperatures are likely to be causes of increasing level of salinity. When crops are stressed, they fail to reach their genetic potential, which is linked to productivity [71]. However, it must be emphasized that the quantitative yield of the crop may not be in line with the qualitative characteristics of the plants. This situation was clearly the subject of discussion in the sub-sections, as was the case with secondary metabolites. In order to meet the requirements for the welfare of both humans and animals, a deal of attempts have been put into action. The attempts have included the application and adaptation of agro-technologies, the use of chemicals and soil amendments, breeding and genetic approaches and the enhancement of soil diversity and activity (as well-reviewed by Ondrasek et al. [68]). Even though the available approaches have also resulted in bolstered plant growth and improved plant health, the search for novel, cost-effective and environmentally friendly approaches to stress management in plants is still ongoing. Herewith the current book chapter, we have compiled the application of nanomaterials (nanofertilizer, nano-structured/based growth promoters) in salt-stress submitted plants. We have focused on the interaction between silver nanoparticles and salt stress. Briefly, the approaches used for the synthesis of silver nanoparticles formed the first part of the chapter. The second section was based on the reports on the use of nanoparticles to mitigate salt stress.

1.2 Synthesis Approaches of Silver Nanoparticles

Nano, represents one billionth of a meter, with a size range of 1–100 nm. At the nanoscale, these structures encompass a wide range of applications, incorporating various new features, both physical and chemical [5]. Due to their unique properties such as physicochemical properties, small size and high monodispersity, nanoparticles are utilized in various fields such as targeted drug delivery, cancer therapy, optical devices, electrochemistry, catalysis, water treatment, food control, textile industry, cosmetic, agriculture and forestry sectors [15, 25, 42–46, 66, 73, 75, 83, 88]. While there are different approaches to the classification of nanoparticles, generally, they are classified as ceramic or oxide NPs, metal NPs, carbon-based NPs, polymer NPs and semiconductor NPs, depending on the material they contain. Among these, metal NPs, especially silver, copper and zinc, are a remarkable class that can be considered quite new [31].

Silver has been used for medical purposes since ancient times. Initially, it was used as a container for transporting or storing water during wars by kings. Subsequently, it was utilized as a blood purifier and for the treatment of skin diseases and epilepsy [17]. With advancements in technology and science today, there are countless possibilities for applications. The combination of the natural properties of metals with the superior characteristics of nanostructures has led to the emergence of nanomaterials with novel features, finding a broad scope of applications [18]. Silver nanoparticles (AgNPs) exhibit biocidal activity by slowing down Ag^{2+} release and interacting with proteins and enzymes, and by suppressing DNA replication, making it difficult for bacteria to develop resistant strains. Moreover, it is a good option for the production of antibacterial materials because of its large surface area, allowing for the absorption of pathogens [10]. According to Web of Science data, a total of 111,517 articles were found through the "silver nanoparticles" search at all times. However, 18,280 articles containing the keywords "antimicrobial activity", 12,405 articles "antimicrobial activity" and 615 articles "antiviral activity" were recorded (30 January, 2024). Antibacterial mechanism suggested in the majority of studies, AgNPs exhibit antibacterial effects by interacting with bacterial DNA plasmids to hinder bacterial reproduction or by interacting with ribosomes to inhibit protein synthesis. Additionally, they also interact with the bacterial cell wall, leading to the lysis of the cell membrane [2].

AgNPs find application in various antibacterial healthcare products such as dressings, bandages, respiratory tubes, bone cements, and implantable devices [79]. In addition, it is preferred as an antifungal and antibacterial material such as packaging and storage package in the food industry [11] and as an antibacterial fabric coating in the textile industry [87]. AgNPs are a suitable candidate as an eco-friendly pesticide in agriculture [33] and in environmental protection [70], remediation [8] and wastewater degradation [23]. Other than these, it has application potential in a wide range of areas such as drug delivery applications, cancer, optical receptors, electronics and biomedicine [19, 36, 38, 86]. The optimization of silver nanomaterials for these applications relies on specific physical, optical and chemical properties, which are considered crucial factors [76].

The general methodologies related to the synthesis of AgNPs are the top-down and bottom-up approaches. These two approaches combine synthesis methods such as physical, chemical, biological and hybrid techniques [90]. The top-down approach involves reducing or transforming large-scale materials into smaller desired sizes through methods such as grinding, crushing and ball milling. However, challenges may arise in this approach, particularly in terms of irregular particle size distribution, uncontrolled geometry, surface chemistry and extended processing durations [78]. Differently, bottom-up approaches enable the fabrication of desired, more complex, and enduring nanoparticles, depending on the reaction kinetics at the molecular level. Compared to the top-down approach, it poses challenges such as cost disadvantages, the need for specialized equipment, and reactor requirements due to involving more complex processes [67]. In the chemical synthesis methods of AgNPs, metal precursors, reducing agents and stabilizing agents are mostly used and are basically produced by reducing silver salts such as $AgNO_3$ and $AgClO_4$. The physical synthesis of AgNPs involves mechanical and vapor-based processes, including condensation and evaporation steps, carried out in a tube furnace under atmospheric pressure. One of the most common physical methods is laser ablation, which involves bombarding silver metal with laser beams to create nanoparticles [65, 90]. However, the synthesis of nanoparticles using physical and chemical methods requires high concentrations of reducing and stabilizing agents, as well as high radiation, and these practices have harmful effects on human health, the environment and the atmosphere. Chemical methods pose a biological risk as they can involve toxic and harmful agents. In addition to being costly and time-consuming, especially physical methods lead to high energy loss due to the required high temperatures, vacuum and pressure conditions [62, 65, 92]. As advances in technology and science have expanded the use of nanomaterials, safe, eco-friendly, economical and sustainable synthesis has become increasingly crucial.

Recent studies have demonstrated that the use of physical and chemical methods in nanoparticle synthesis has been replaced by "green synthesis" due to its associated disadvantages [16, 49, 65, 92]. Green synthesis is a biological method that aims to reduce the environmental impacts of traditional synthesis methods, minimize the use of toxic chemicals and enhance energy efficiency. They typically involve the use of plant extracts, microorganisms, fungi, or other biological agents [72]. The microbial reducing agents commonly used in silver nanoparticle synthesis are bioactive components obtained from bacteria, fungi and algae. These agents reduce and precipitate silver ions, facilitating the formation of silver nanoparticles [29]. They are considered ideal candidates for the synthesis of silver nanoparticles in various sizes (Table 1).

AgNPs have been previously produced at various sizes with antibacterial, antimicrobial, anti-viral, anti-fungal and anti-larvicidal activities, utilizing cost-effective, environmentally friendly and efficient methods [30, 40, 74, 87]. Here, the microbial reducing agents in the synthesis of AgNPs can be DNA, genes, antigens, enzymes, proteins, or cell ligands [2, 63]. Biological synthesis of AgNPs was found to be

Table 1 Bacterial-, fungal- and algal-mediated biological synthesis of AgNPs using $AgNO_3$

Producer organism	Size (nm)	Shape	Applications	Author(s)
Aspergillus niger *Aspergillus flavus* *Aspergillus fumigatus*	25–175 45–185 5–95	Spherical	Nitrate reductase activity	[98]
Anamorphous Bjerkandera sp. R1	70–90	Spherical	Anti-antimicrobial, effective synthesis	[69]
Colletotrichum sp.	20–50	Polyform shapes including spherical, near to spherical, triangular and hexagonal	Bactericidal activity, action mode on DNA damage	[14]
Lactobacillus acidophilus	10–20	Spherical	Antioxidant, antimicrobial activity, biofilm inhibition	[3]
Bacillus sp.	22–41	Spherical	Anti-fungal activity	[7]
Unicellular ulvophyte sp. MBIC10591	12.1 ± 1.2	Polyform shapes, including spherical and cubic	Anti-tumour and microbial agents	[39]
Botryococcus braunii	40–90	Cubical and spherical	Antimicrobial	[13]
Streptomyces sp.	10–30	Spherical	Antibacterial activity, biofilm inhibition	[37]

between 50 and 100 nm using *Candida albicans* [81], 15.5 nm using *Aspergillus versicolor* [98] and 5–30 nm using *Bacillus safensis* [59].

The use of plants for AgNP synthesis is an attention-grabbing production method, offering a rapid, non-toxic, inexpensive and single-step technique for biosynthetic processes [97]. The step-by-step process of biosynthesis of AgNPs by plants typically involves the preparation of plant extract, addition of silver salts, reduction of silver ions by plant bioactive components, nanoparticle formation and separation and characterization stages [27]. The necessary bio-reducing phytochemicals for the formation of NPs include vitamin C, alcohols, flavonoids, carbohydrates, alkaloids and glycosides [2, 29]. For instance, in the case of *Panax ginseng*, the biomolecules responsible for reducing Ag ions are proteins, carbohydrates, triterpenes, saponins and phenolic compounds [96] whereas for *Azadirachta indica*, they are flavonoids and terpenoids [6].

In the synthesis of AgNPs, synthesis conditions play a crucial role. Changes in the components during the synthesis process affect the morphology, size, shape, activity and consequently, the applicability of the synthesized product. These parameters include pH, temperature, plant extracts concentration, silver metal salt concentration and reaction time [29, 60, 95].

The most common techniques used in the characterization of silver nanoparticles include UV-visible, infra-red spectroscopy (FTIR) spectrum, dynamic light scattering

(DLS) spectroscopic techniques, scanning electron microscopy (SEM), transmission electron microscopy (TEM), Energy Dispersive X-ray Analysis (EDX), X-ray diffraction (XRD) spectrum, X-Ray photoelectron spectroscopy (XPS), selected area electron diffraction (SAED) pattern techniques [16, 28].

1.3 Silver Nanoparticles and Salt Stress Studies

Along with the salt-stress mediated perturbations in redox-stability of the plants, a series of state-of-arts concerned with ionic and osmotic balance, impairment in photosynthesis machinery, increases in values of stress-associated attributes, stunted plant growth and consequently decreased crop productivity has been reported and reviewed in large number documents [45–47, 48, 50, 77, 82, 94]. However, by supplying the plants with silver nanoparticles, the harmful effects caused by salt stress can be buffered to a considerable extent [99]. We have detailed the application of silver nanoparticles to salt-stressed plants, demonstrating the improved growth and beneficial properties of salt-stressed plants (Table 1). Concerning the signatures of salt stress, ionic imbalance or toxicity is one of the critical endpoints of salt stress. Silver treatments to the salt-submitted plants exhibited substantial improvements in chloride and potassium contents, ultimately maintaining a ratio of Na/K (Fig. 1, Table 2). Abasi et al. [1] highlighted that a continuous application of silver nanoparticles could be a promising option to buffer the damage of salt stress, as the silver nanoparticles are more stable in low saline waters. As a secondary effect of the toxicity of the ions, the reactive oxygen species are more pronounced in the stressed plants, resulting in oxidative damage. Oxidative damage attributes such as hydrogen peroxide, thiobarbituric acid reactive substances and percentage of electrolyte leakage were buffered in stressed plants after application of silver nanoparticles in wheat plants [93]. Up to our best literature survey, we have recorded a limited number of studies concerned with silver nanoparticles as a solution to allay the damage triggered by salt stress. Out of the documents available, most of them are based on the physiological and biochemical responses of the plants. However, molecular responses-related reported are very rare. Of the molecular reports, Almutairi [12] examined the changes in tomato plants after exposure to the combination of salt (150 and 100 mM) and silver nanoparticles (0.05, 0.5, 1.5, 2 and 2.5 mg L^{-1}). In this regard, the seeds were first primed with silver nanoparticles and were then allowed to germination under salt stress. As expected, plant performance was markedly inhibited but the inhibitory impact, in an extent, was relieved by exposure to silver nanoparticles. According to semi-quantitative RT-PCR, four genes including AREB, MAPK2, P5CS and CRK1 were up-regulated in salt-stressed-plants by silver nanoparticles. On the other hand, TAS14, DDF2 and ZFHD1 were down-regulated in response to AgNPs under salt stress.

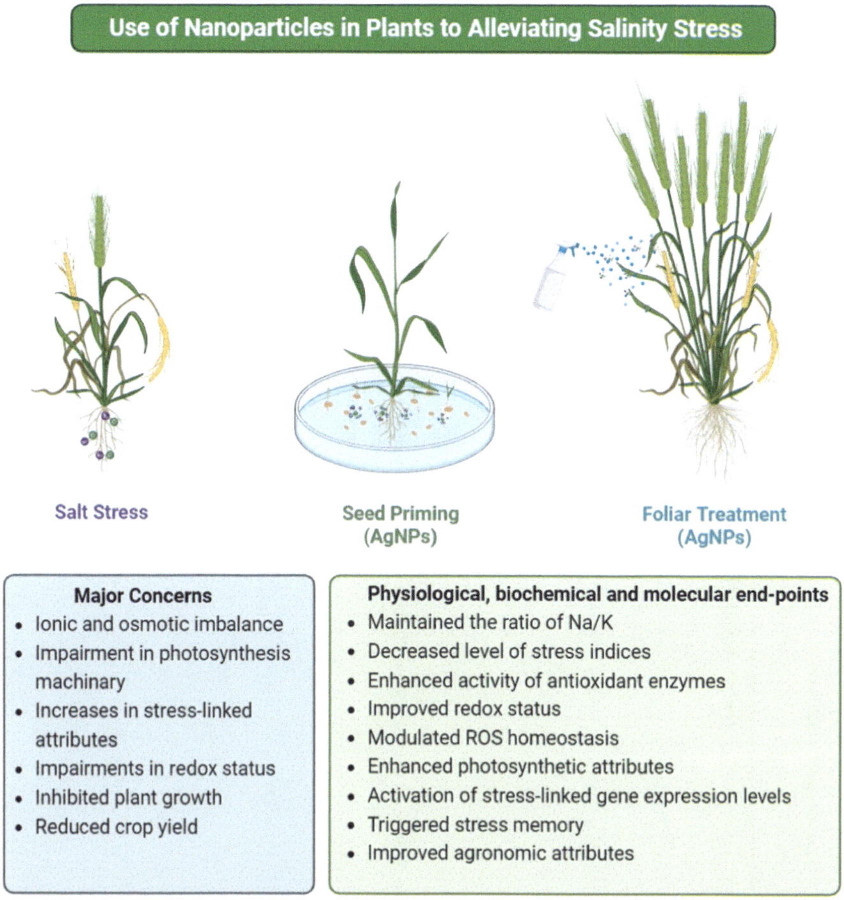

Fig. 1 Use of nanoparticles in plants to alleviate salinity stress (modified according to Junedi et al. [53])

2 Conclusion and the Way Forward: Outstanding Questions

In conclusion, silver nanoparticles can be considered as alternative options in mitigation of the salt stress effects but they exhibit affirmative behavior in a concentration manner. In this chapter, rather than focusing on the phytotoxic effects of silver nanoparticles, we have concentrated on their beneficial effects. Reports available have shown that the silver nanoparticles can shape the biochemical and physiological status of the plants, which in turn are translated into improved plant health and crop productivity. However, the modus operandi of the silver nanoparticles is still in question and deserves further investigation. In this context, We have suggested some outstanding issues to be addressed in future plant science reports, as given below.

Table 2 Physiological and chemical response of the plants treated with silver nanoparticles and exposed to salt stress

Nanoparticle	Plant	AgNPs form, concentration and mode of application	Np shape and size	Salt level	Experimental conditions	NPs effects on salt-stressed plants	Author(s)
AgNPs	Maize (*Zea mays* L.)	Foliar application with 1 mM Cit-AgNPs and IAA	NA	0, 40 and 80 mM NaCl	35–40 °C temperature in the month of June and 23–30 °C in the month of September along with 513 mm annual rainfall. 2 kg silt and soil (1:2) in triplicates in the green house condition	Photosynthetic pigments amplified by 35–63%, Total osmolytes were augmented by 39–68% antioxidant enzymes (SOD and POD) were boosted by 42–57 and 37–62% leaf, root and shoot fresh and dry weight was improved by 52–74, 43–69 and 36–79%	[4]
AgNPs	Tomato (*Solanum lycopersicum*)	0.05, 0.5, 1.5, 2 and 2.5 mg L^{-1}	NA, 99.99%, 20 nM	0, 150 and 200 mM NaCl	Seeds incubated in petri dishes at room temperature. Findings were collected 8 days after germination	Germination percentage, germination rate, root length and seedling fresh and dry weight were improved. Salt stress genes; AREB, MAPK2, P5CS and CRK1 were up-regulated, TAS14, DDF2 and ZFHD1 were down-regulated	[12]

(continued)

Table 2 (continued)

Nanoparticle	Plant	AgNPs form, concentration and mode of application	Np shape and size	Salt level	Experimental conditions	NPs effects on salt-stressed plants	Author(s)
AgNPs	Garlic callus	0, 1, and 2 mg L^{-1}	NA	0, 25 and 50 mmol L^{-1} NaCl	In-vitro culture, incubated 25 ± 2 °C	Increase allicin, di-allyldisulfide and vinyldithiin content in culture	[9]
AgNPs	Spinach (*Spinacia oleracea* L.)	20, 40, 60 ppm	NA	4.0, 6.0 and 10.0 dS m^{-1}	Seedlings were then grown in a hydroponic system. Each uniform plant was moved sterile perlite and Hoagland's solution at greenhouse conditions (day temperature, humidity and light intensity were (24.6 ± 0.039 °C), (51% ± 1.48)	Ag NPs negatively impacts stomatal conductance and chlorophyll content when it was added under saline conditions and didn't improve physiological parameters	[20]
AgNPs	Bright Pixi	Priming 100 ppm	7–25 nm	600 mM NaCl	Planted into plastic pots 16 cm diameter and 2 dm^3 capacity into TS1 medium with pH 6.4, an electrical conductivity (EC) of 0.54 mS cm^{-1} and the following composition (mg dm^{-3}): N–NO_3_182; P—131, K—402, Ca—1646, Mg—172, Cl—18	Positive effect on plant height, fresh bulb weight, number of scales per bulb and bulb diameter. Decrease Cu, Mn and Zn content. Increased Na content. Similar effect (with NaCl) on Chl content	[21]

(continued)

Table 2 (continued)

Nanoparticle	Plant	AgNPs form, concentration and mode of application	Np shape and size	Salt level	Experimental conditions	NPs effects on salt-stressed plants	Author(s)
AgNPs	Maize (*Zea mays* L.)	Priming 40 mg/L	17−39 nm	50, 100 mM NaCl	Climate incubator (25 °C, 70% relative humidity and darkness)	Increased seed germination rate, seedling vigor, root lenght, root dry weight. Induced a transcriptomic shift. Stress memory triggered. Modulated ROS homeostasis	[24]
AgNPs	Wheat (*Triticum aestivum* L.)	AgNPs via (*Chrysthanthemum cornarium* leaf aqueous extract) priming 0, 20, 40 and 80 mg L^{-1}	15–30 nm	150 nM NaCl	NA	Caused toxicity to meristematic root cells (mitotic index, chromosomal abnormalities) and reduction in seedling growth	[32]
AgNPs	*Arabidopsis thaliana*	MWCNTs AgNPs 0.02 mg/mL	20, 40, and 80 nm	100 mM NaCl	In-vitro condition. Growth chamber at 22 °C, 70% relative humidity and 200 $\mu M \times m^2/s$ of cool white fluorescence illumination (12 h light/12 h dark). Plants treated at 8–12 leaf stage	Lower impact on gene regulation. repressed transcriptional responses to microbial pathogens, resulting in increased bacterial colonization during an experimental infection	[34]

(continued)

Table 2 (continued)

Nanoparticle	Plant	AgNPs form, concentration and mode of application	Np shape and size	Salt level	Experimental conditions	NPs effects on salt-stressed plants	Author(s)
AgNPs	*Thymus vulgaris* L. and *Thymus daenensis* Celak	0, 10, 20 and 30 mL	NA	0, 100, 200 and 300 nM NaCl	Germinated 26 °C for 16 h photoperiod, 8 h darkness and the relative humidity of 70%	Increased germination percent, germination rate and root length	[35]
AgNPs	Bitter Vetch (*Vicia ervilia* L.)	Polyvinylpyrrolidone-coated AgNPs 0, 5, 10, 15 and 20 ppm	99.99% purity, 20 nm	0, 4, 8, 12 16 and 20 dS/m NaCl	During germination stage (25 ± 1 °C) Growth in pots	Improved germination parameter, plant root length, shoot length, root fresh weight, shoot dry weight, root dry weight and shoot fresh weight	[41]
AgNPs	Onion (*Allium cepa* L.)	5 ppm AgNPs and PGPR (*Bacillus pumilus* and *Pseudomonas moraviensis*)	NA	50 mM NaCl	Broth culture in rhizosphere soil. Pots were filled with soil and sand mixture (3:1) and 1 mL of inocula was given to each seedling, having two seedlings per pot	Increased the total chlorophyll and carotenoids contents protein content of bulb, sugar and proline content, and bulb weight. Decreased the leaf flavonoids but had significant increase in the bulb flavonoid contents	[52]

(continued)

Table 2 (continued)

Nanoparticle	Plant	AgNPs form, concentration and mode of application	Np shape and size	Salt level	Experimental conditions	NPs effects on salt-stressed plants	Author(s)
AgNPs	Linseed (*Linum usitatissimum* L.)	AgNPs (isolated from *Rosmarinus officinalis*) and *Comamonas testosteroni*	75 nm	0, 25, 50, and 100 mM NaCl	20–25 °C in a greenhouse enlightened with regular light. Watered with 200 mL water twice a week until growth	Increased; chlorophyll content, soluble sugars, proteins, and proline, SOD, CAT, POD, APX, and G.R. activity. Decreased H_2O_2 and MDA	[54]
AgNPs	Pearl millet (*Pennisetum glaucum* L.)	Seed priming with AgNPs 0, 10, 20 and 30 mM	99.5% purity, 10.49 density, 50–100 nm	0, 120 and 150 mM NaCl	Greenhouse condition. 31/25 °C day and night temperature, 43 and 84% relative humidity, 12 h photoperiod, ~864 μmol m^{-2} s^{-1} photon flux density. Seeds soaked in prepared NPs solutions for 20 h in the dark at room temperature. Per pot having 5.0 kg soil containing sand, silt and clay of 36, 41 and 23%. Soil pH 7.48, EC 1.99 dS m^{-1} and organic matter 1.11%. Zn, Cu, Fe, Mn and B were present in the soil as 1.09, 0.74, 3.2, 7.5 and 0.9 mg/kg. Harvested 40 days after sowing	Improved plant height, relative water content, fresh and dry weight, SOD and POD activity, total phenolic and flavonoids content. Decreased proline, H_2O_2, MDA, Na^+ and Na^+/K^+ ratio	[57]

(continued)

Table 2 (continued)

Nanoparticle	Plant	AgNPs form, concentration and mode of application	Np shape and size	Salt level	Experimental conditions	NPs effects on salt-stressed plants	Author(s)
AgNPs	Pearl millet (*Pennisetum glaucum* L.) Tifleaf 3	Seed priming with AgNPs 0, 10, 20 and 30 mM	99.5% purity, 10.49 density, 50–100 nm	0, 120, and 150 NaCl	Greenhouse condition. 31/25 °C day and night temperature, 43 and 84% relative humidity, 12 h photoperiod, ~864 μmol m^{-2} s^{-1} photon flux density. Harvested 70 days after sowing	Reduced Na and Cl uptake. Maintained the ionic balance of cell (Na^+, K^+ and Na^+/K^+ ratio) Increased plant height, yield, and photosynthesis, SOD and CAT. Decreased Proline, H_2O_2, MDA	[55]
AgNPs	Pearl millet (*Pennisetum glaucum* L.) Tifleaf 3	AgNPs 0, 20 mg/L	99.5% purity, 10.49 density, 50–100 nm	0, 250 mM NaCl	Day and night conditions were set at 14 h/26 °C and 10 h/22 °C. AgNPs treatment were given to plants by dissolving in distilled water through sonication and then given to plants (13 days after sowing) 1 h after treatment leaf sample collected for RNA-Seq	6016 Differentially expressed transcripts (DETs) detected. Among DETs 494 upregulated, 231 downregulated with AgNPs + NaCl. MAPK signaling pathway, biosynthesis of secondary metabolites, and plant hormonal signal transduction pathway were enriched. Improved the morphology, physiology, biochemistry, and gene expression pattern	[56]

(continued)

Table 2 (continued)

Nanoparticle	Plant	AgNPs form, concentration and mode of application	Np shape and size	Salt level	Experimental conditions	NPs effects on salt-stressed plants	Author(s)
AgNPs	*Arabidopsis thaliana*	AgNPs	20 nm	NA	NA	111 Genes were unique in AgNPs and they were enriched in three biological functions: response to fungal infection, anion transport, and cell wall/plasma membrane related. 60 AgNP-specific genes that are not affected/regulated by stressors	[58]
AgNPs	Wheat (*Triticum aestivum* L.)	Priming (0, 2, 5, and 10 mM) 24 h. Isolated from *Aspergillus terreus* MALEX	Spherical-shaped 15–19 nm	150 mM NaCl	Greenhouse experiment. Irrigated 70% of soil water holding capacity. Fertilized with $MgCO_3$, P, NH_4NO_3 and H_3BO_3. At the fourth leaf, irrigated 25 mM NaCl solution every day until the final concentration (150 mM). 26 ± 2/18 ± 2 °C day and night temperature	Increased total chlorophyll, MDA, fresh and dry biomass, soluble sugars and proline contents, POD. Decreased H_2O_2, CAT and MDA	[61]

(continued)

Table 2 (continued)

Nanoparticle	Plant	AgNPs form, concentration and mode of application	Np shape and size	Salt level	Experimental conditions	NPs effects on salt-stressed plants	Author(s)
AgNPs	*Satureja hortensis* L.	Priming, 0, 40, 60, and 80 ppm	99.80% purity, NA	0, 60, 90, and 120 mM L^{-1} NaCl	Germinated growth chamber at 24 ± 1 °C, 70% relative humidity. Growth greenhouse conditions. 17 h light periods and 200 μmol quanta m^{-2} s^{-1} light intensity, 25 ± 1/18 ± 1 °C day/night, irrigated Hougland solution. Salt stress applied 38 days after	Improved germination percentage, seedling growth, shoot and stem length	[64]
AgNPs	Roselle (*Hibiscus sabdarifa*)	Synthesized aqueous leaf extract of *Hibiscus sabdarifa* 0, 15, 30, 45, 60, 75, and 90 mg/L	~8 nm	0, 15, 30, 45, 60, 90, 120, 150, and 180 mM NaCl	Germinated at 20 ± 1 °C. Growth in pots. A fresh solution of nutrients (150 mL) was poured into each pot every other day. To prevent plant shock, the salt was applied four times and the applications were distributed evenly over a total period of 10 days	Increased seed germination, plant height, leaf, and root fresh and dry weights, as well as relative water content (RWC). Enhancing CAT, APX, POD, and SOD activity. Increase amounts of proline, soluble sugars and carbohydrates. Improved flavonoid and anthocyanin content. Higher expression CHS, F3H, and ANS genes	[80]

(continued)

Table 2 (continued)

Nanoparticle	Plant	AgNPs form, concentration and mode of application	Np shape and size	Salt level	Experimental conditions	NPs effects on salt-stressed plants	Author(s)
AgNPs	*Maerua oblongifolia*	Spherical AgNPs 0, 10, 20, an 30 mg/L	9–30 nm	0, 50, and 200 mM NaCl	In-vitro condition. Growth chamber 25 ± 2 °C, 60% relative humidity and 25–30 μmol m^2 s^{-1} SFP for 16 h light, 8 h dark. Harvested after 45 days of culture	Improve growth, shoots, photosynthetic pigment, antioxidant enzyme activity. Decreased H_2O_2, MDA, proline, soluble sugar, total phenolics and flavonoids	[84]
AgNPs	*Chenopodium Quinoa* (Q6 Line)	25, 50, and 75 mg/L	20.27 nm	0, 25, 50, 75, 100, and 200 mM NaCl	In-vitro condition. Growth room condition (16 h light, 8 h dark 24 ± 1 °C). Data collected after 4 weeks of culturing	Improved shoot length, leaf number, plant fresh and dry weight, chlorophyll, nitrogen content. Decreased proline content	[85]

(continued)

Table 2 (continued)

Nanoparticle	Plant	AgNPs form, concentration and mode of application	Np shape and size	Salt level	Experimental conditions	NPs effects on salt-stressed plants	Author(s)
AgNPs	Rice (*Oryza sativa* L.)	Seed priming 0.1, 1, 10, 100 and 1,000 mg/L	20, 30–70, 70–120 and 150 nm	85 mM NaCl	Controlled greenhouse conditions. After 14 days salt stress treated. Harvested 7 days after treatments	Improve antioxidant enzyme activities, GPX, GR, APX, SOD, CAT. Decreased H_2O_2	[89]
AgNPs	Maize (*Zea mays* L.) Pahari white	Foliar treatment, spherical Nit-AgNPs and IAA	90 nm	40 and 80 mM NaCl	Seeds were propagated in pots. Silt and soil (1:2), EC: 3.09–5.12, pH: 6.8–7.3, Moisture: 4–16%	Improved shoot and root length, fresh and dry weight, number and fresh weight of leaves. Increased chlorophyll content, antioxidant enzyme activity	[91]

(continued)

Table 2 (continued)

Nanoparticle	Plant	AgNPs form, concentration and mode of application	Np shape and size	Salt level	Experimental conditions	NPs effects on salt-stressed plants	Author(s)
AgNPs	Wheat (*Triticum aestivum* L.)	Foliar treatment AgNPs, 30, 40, and 50 µg/mL	27–33 (~26) nm	0 and 100 mM NaCl	The randomized pot experiment proceeded in the natural environmental conditions with characterized agro-climatic conditions of 400 mm annual rainfall in the year 2019	Increased the average number of seed germination, germination rate, and germination index. Decreased H_2O_2, ABA, TBARS, Na^+, Cl^- content. Increased chlorophyll content, SOD, APX, GR, and GPX activity, NR, NiR, and content of N	[93]

(1) What are the specific mechanisms that occur after exposure to silver nanoparticles in salt-stressed plants at the physiological and molecular level? (2) are the mechanisms common in all stress types? (3) Do the responses of the plants differ by the biotic or abiotic stress factors? (4) How much of the potential to mitigate the oxidative damage of salt stress is influenced by the physicochemical properties of the nanoparticles? (5) what are the long term effects of the silver nanoparticles on stress memory of the plants? (6) After treatment with silver nanoparticles under either no stress or salt stress, what kind of epigenetic regulations are involved in plants? (7) it is likely to be experiencing different stressors at the same time and so, what will be the interactions between the silver nanoparticles and other stress factors (i.e., drought and heat stress) in salt-affected agro-ecosystems?

References

1. Abasi F, Raja NI, Mashwani ZUR, Amjad MS, Ehsan M, Mustafa N, Proćków J (2022) Biogenic silver nanoparticles as a stress alleviator in plants: a mechanistic overview. Molecules 27(11):3378
2. Abdelghany TM, Al-Rajhi AMH, Al Abboud MA et al (2018) Recent advances in green synthesis of silver nanoparticles and their applications: about future directions a review. BioNanoSci 8:5–16
3. Abishad P, Vergis J, Unni V, Ram V, Niveditha P, Yasur J, Rawool DB (2022) Green synthesized silver nanoparticles using *Lactobacillus acidophilus* as an antioxidant antimicrobial and antibiofilm agent against multi-drug resistant enteroaggregative *Escherichia coli*. Probiotics Antimicrob Proteins 14(5):904–914
4. Afridi GM, Ullah N, Ullah S, Nafees M, Khan A, Shahzad R, Fahad S et al (2023) Modulation of salt stress through application of citrate capped silver nanoparticles and indole acetic acid in maize. Plant Physiol Biochem 201:107914
5. Ahire S, Bachhav AA, Pawar TB, Jagdale BS, Patil AV, Koli PB (2022) The augmentation of nanotechnology era: a concise review on fundamental concepts of nanotechnology and applications in material science and technology. Results Chem 100633
6. Ahmed S, Mudasir S, Babu A, Swami L, Ikram S (2016) Green synthesis of silver nanoparticles using Azadirachta indica aqueous leaf extract. J Radiat Res Appl Sci 9:1–7
7. Ajaz S, Ahmed T, Shahid M, Noman M, Shah AA, Mehmood MA, Li B et al (2021) Bioinspired green synthesis of silver nanoparticles by using a native *Bacillus* sp., strain AW1-2: characterization and antifungal activity against *Colletotrichum falcatum* Went. Enzyme Microb Technol 144:109745
8. Akhil T, Bhavana V, Ann Maria CG, Nidhin M (2023) Role of biosynthesized silver nanoparticles in environmental remediation: a review. Nanotechnol Environ Eng 8(3):829–843
9. Al-Taie AAG, Aboohanah MA, Issa FH (2021) Effect of silver nanoparticles in stimulating some active compounds in garlic callus under salt stress in vitro. In: IOP conference series: earth and environmental science, vol 923, No 1. IOP Publishing, p. 012023
10. Alavi M, Hamblin MR (2023) Antibacterial silver nanoparticles: effects on bacterial nucleic acids. Cell Mol Biomed Rep 3(1):35–40
11. Ali M, Ullah M, Khan MM, Rasheed A, Khan M, Aswad S, Zaman MQ (2023) Role of silver nanoparticles (AgNPs) in food surface packaging: effects of AgNPs on packaging. Biol Sci-PJSIR 66(1):65–73
12. Almutairi ZM (2016) Influence of silver nano-particles on the salt resistance of tomato (*Solanum lycopersicum*) during germination. Int J Agric Biol 18(2)

13. Arya A, Vaibhav M, Singh CT (2019) Green synthesis of silver nanoparticles from green algae (*Botryococcus braunii*) and its catalytic behavior for the synthesis of benzimidazoles. Chem Data Collect 20:100190
14. Azmath P, Baker S, Rakshith D, Satish S (2016) Mycosynthesis of silver nanoparticles bearing antibacterial activity. Saudi Pharm J 24:140–146
15. Bahrami B, Hojjat-Farsangi M, Mohammadi H, Anvari E, Ghalamfarsa G, Yousefi M, Jadidi-Niaragh F (2017) Nanoparticles and targeted drug delivery in cancer therapy. Immunol Lett 190:64–83
16. Barabadi H, Mojab F, Vahidi H, Marashi B, Talank N, Hosseini O, Saravanan M (2021) Green synthesis characterization antibacterial and biofilm inhibitory activity of silver nanoparticles compared to commercial silver nanoparticles. Inorg Chem Commun 129:108647
17. Barillo DJ, Marx DE (2014) Silver in medicine: a brief history BC 335 to present. Burns 40:S3–S8
18. Ben-Shahar Y, Stone D, Banin U (2023) Rich landscape of colloidal semiconductor–metal hybrid nanostructures: synthesis synergetic characteristics and emerging applications. Chem Rev 123(7):3790–3851
19. Bouafia A, Laouini SE, Ahmed AS, Soldatov AV, Algarni H, Feng Chong K, Ali GA (2021) The recent progress on silver nanoparticles: synthesis and electronic applications. Nanomaterials 11(9):2318
20. Bsoul EY, Othman SA, Al-Ghzawi ALA, Massadeh MI (2023) Effect of silver nanoparticles on growth and physiological responses of spinach (*Spinacia oleracea* L.) under salt stress. Jordan J Biol Sci 16(1)
21. Byczyńska A, Zawadzińska A, Salachna P (2023) Colloidal silver nanoparticles enhance bulb yield and alleviate the adverse effect of saline stress on lily plants. J Ecol Eng 24(6)
22. Cebeci E, Boyacı HF, Kıran S, Ellialtıoğlu ŞŞ (2024) Assessment results of salinity stressed F2 population originated from interspecific hybridization of eggplant with wild relative Solanum incanum L. Hortic Stud 41(2):50–59
23. Chandhru M, Rani SK, Vasimalai N (2020) Reductive degradation of toxic six dyes in industrial wastewater using diaminobenzoic acid capped silver nanoparticles. J Environ Chem Eng 8(5):104225
24. Chen S, Liu H, Yangzong Z, Gardea-Torresdey JL, White JC, Zhao L (2023) Seed priming with reactive oxygen species-generating nanoparticles enhanced maize tolerance to multiple abiotic stresses. Environ Sci Technol 57(48):19932–19941
25. Chiari-Andréo BG, de Almeida-Cincotto MGJ, Oshiro Jr JA, Taniguchi CYY, Chiavacci LA, Isaac V, LB (2019) Nanoparticles for cosmetic use and its application. In: Nanoparticles in pharmacotherapy. William Andrew Publishing, pp 113–146
26. De Clercq WP, Van Meirvenne M (2005) Effect of long-term irrigation application on the variation of soil electrical conductivity in vineyards. Geoderma 128(3–4):221–233
27. Das CA, Kumar VG, Dhas TS, Karthick V, Govindaraju K, Joselin JM, Baalamurugan J (2020) Antibacterial activity of silver nanoparticles (biosynthesis): a short review on recent advances. Biocatal Agric Biotechnol 27:101593
28. Dawadi S, Katuwal S, Gupta A, Lamichhane U, Thapa R, Jaisi S, Parajuli N et al (2021) Current research on silver nanoparticles: synthesis characterization and applications. J Nanomater 2021:1–23
29. Dhaka A, Mali SC, Sharma S, Trivedi R (2023) A review on biological synthesis of silver nanoparticles and their potential applications. Results Chem 101108
30. Dridi R, Essghaier B, Hannachi H, Khedher GB, Chaffei C, Zid MF (2022) Biosynthesized silver nanoparticles using *Anagallis monelli*: evaluation of antioxidant activity antibacterial and antifungal effects. J Mol Struct 1251:132076
31. Ealia SAM, Saravanakumar MP (2017) A review on the classification characterisation synthesis of nanoparticles and their application. In: IOP conference series: materials science and engineering, vol 263, No 3. IOP Publishing, p 032019
32. El-Badan GE, Abou-Zeid HM (2023) Impact of biogenic silver nanoparticles on some physiological attributes mitotic index and chromosomal abnormalities of wheat (*Triticum aestivum*) under salt stress. Intl J Agric Biol 29:55–64

33. Eswaran SG, Mamat MHB, Vasimalai N (2023) Facile ultrasonication-assisted synthesis of purpald-functionalized silver nanoparticles for the rapid spectrophotometric detection of acetamiprid pesticide in food and environmental samples. J Mol Liq 122425
34. García-Sánchez S, Bernales I, Cristobal S (2015) Early response to nanoparticles in the Arabidopsis transcriptome compromises plant defence and root-hair development through salicylic acid signalling. BMC Genomics 16(1):1–17
35. Ghavam M (2018) Effect of silver nanoparticles on seed germination and seedling growth in *Thymus vulgaris* L., and *Thymus daenensis* Celak under salinity stress. J Rangel Sci 8(1):93–100
36. Gherasim O, Puiu RA, Bîrcă AC, Burduşel AC, Grumezescu AM (2020) An updated review on silver nanoparticles in biomedicine. Nanomaterials 10(11):2318
37. Goel N, Ahmad R, Singh R, Sood S, Khare SK (2021) Biologically synthesized silver nanoparticles by Streptomyces sp., EMB24 extracts used against the drug-resistant bacteria. Bioresour Technol Rep 15:100753
38. Gomes HI, Martins CS, Prior JA (2021) Silver nanoparticles as carriers of anticancer drugs for efficient target treatment of cancer cells. Nanomaterials 11(4):964
39. Hamida RS, Ali MA, Alkhateeb MA, Alfassam HE, Momenah MA, Bin-Meferij MM (2022) Algal-derived synthesis of silver nanoparticles using the Unicellular ulvophyte sp., MBIC10591: optimisation characterisation and biological activities. Molecules 28(1):279
40. Hasanin M, Elbahnasawy MA, Shehabeldine AM, Hashem AH (2021) Ecofriendly preparation of silver nanoparticles-based nanocomposite stabilized by polysaccharides with antibacterial antifungal and antiviral activities. Biometals 34:1313–1328
41. Hojjat SS, Mozumder C, Bora T, Hornyak GL (2019) Polyvinylpyrrolidone-coated silver nanoparticle mitigation of salinity on germination and seedling parameters of bitter vetch (*Vicia ervilia* L.) plants. Nanotechnol Russ 14:582–587
42. Husen A, Iqbal M, Aref IM (2016) IAA-induced alteration in growth and photosynthesis of pea (*Pisum sativum* L.) plants grown under salt stress. J Environ Biol 37:421–429
43. Husen A, Iqbal M, Khanum N, Aref IM, Sohrab SS, Meshresa G (2019) Modulation of salt-stress tolerance of Niger (*Guizotia abyssinica*), an oilseed plant, by application of salicylic acid. J Environ Biol 40:94–104
44. Husen A (2022) Engineered nanomaterials for sustainable agricultural production, soil improvement and stress management. Elsevier Inc., Cambridge, USA. https://doi.org/10.1016/C2021-0-00054-7
45. Husen A (2023) Nanomaterials and nanocomposites exposures to plants. In: Response, interaction, phytotoxicity and defense mechanisms. Springer Nature Singapore Pte Ltd., Singapore. https://doi.org/10.1007/978-981-99-2419-6
46. Husen A, Bachheti RK, Bachheti A (2023) Current trends in green nano-emulsions. In: Food, agriculture and biomedical sectors. Springer Nature Singapore Pte Ltd., Singapore
47. Husen A, Iqbal M, Sohrab SS, Ansari MKA (2018) Salicylic acid alleviates salinity-caused damage to foliar functions, plant growth and antioxidant system in Ethiopian mustard (*Brassica carinata* A. Br.). Agric Food Secur 7(44):1–14
48. Hussein M, Embiale A, Husen A, Aref IM, Iqbal M (2017) Salinity-induced modulation of plant growth and photosynthetic parameters in faba bean (*Vicia faba*) cultivars. Pak J Bot 49:867–877
49. Ijaz I, Gilani E, Nazir A, Bukhari A (2020) Detail review on chemical physical and green synthesis classification characterizations and applications of nanoparticles. Green Chem Lett Rev 13(3):223–245
50. Isayenkov S (2012) Physiological and molecular aspects of salt stress in plants. Cytol Genet 46(5):302–318
51. Ivushkin K, Bartholomeus H, Bregt AK, Pulatov A, Kempen B, De Sousa L (2019) Global mapping of soil salinity change. Remote Sens Environ 231:111260
52. Jahangir S, Javed K, Bano A (2020) Nanoparticles and plant growth promoting rhizobacteria (PGPR) modulate the physiology of onion plant under salt stress. Pak J Bot 52(4):1473–1480
53. Junedi MA, Mukhopadhyay R, Manjari KS (2023) Alleviating salinity stress in crop plants using new engineered nanoparticles (ENPs). Plant Stress 100184

54. Khalofah A, Kilany M, Migdadi H (2021) Phytostimulatory influence of *Comamonas testosteroni* and silver nanoparticles on *Linum usitatissimum* L., under salinity stress. Plants 10(4):790
55. Khan I, Awan SA, Raza MA, Rizwan M, Tariq R, Ali S, Huang L (2021) Silver nanoparticles improved the plant growth and reduced the sodium and chlorine accumulation in pearl millet: a life cycle study. Environ Sci Pollut Res 28:13712–13724
56. Khan I, Awan SA, Rizwan M, Akram MA, Zia-ur-Rehman M, Wang X, Huang L (2023) Physiological and transcriptome analyses demonstrate the silver nanoparticles mediated alleviation of salt stress in pearl millet (*Pennisetum glaucum* L.). Environ Pollut 318:120863
57. Khan I, Raza MA, Awan SA, Shah GA, Rizwan M, Ali B, Huang L (2020) Amelioration of salt induced toxicity in pearl millet by seed priming with silver nanoparticles (AgNPs): the oxidative damage antioxidant enzymes and ions uptake are major determinants of salt tolerant capacity. Plant Physiol Biochem 156:221–232
58. Kohan-Baghkheirati E, Geisler-Lee J (2015) Gene expression protein function and pathways of *Arabidopsis thaliana* responding to silver nanoparticles in comparison to silver ions cold salt drought and heat. Nanomaterials 5(2):436–467
59. Lateef A, Adelere IA, Gueguim-Kana EB, Asafa TB, Beukes LS (2015) Green synthesis of silver nanoparticles using keratinase obtained from a strain of *Bacillus safensis* LAU 13. Int Nano Lett 5:29–35
60. Menichetti A, Mavridi-Printezi A, Mordini D, Montalti M (2023) Effect of size shape and surface functionalization on the antibacterial activity of silver nanoparticles. J Funct Biomater 14(5):244
61. Mohamed AKS, Qayyum MF, Abdel-Hadi AM, Rehman RA, Ali S, Rizwan M (2017) Interactive effect of salinity and silver nanoparticles on photosynthetic and biochemical parameters of wheat. Arch Agron Soil Sci 63(12):1736–1747
62. Mughal SS, Hassan SM (2022) Comparative study of AgO nanoparticles synthesize via biological chemical and physical methods: a review. Am J Mater Synth Process 7(2):15–28
63. Mustapha T, Misni N, Ithnin NR, Daskum AM, Unyah NZ (2022) A review on plants and microorganisms mediated synthesis of silver nanoparticles role of plants metabolites and applications. Int J Environ Res Public Health 19(2):674
64. Nejatzadeh F (2021) Effect of silver nanoparticles on salt tolerance of *Satureja hortensis* L., during in vitro and in vivo germination tests. Heliyon 7(2)
65. Nguyen NPU, Dang NT, Doan L, Nguyen TTH (2023) Synthesis of silver nanoparticles: from conventional to 'modern' methods: a review. Processes 11(9):2617
66. Noël JM, Lemineur JF (2021) Optical microscopy to study single nanoparticles electrochemistry: from reaction to motion. Curr Opin Electrochem 25:100647
67. De Oliveira PF, Torresi RM, Emmerling F, Camargo PH (2020) Challenges and opportunities in the bottom-up mechanochemical synthesis of noble metal nanoparticles. J Mater Chem A 8(32):16114–16141
68. Ondrasek G, Rathod S, Manohara KK, Gireesh C, Anantha MS, Sakhare AS, Horvatinec J et al (2022) Salt stress in plants and mitigation approaches. Plants 11(6):717
69. Osorio-Echavarría J, Osorio-Echavarría J, Ossa-Orozco CP, Gómez-Vanegas NA (2021) Synthesis of silver nanoparticles using white-rot fungus *Anamorphous Bjerkandera* sp., R1: influence of silver nitrate concentration and fungus growth time. Sci Rep 11(1):3842
70. Palani G, Trilaksana H, Sujatha RM, Kannan K, Rajendran S, Korniejenko K, Uthayakumar M et al (2023) Silver nanoparticles for waste water management. Molecules 28(8):3520
71. Panahirad S, Morshedloo MR, Ali S, Hano C, Kulak M (2023), Secondary metabolites and their potential roles in plant tolerance against abiotic and biotic stress. Plant Stress 100292
72. Pandit C, Roy A, Ghotekar S, Khusro A, Islam MN, Emran TB, Bradley DA et al (2022) Biological agents for synthesis of nanoparticles and their applications. J King Saud Univ-Sci 34(3):101869
73. Pasinszki T, Krebsz M (2020) Synthesis and application of zero-valent iron nanoparticles in water treatment environmental remediation catalysis and their biological effects. Nanomaterials 10(5):917

74. Patel HV, Mangrola AV, Bariya HS, Patel JD (2020) Antibacterial and larvicidal activity of biologically synthesized silver nanoparticles from *Bambusa arundinaceae* leaves extract. Asian J Biol Life Sci 9(1):43
75. Prakash J, Vignesh K, Anusuya T, Kalaivani T, Ramachandran C, Sudha RR, Devanand Venkatasubbu G et al (2019) Application of nanoparticles in food preservation and food processing. J Food Hyg Saf 34(4):317–324
76. Pryshchepa O, Pomastowski P, Buszewski B (2020) Silver nanoparticles: Synthesis investigation techniques and properties. Adv Coll Interface Sci 284:102246
77. Rasool S, Hameed A, Azooz MM, Siddiqi TO, Ahmad P (2013) Salt stress: causes types and responses of plants. In: Ecophysiology and responses of plants under salt stress, pp 1–24
78. Ratan ZA, Haidere MF, Nurunnabi MD, Shahriar SM, Ahammad AS, Shim YY, Cho JY et al (2020) Green chemistry synthesis of silver nanoparticles and their potential anticancer effects. Cancers 12(4):855
79. Roy S (2023) Anti-microbial activities of nano-polymers for biomedical applications. In: Assessment of polymeric materials for biomedical applications. CRC Press, pp 101–120
80. Sadat-Hosseini M, Naeimi A, Boroomand N, Aalifar M, Farajpour M (2022) Alleviating the adverse effects of salinity on Roselle plants by green synthesized nanoparticles. Sci Rep 12(1):18165
81. Saminathan K (2015) Biosynthesis of silver nanoparticles from dental caries causing fungi *Candida albicans*. Int J Curr Microbiol App Sci 4(3):1084–1091
82. Serrano R, Rodriguez-Navarro A (2001) Ion homeostasis during salt stress in plants. Curr Opin Cell Biol 13(4):399–404
83. Shahidi S (2021) Magnetic nanoparticles application in the textile industry: a review. J Ind Text 50(7):970–989
84. Shaikhaldein HO, Al-Qurainy F, Nadeem M, Khan S, Tarroum M, Salih AM, Alkahtani J et al (2022) Assessment of the impacts of green synthesized silver nanoparticles on *Maerua oblongifolia* shoots under in vitro salt stress. Materials 15(14):4784
85. Shibli R, Mohusaien R, Abu-Zurayk R, Qudah T, Tahtamouni R (2022) Silver nanoparticles (Ag NPs) boost mitigation powers of *Chenopodium Quinoa* (Q6 Line) Grown under in-vitro salt stressing conditions. Water 14(19):3099
86. Singh KR, Natarajan A, Pandey SS (2023) Bioinspired multifunctional silver nanoparticles for optical sensing applications: a sustainable approach. ACS Appl Bio Mater 6(11):4549–4571
87. Srikhao N, Ounkaew A, Srichiangsa N, Phanthanawiboon S, Boonmars T, Artchayasawat A, Kasemsiri P et al (2023) Green-synthesized silver nanoparticle coating on paper for antibacterial and antiviral applications. Polym Bull 80(9):9651–9668
88. Taghiyari HR, Morrell JJ, Husen A (2023) Emerging nanomaterials for forestry and associated sectors: an overview. In: Taghiyari HR, Morrell JJ, Husen A (eds) Emerging nanomaterials. Springer Nature Switzerland AG, Cham, Switzerland, pp 1–16. ISBN: 978-3-031-17377-6. https://doi.org/10.1007/978-3-031-17378-3_1
89. Thuesombat P, Hannongbua S, Ekgasit S, Chadchawan S (2016) Effects of silver nanoparticles on hydrogen peroxide generation and antioxidant enzyme responses in rice. J Nanosci Nanotechnol 16(8):8030–8043
90. Tripathy S, Rodrigues J, Shimpi NG (2023) Top-down and bottom-up approaches for synthesis of nanoparticles, nanobiomaterials: perspectives for medical applications in the diagnosis and treatment of diseases, vol 145, pp 92–130
91. Ullah S, Ali S, Binte Abid A, Nafees M (2022) Modulating response of *Zea mays* to induced salinity stress through application of nitrate mediated silver nanoparticles and indole acetic acid. Microsc Res Tech 85(3):1135–1145
92. Vishwanath R, Negi B (2021) Conventional and green methods of synthesis of silver nanoparticles and their antimicrobial properties. Curr Res Green Sustain Chem 4:100205
93. Wahid I, Kumari S, Ahmad R, Hussain SJ, Alamri S, Siddiqui MH, Khan MIR (2020) Silver nanoparticle regulates salt tolerance in wheat through changes in ABA concentration ion homeostasis and defense systems. Biomolecules 10(11):1506

94. Yang Y, Guo Y (2018) Unraveling salt stress signaling in plants. J Integr Plant Biol 60(9):796–804
95. Yaqoob AA, Umar K, Ibrahim MNM (2020) Silver nanoparticles: various methods of synthesis size affecting factors and their potential applications: a review. Appl Nanosci 10:1369–1378
96. Yugay Y, Rusapetova T, Mashtalyar D, Grigorchuk V, Vasyutkina E, Kudinova O, Shkryl Y et al (2021) Biomimetic synthesis of functional silver nanoparticles using hairy roots of Panax ginseng for wheat pathogenic fungi treatment. Colloids Surf, B 207:112031
97. Zhao X, Xu X, Ai C, Yan L, Jiang C, Shi J (2022) Advantages of silver nanoparticles synthesized by microorganisms in antibacterial activity. In: Green synthesis of silver nanomaterials. Elsevier, pp 571–586
98. Zomorodian K, Pourshahid S, Sadatsharifi A, Mehryar P, Pakshir K, Rahimi MJ, Arabi Monfared A (2016) Biosynthesis and characterization of silver nanoparticles by aspergillus species. Biomed Res Int 2016:5435397
99. Zulfiqar F, Ashraf M (2021) Nanoparticles potentially mediate salt stress tolerance in plants. Plant Physiol Biochem 160:257–268

Stress Responses of Wetland and Aquatic Plants to Silver Nanoparticles

Bushra Ahmad, Arshad Jamil, Sana Shabir, and Ruqia Bibi

Abstract Nanotechnology is an emerging approach gaining attention in the design, production, and manipulation of tiny particles ranging from 1 to 100 nm in size. Among various metal nanoparticles, Silver Nanoparticles (AgNPs) are of utmost importance due to their distinctive capabilities which can play a part as antimicrobial agents, biosensor materials, cosmetics, composite fibers, etc. As previous research concluded, numerous biotic and abiotic factors perform principal roles in evaluating the transportation, toxicity profile, and mechanism of action of AgNPs in water and soil. AgNPs when mixed into land and hydrated environments set serious health issues and environmental hazards leading to instigating stress responses in aquatic and wetland plants. These include poor growth of plant, low quality, and yield of crop due to stress responses induced by AgNPs.

Keywords Silver nanoparticles · Stress response · Biotic factor · Wetland plants · Toxicity

1 Introduction

Nanotechnology is emerging research gaining attention in design, production, and manipulation of tiny particles (1–100 nm). Nanoparticles (NPs) have numerous applications in various domains including food, cosmetics, health, environment, optics, electronics, biomedical sciences, agricultural and forestry sectors, chemical and space industries, pharmaceutical industry, catalysis, single electron transistors, emitters, nonlinear optical devices, light, and photo-electrochemical devices, etc. [5,

B. Ahmad (✉) · S. Shabir · R. Bibi
Department of Biochemistry, Shaheed Benazir Bhutto Women University, Peshawar 29000, Pakistan
e-mail: bushraahmad@sbbwu.edu.pk

A. Jamil
Department of Plant Breeding and Genetics, University of Agriculture, Dera Ismail Khan 25000, Pakistan

A. Husen (ed.), *Plant Response to Silver Nanoparticles*, Smart Nanomaterials Technology, https://doi.org/10.1007/978-981-97-7352-7_12

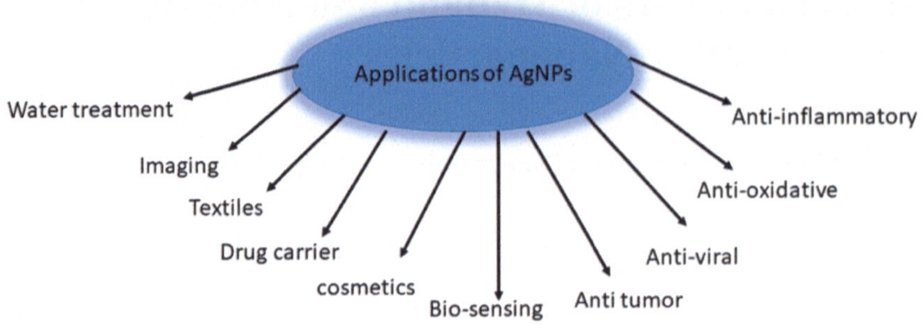

Fig. 1 Applications of AgNP

18–25, 27, 30, 33, 42, 46, 47] (Bachheti et al. 2024). It is designated as a multidisciplinary domain emerging from the tentative use of NPs in biological systems, e.g., biology, chemistry, biochemistry, medicine, engineering, and physics.

Silver NPs (AgNPs) are of utmost significance owing to their unique properties with antimicrobial potentials, biosensor supplies, composite fibers, cosmetic products, cryogenic super-conducting materials, and electronic devices. Silver NPs have been synthesized and stabilized by various physical and chemical methods [28]. Silver nanoparticles (AgNPs) are serving as potent candidates in medical and health domain due to their antimicrobial characteristics. AgNPs may also pose risks as they can significantly control the growth of nitrifiers by approximately 86% (Choi et al. 2008), and 0.01 mg L^{-1} AgNPs altered the microbial composition in waters after exposure of 5 days (Fig. 1).

Being a substantial constituent of the ecosystem, plants also practice the stress-mediated adverse effects of AgNPs. As a result, stress instigates the excessive production of reactive oxygen species (ROS), ultimately affecting seed germination, photosynthesis in leaves, and root growth (Stefanic et al. 2018). A number of research investigations revealed that various biotic and abiotic components often perform major roles in evaluating the toxicity behavior, transport, and fate of AgNPs in water and soil system (Prasher et al. 2022). Moreover, AgNPs released into terrestrial and marine surroundings cause serious health issues and environmental hazards. Synthesis of AgNPs is carried out using various methods. Generally, standard physical and chemical methodologies seem to be very high-priced and risky [6]. Remarkably, biologically synthesized AgNPs are believed to have appreciably high yield, stability, and solubility [6]. Furthermore, it is concluded that biological methods seem to be rapid, simple, and non-toxic [6].

Wetlands and aquatic ecosystems play a crucial role in maintaining environmental balance and supporting diverse plant and animal life. However, the increasing use of nanomaterials, such as silver nanoparticles (AgNPs), has raised concerns about their potential impact on these delicate ecosystems. This chapter explores the stress responses of wetland and aquatic plants to AgNPs, shedding light on the intricate interactions between nanomaterials and plant physiology.

2 Silver Nanoparticles in the Environment

AgNPs are broadly utilized in numerous customer products, including textiles, electronics, and medical devices, leading to their release into water reservoirs. The unique properties that make AgNPs desirable in these applications, such as antimicrobial activity, also raise questions about their potential ecological consequences. Understanding how wetland and aquatic plants react to silver nanoparticle exposure is needed for evaluating the risks accompanying their environmental presence. Further, the nanobiotechnology also serves as a significant technique to develop clean, nontoxic, as well as eco-friendly procedures to manufacture metallic NPs having the ability to reduce metals involved in various metabolic pathways [6, 17, 34, 45].

3 Wetland Plant

Wetland plants are hydrophytes (hydro = water, phyte = plant). Wetland plants are usually believed to grow in all wetlands, or where soils are waterlogged or saturated for anaerobic conditions to be happened in the root zone. These plants grow in water or on soil that however occasionally is deficient in oxygen due to too much water content. Hydrophytes have physiological, morphological, and reproductive adaptations that make them flourish in saturated soils where non-hydrophytes (upland plants) cannot survive. Even though wetland plants are considered by their capability to restrain wet places, they represent a diverse accumulation of species with different adaptative features and ecological tolerances. Some of the adaptative features acquired by wetland plant species are similarly found in relevant terrestrial species; however, many characteristics are distinctive, or if common, have attained a high degree of specialization [9].

4 Aquatic Plants

Any kind of plant which exists and grows in water is termed an aquatic plant. They can be found in all varieties of water including freshwater and seawater. They are also known as hydrophytes or macrophytes to differentiate them from microphytes and algae. A macrophyte can grow in water as merged, submerged, or floating body. In rivers and lakes macrophytes spread as cover for fish, or they may act as substrate for aquatic invertebrates, yield oxygen, and can be a food source for fish [9]. Aquatic plants demand special adaptative features for staying submerged in water. The prominent feature is the occurrence of lightweight packing cells, and aerenchyma, but floating and finely dissected leaves are also frequent [7, 26, 39]. They are a familiar element of wetlands. Aquatic plants serve as a food source, carbon dioxide removal and oxygen yield by photosynthesis. Aquatic plants also give numerous items for use. Some of these include rice, blueberries, cranberries, fiber for rope, herbs, medicinal phytoconstituents, and decorative items such as flowers and colorful berries and fruits for ornamental purposes.

5 Stress Responses

Stress can be identified as any exterior and interior factor that reduces the rate of photosynthesis and decreases the energy conversion capacity of a plant to a relevant product []. Stress responses of a plant are diverse including alterations in gene expression, growth rates, metabolism, yield, etc. Plants show stress due to their response to various environments. Still, bearing a certain stress by stress-tolerant plant species may cause resistance with the passage of time [12]. Plants that live in rivers are exposed to several types of environmental stresses including flooding, waves, high salt levels, and low soil oxygen.

6 Abiotic Stress

Abiotic stress is supposed to be an environmental factor that plants face, consequently due to various chemical or physical stress []. Plants are exposed to many different abiotic stresses, which pose different impacts on crop yield globally. These involve salt, drought, heat, cold, toxins, etc. (Fig. 2).

Abiotic stress can be explained as how lifeless components destructively influence living in a distinct environment. Salts, heavy metals, drought, harsh temperatures, etc., are considered a few possible stresses that are critically unavoidable globally. Owing to climate change mechanisms, plants are becoming vulnerable to abiotic stresses. Climate change due to global warming phenomenon is complemented by a prominent rise in the incidence and harshness of drought, heat waves, and rest of the abiotic stresses, e.g., Salinity, flooding, and freezing [49]. Over and above natural phenomena, human-caused agitations of the environment, such as enhanced industrialization, vast mining, and intensive agriculture, tangled with growing inhabitants and fast urbanization, have caused devastations in global warming, thus incidentally instigating abiotic stress and the severe adulteration of the vital life on the earth []. Plants usually tolerate the abiotic stress [50]. Stress responses are multifaceted, integrated routes, linear routes, integrating several reactions, and interactions with extra cofactors or signaling molecules to instigate certain responses to a stimulus []. The parts of plants that are bearing stress will reveal the plant response. Besides this, the complication of the response can be significantly changed by the strength and extent of the stress [8]. Plants trigger early stress-signaling in reply to abiotic stresses to respond to stress reactions and boost tolerance [35, 36]. When a plant cell reveals stress, second messengers, e.g., calcium, phospholipids, ROS, and nitric oxide (NO), various protein kinases, transmit and amplify the signal [].

Fig. 2 Abiotic stress responses

7 Biotic

Biotic stress encompass biological incidence, e.g., illnesses, insects, and other pests that are exposed to crops. Some stresses trigger plant injuries. These plants usually face various metabolical issues. Biotic stress in plants is instigated by living beings, e.g., viruses, fungus, bacteria, nematodes, insects, weeds, etc. The agents that induce biotic stress make their host deficient in nutrients, ultimately leading to plant death. Due to pre- and post-harvest damages, biotic stress can become terrible for plants (Fig. 3).

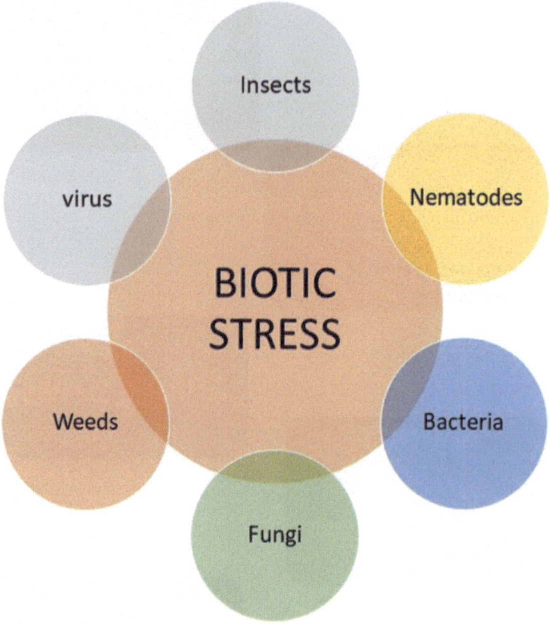

Fig. 3 Biotic stress responses

8 Stress Responses of Wetland and Aquatic Plants to AgNPs

Nanotechnology is a recently evolved field that progressed during past years hence executed in a variety of industrial uses, which caused a predictable release of nanomaterials obsessed by the environmental system. AgNPs are among the most familiar nanoformulations in numerous fields, markedly in farming. Green inheritance is the chief contributor to the ecosystem and the utmost significant food source for manhood; consequently, considering the influences of AgNPs on the growth and development of plants is key for the estimation of environmental hazards on food security and human health. The fruitful influences of silver nanoparticles are examined in several economy-based sectors, including the field of agriculture. The scientific novelty of AgNPs in plant biotechnology and agriculture has proved potent efficiency in past decades. It is believed not just to improve seed germination and plant growth, but furthermore boost the energy productivity of photosynthesis. AgNPs play a central role in the domain of agriculture through numerous approaches that are necessary to ensure food security and enhance crop yield. Likewise, these AgNPs can also be used as nano-pesticides, keeping appropriate dosage to the desired plants with no excessive pesticides release in the environment. Nanofertilizers gradually provide nutritious moiety to the plants, in this manner inhibiting excessive nutrient shortage. AgNPs are applied for applicable and safe pest management, compelling them a tremendous implement to combat pests securely. They merge either non-biodegradable or edible polymers for effective food packaging. Besides this, AgNPs also have antimicrobial characteristics that safeguard plants from destructive microorganisms. AgNPs are promising tools for efficient seed germination. Numerous prior studies showed their positive impact on the growth of the plant. The suitable dose of AgNPs has appreciably enhanced the growth and crop production. It was concluded in a flurry of literature that AgNPs beneficially enhanced crops with negligible phototoxic consequences on the growth of the plant. On the opposite hand, negative effects were also described in few plant species.

9 Positive Impacts on Growth of the Plant

In agriculture research, AgNPs are reported as antimicrobials to treat bacterial, viral, and fungal infection. It conserves the shelf life of all plant parts and encourages growth and metabolism in plants. The reactions of plants to the AgNPs are different, dependent on the AgNPs exposure interval, plant species, composition, concentration, particle size, and numerous relevant attributes [16].

10 Negative Impacts on Growth of the Plant

The toxicity profile of AgNPs to numerous plants inhabiting land-dwelling ecosystems, notably crops, depends on the size and concentration of these nanoparticles. It decreases seed germination and growth and also influences the length and mass of roots and shoots. Its addition this initiates the defense mechanism at cellular/tissue levels and alters proteomic expression, metabolism, and antioxidant potential. Plant-derived components, e.g., ROS production, chlorophyll, superoxide dismutase activities, elevated glutathione, H_2O_2, proline, ascorbate, carotenoid, etc., were either enhanced or diminished following AgNPs exposures. These activities initiate the inhibition of photosynthesis, random morphological variations, etc. [4].

11 AgNPs as Nano-fertilizers

The consumption of AgNPs as nano-fertilizer is thought to be a highly fruitful tactic to sufficiently manage phytonutrient distribution. Extreme fertilizers were wasted in the field terribly disturbing the surrounding environment. Nanoparticles become attached to the fertilizers and distribute them properly to the plants. Hence, the overuse of fertilizer was reduced in the field. The manufacturing of Onion-derived nano fertilizers was reported to be efficient for tomato and brinjal plants. It assists the plant in enhancing the absorption of valuable nutrients from the soil []. The nano-biofertilizer mitigated the farm management costs, environmental contamination, and useless chemical-based fertilizers in the field [13].

12 AgNPs as Nano-pesticides

Pests always remain an evolving hazard for the agriculture sector, e.g., reduced yield and poor quality of crops, along with increased use of synthetic pesticides for plants, have destructive effect on the environment. Nano pesticide is a recently developed tool to control pests with agricultural concern. AgNPs serve as potent pesticides ultimately helping to control pest-related issues. The nanocides (nanotechnology-based pesticides) are significantly important owing to their eco-friendly relationship to the environment and negligible effect on non-targeted insects. It enhances the nutritional value of the plants and gives defense against pests. Thus, persistent use of fertilizer is lowered in typical farming. It also eliminates detrimental microbial species in soil. The use of foliar spray is favorable in preventing rot, fungi, molds, and several other infectious diseases of plants [3]. In another study, AgNPs were manufactured from *Ulva lactuca* that have the capability of reduction of silver nitrate to AgNPs, and the appearance of brown color approves their occurrence in the solution. The anti-microbial efficacy of biologically synthesized pesticides proved their protective role against microbes, triggering numerous crop illnesses [1].

13 Role of AgNPs in Crops and Foods Protection

Plants are extremely influential in maintaining a healthy life, but they obtain less importance in health literature. This green inheritance is the chief source of nutrition for livestock and provides approximately 80% of human nutrition. Pest invasion and plant diseases compromise plant accessibility and protection for human and animal utilization. Worldwide, damages of critical staple crops make up 30% [37]. Nanotechnology is a significantly appreciable and secure practice, giving healthier food and enlightening precision farming (Gruere 2021). They can eradicate risky microbial species from hydroponics, soil, and plants by protecting plants in various ways, for instance appearing as nano-antifungal nano-antiviral, antibacterial.

14 Conclusion

As the use of nanomaterials grows, accepting their impact on wetlands and aquatic ecosystems becomes progressively more important. This chapter provides a comprehensive overview of the stress responses of wetland and aquatic plants to silver nanoparticles, offering valuable insights for researchers, policymakers, and environmentalists working toward sustainable solutions for nanoparticle-related challenges in aquatic environments.

References

1. Amin HH (2020) Biosynthesized silver nanoparticles using Ulva lactuca as a safe synthetic pesticide (in vitro). Open Agric 5(2020):291–299. https://doi.org/10.1515/opag-2020-0032
2. Amulyavichus A, Daugvila A, Davidonis R, Sipavichus C (1998) Study of chemical composition of nanostructural materials prepared by laser cutting of metals. Fiz Met Metalloved 85:111–117
3. Babu M, Devi V, Ramakritinan CM, Umarani R, Taredahalli N, Kumaraguru AK (2013) Application of biosynthesized silver nanoparticles in agricultural and marine pest control. Curr Nanosci 10:374–381. https://doi.org/10.2174/157341371130966601O3
4. Budhani S, Egboluche NP, Arslan Z, Yu H, Deng H (2019) Phytotoxic effect of silver nanoparticles on seed germination and growth of terrestrial plants. J Environ Sci Health Part C Environ Carcinog Ecotoxicol Rev 37:330–355. https://doi.org/10.1080/10590501.2019.1676600
5. Chawla S, Singh S, Husen A (2023) Smart nanomaterials targeting pathological hypoxia. Springer Nature Singapore Pte Ltd., Singapore. ISBN: 978-981-99-1717-4. https://doi.org/10.1007/978-981-99-1718-1
6. Colvin VL, Schlamp MC, Alivisatos A (1994) Light emitting diodes made from cadmium selenide nanocrystals and a semiconducting polymer. Nature 370:354–357
7. Cook CDK (ed) (1974) Water plants of the world. Dr. W. Junk Publishers, The Hague. ISBN: 90-6193-024-3
8. Cramer GR, Urano K, Delrot S, Pezzotti M, Shinozaki K (2011) Effects of abiotic stress on plants: a systems biology perspective. BMC Plant Biol 11:163
9. Cronk JK, Fennessy MS (2014) Macrophytes as Indicators of freshwater marshes in Florida. In: Encyclopedia of inland waters, p 200928
10. Deepak V, Umamaheshwaran PS, Guhan K, Nanthini RA, Krithiga B, Jaithoon NM, Gurunathan S (2011) Synthesis of gold and silver nanoparticles using purified URAK. Colloid Surface B 86:353–358. https://doi.org/10.1016/j.colsurfb.2011.04.019
11. Elsupikhe RF, Shameli K, Ahmad MB, Ibrahim NA, Zainudin N (2015) Green sonochemical synthesis of silver nanoparticles at varying concentrations of κ-carrageenan. Nanoscale Res Lett 10:302. https://doi.org/10.1186/s11671-015-0916-1
12. Fahad S, Bajwa AA, Nazir U, Anjum SA, Farooq A, Zohaib A et al (2017) Crop production under drought and heat stress: plant responses and management options. Front Plant Sci 29:01147
13. Gosavi V, Daspute A, Patil A, Gangurde A (2020) Synthesis of green nanobiofertilizer using silver nanoparticles of Allium cepa extract short title: green nanofertilizer from Allium cepa S.W. IJCS. 0.22271/chemi.2020.v8.i4q.9854
14. Gurav AS, Kodas TT, Wang LM, Kauppinen EI, Joutsensaari J (1994) Generation of nanometer-size fullerene particles via vapor condensation. Chem Phys Lett 218:304–308. https://doi.org/10.1016/0009-2614(93)E1491-X
15. Guzman-Baez GA, Trejo-Tellez LI, Ramírez-Olvera SM, Salinas-Ruiz J, Bello-Bello JJ, Alcantar-Gonzalez G, Hidalgo-Contreras JV, Gómez-Merino FC Gruère GP (2012) Implications of nanotechnology growth in food and agriculture in OECD countries. Food Pol 37:191–198. https://doi.org/10.1016/j.foodpol.2012.01.001
16. Guzman-Baez GA, Trejo-Tellez LI, Ramírez-Olvera SM, Salinas-Ruiz J, Bello-Bello J, Alcantar-Gonzalez G, Hidalgo-Contreras JV, Gomez-Merino FC (2021) Silver nanoparticles increase nitrogen, phosphorus, and potassium concentrations in leaves and stimulate root length

and number of roots in tomato seedlings in a hormetic manner. Dose Response 19:1–15. https://doi.org/10.1177/15593258211044576
17. Hamilton JF, Baetzold R (1979) Catalysis by small metal clusters. Science 205:1213–1220
18. Husen A (2019) Natural product-based fabrication of zinc oxide nanoparticles and their application. In: Husen A, Iqbal M (eds) N*anomaterials* and plant potential. Springer International Publishing AG, Cham, pp 193–291. https://doi.org/10.1007/978-3-030-05569-1_7
19. Husen A (2022) Engineered nanomaterials for sustainable agricultural production, soil improvement and stress management. Elsevier Inc., Cambridge, MA, USA. https://doi.org/10.1016/C2021-0-00054-7
20. Husen A (2023) Secondary metabolites based green synthesis of nanomaterials and their applications. Springer Nature Singapore Pte Ltd., Singapore. https://doi.org/10.1007/978-981-99-0927-8
21. Husen A (2023) Nanomaterials from agricultural and horticultural products. Springer Nature Singapore Pte Ltd., Singapore. https://link.springer.com/book/9789819934348
22. Husen A (2023) Nanomaterials and nanocomposites exposures to plants. In: Response, interaction, phytotoxicity and defense mechanisms. Springer Nature Singapore Pte Ltd., Singapore. https://doi.org/10.1007/978-981-99-2419-6
23. Husen A, Bachheti RK, Bachheti A (2023) Current trends in green nano-emulsions. In: Food, agriculture and biomedical sectors. Springer Nature Singapore Pte Ltd., Singapore
24. Husen A, Iqbal M (2019) Nanomaterials and plant potential. Springer International Publishing AG, Cham, Switzerland. https://doi.org/10.1007/978-3-030-05569-1
25. Husen A, Siddiqi KS (2023) Advances in smart nanomaterials and their applications. Elsevier Inc., Cambridge, MA, USA. https://doi.org/10.1016/C2021-0-02202-1
26. Hutchinson GE (1975) A treatise on limnology, 3, Limnological botany. John Wiley, New York
27. Jin-Chul K, Madhusudhan A, Husen A (2021) Smart nanomaterials in biomedical applications. Springer Nature Switzerland AG, Cham, Switzerland. https://doi.org/10.1007/978-3-030-84262-8
28. Klaus-Joerger T, Joerger R, Olsson E, Granqvist CG (2001) Bacteria as workers in the living factory: metal-accumulating bacteria and their potential for materials science. Trends Biotechnol 19:15–20
29. Kruis FE, Fissan H, Rellinghaus B (2000) Sintering and evaporation characteristics of gas-phase synthesis of size-selected PbS nanoparticles. Mater Sci Eng B 69:329–334. https://doi.org/10.1016/S0921-5107(99)00298-6
30. Madhusudhan A, Purohit SD, Prasad R, Husen A (2024) Functional smart nanomaterials and their Theranostics approaches. Springer Nature Singapore Pte Ltd., Singapore
31. Magnusson MH, Deppert K, Malm JO, Bovin JO, Samuelson L (1999) Size-selected gold nanoparticles by aerosol technology. Nanostruct Mater 12:45–48. https://doi.org/10.1016/S0965-9773(99)00063-X
32. Mallick K, Witcomb MJ, Scurrell MS (2004) Polymer stabilized silver nanoparticles: a photochemical synthesis route. J Mater Sci 39:4459–4463. https://doi.org/10.1023/B:JMSC.0000034138.80116.50
33. Mandal AK, Ghorai S, Husen A (2023) Functionalized smart nanomaterials for point-of-care testing. Springer Nature Singapore Pte Ltd., Singapore
34. Mansur HS, Grieser F, Marychurch MS, Biggs S, Urquhart RS, Furlong D (1995) Photoelectrochemical properties of 'q-state' cds particles in arachidic acid langmuir-blodgett films. J Chem Soc Faraday Trans 91:665–672
35. Mohanta TK, Bashir T, Hashem A, Abd_Allah EF, Khan AL, Al-Harrasi AS (2018) Early events in plant abiotic stress signaling: interplay between calcium, reactive oxygen species and phytohormones. J Plant Growth Regul 37:1033–1049
36. Molassiotis A, Fotopoulos V (2011) Oxidative and nitrosative signaling in plants: two branches in the same tree? Plant Signal Behav 6:210–214
37. Rizzo DM, Lichtveld M, Mazet JAK, Togami E, Miller SA (2021) Plant health and its effects on food safety and security in a one health framework: four case studies. One Heal. Outlook 3. https://doi.org/10.1186/s42522-021-00038-7

38. Schmidt-Ott A (1988) New approaches to in situ characterization of ultrafine agglomerates. J Aerosol Sci 19:553–563. https://doi.org/10.1016/0021-8502(88)90207-8
39. Sculthorpe CD (1967) The biology of aquatic vascular plants. Reprinted 1985 Edward Arnold, by London
40. Shameli K, Ahmad MB, Yunus WM, Rustaiyan A, Ibrahim NA, Zargar M, Abdollahi Y (2010) Green synthesis of silver/montmorillonite/chitosan bionanocomposites using the UV irradiation method and evaluation of antibacterial activity. Int J Nanomed 5:875–887. https://doi.org/10.2147/IJN.S13632
41. Shameli K, Ahmad MB, Yunus WMZW, Ibrahim NA, Gharayebi Y, Sedaghat S (2010) Synthesis of silver/montmorillonite nanocomposites using γ-irradiation. Int J Nanomed 5:1067–1077. https://doi.org/10.2147/IJN.S15033
42. Taghiyari HR, Morrell JF, Husen A (2023) Emerging nanomaterials. In: Opportunities and challenges in forestry sectors. Springer Nature Switzerland AG, Cham, Switzerland. https://doi.org/10.1007/978-3-031-17378-3
43. Tao A, Sinsermsuksakul P, Yang P (2006) Polyhedral silver nanocrystals with distinct scattering signatures. Angew Chem Int Ed 45:4597–4601. https://doi.org/10.1002/anie.200601277
44. Tsuji M, Hashimoto M, Nishizawa Y, Kubokawa M, Tsuji T (2005) Microwave-assisted synthesis of metallic nanostructures in solution. Chem Eur J 11:440–452. https://doi.org/10.1002/chem.200400417
45. Wang YG, Herron N (1991) Nanometer-sized semiconductor clusters: materials synthesis, quantum size effects, and photophysical properties. J Phys Chem 95:525–532
46. Wassie AT, Ahmed IN, Bachheti A, Husen A, Bachheti RK (2024) Plant extract-assisted green synthesis and characterization of copper nanocomposite. In: Bachheti RK, Bachheti A, Husen A (eds) Metal and metal-oxide based nanomaterials. Smart nanomaterials technology. Springer, Singapore, pp 43–66. https://doi.org/10.1007/978-981-99-7673-7_3
47. Wassie AT, Pandey DP, Bachheti A, Husen A, Deepti Singh Vashishth, Bachheti RK (2024) Use of green synthesized platinum nanoparticles for biomedical applications. In: Bachheti RK, Bachheti A, Husen A (eds) Metal and metal-oxide based nanomaterials. Smart nanomaterials technology. Springer, Singapore, pp 271–285 https://doi.org/10.1007/978-981-99-7673-7_13
48. Wiley B, Sun Y, Mayers B, Xia Y (2005) Shape-controlled synthesis of metal nanostructures: the case of silver. Chemistry 11:454–463. https://doi.org/10.1002/chem.200400927
49. Zandalinas SI, Fritschi FB, Mittler R (2021) Global warming, climate change, and environmental pollution: recipe for a multifactorial stress combination disaster. Trends Plant Sci 26:588–599
50. Zhang H, Zhu J, Gong Z, Zhu JK (2022) Abiotic stress responses in plants. Nat Rev Genet 23:104–119

Effects of Silver Nanoparticles on Wetland Plant and Soil Systems

Zahra Tariq, Tabinda Shahzadi, Hijab Fatima, Irha Mujeeb, Sara Maham, Areej Imran, M. Iftikhar Hussain, Fatima, and Sumaira Anjum

Abstract Wetland plants, commonly known as hydrophytes, are present in the areas flooded with water or having saturated anaerobic conditions. The corresponding soil systems of different plant species differ with respect to the physiological and morphological conditions. A diverse range of parameters including the hydrology, temperature fluctuations, soil characteristics, light conditions, oxygen availability, redox potential, pH, ion solubility, microbiome, and nutrient cycling influence the growth of wetland plants. The maintenance of the optimum conditions is crucial in investigating the important processes occurring in the soil–plant systems. Latest developmental approaches have allowed the researchers to investigate the role of important nano-technological applications in regulating the morpho-physiological conditions. The role of metallic nanoparticles, most importantly silver nanoparticles (AgNPs) has been thoroughly studies in the terrestrial plants. However, not much research have been conducted on wetland plants and soil systems. Rigorous studies are being conducted to demonstrate the role of AgNPs in different wetland plant species along with their impact on the soil system especially the inhabited microbial community. Having both positive and negative impacts, AgNPs have been found to play important roles in the nutrient uptake and translocation, bioaccumulation and phytoremediation, growth and development, and seed germination and fruit production in different wetland plant species. However, massively induced phytotoxicity has also been found with the potential application of AgNPs. This book chapter focuses on the diverse effects on AgNPs on different species of the wetland plants.

Keywords Wetland plants · Hydrophytes · Nutrients · Silver nanoparticles · Growth · Bioaccumulation · Phytotoxicity

---

Z. Tariq · T. Shahzadi · H. Fatima · I. Mujeeb · S. Maham · A. Imran · S. Anjum (✉)
Department of Biotechnology, Kinnaird College for Women, 92-Jail Road, Lahore 54000, Pakistan
e-mail: Sumaira.anjum@kinnaird.edu.pk

M. I. Hussain
Department of Plant Biology and Soil Science, Universidad de Vigo, Campus As Lagoas Marcosende, 36310 Vigo, Spain

Fatima
School of Biochemistry and Biotechnology, University of the Punjab, Lahore 52254, Pakistan

A. Husen (ed.), *Plant Response to Silver Nanoparticles*, Smart Nanomaterials Technology, https://doi.org/10.1007/978-981-97-7352-7_13

1 Introduction

Wetland plants are defined as plant species that grow in the wetlands and are commonly termed as hydrophytes. These plants grow in or on the water; requiring the soils to be either completely flooded or saturated for the anaerobic conditions to be developed in the root zones [26]. The wetlands could be classified as bogs and fens based on the soil type and as marshes and swaps based on the soil nutrient contents [16]. Although wetlands have many common features with other aquatic and terrestrial ecosystems, yet, the distinctive features of the wetland ecosystem are the anaerobic soil and wetland plants [90, 132]. The evolution of wetland plants has allowed the development of structurally and physiologically adaptive features to survive in anaerobic soil and water-dominated environments [43, 116].

The plant species inhabited in different types of wetlands could be classified as floating, floating-leaved, emergent, or submerged [135]. Other terms used for hydrophytes are macrophytes (visible to the naked eye), linnophytes (freshwater plants), helophytes (emergent plants), amphiphytes (plants growing in both land and water), and amphibious species [17, 26]. The growth and development of these plants are based on distinctive environmental characteristics including hydrology, temperature fluctuations, soil characteristics, light conditions, oxygen availability, redox potential, pH, ion solubility, and nutrient cycling [12, 62, 79, 100, 114]. All these species play significant roles in biodiversity maintenance, water purification, carbon sequestration, nutrient accumulation, and flood mitigation [114].

Soil is the chief nutrient reservoir for wetland plants playing a significant role in the vegetation processes including nutrient uptake and interaction with agents of disturbance or stress [15, 123]. The fluctuating moisture, gaseous, and nutrient contents of varying soil systems are crucial for the growth and development of the plant species in the wetlands [32]. Of particular importance for the survival of the plants are the nutrient cycling processes involving ammonification, nitrification, denitrification, methanogenesis, and methanotrophy which are largely dependent on the microbial ecology and wetland biogeochemistry (including soil systems and wetland ecosystems) [44, 109]. Contrastingly, the plants contribute to the soil through their mineral sediment deposition and retention properties and by organic matter contributions to soil volume [18].

The interplay of plant species and the soil microbiota plays a significant role in nutrient cycling, biomass production, and other biogeochemical processes including the release of oxygen and organic compounds from rhizosphere and regulating methane to contribute to plant growth and development [44, 76, 100, 108, 123]. Different plants might respond differently to the available nutrients, metals, enzymes, and pH. A number of plant species might be hyperaccumulators with adaptive features to protect the plants from higher metal contents while others may have increased root exudates in lower nutrient availability [92, 100, 142]. Similarly, the amount and composition of the organic acids released by plants may also differ significantly depending on the enzymes and the nutrients provided in the given wetland and soil ecosystems [142].

The potentially stressful wetland environment may pose varying threats to a number of plant species [92]. Different treatment approaches have been adopted over the years to investigate the growth and developmental characteristics of different wetland plants. Scientists have found the effective use of additives such as fertilizers, nitrogen, phosphorus, potassium, and antibiotics to be beneficial in the growth and development of these plants [128]. Although, the impact of these additives has been found to be slow, or less effective by various other researchers, advanced approaches have been developed especially using non-anthropogenic sources of introducing nutrients in the plant–soil systems [46, 66].

Over past few years, nanotechnological approaches have been developed and are being implemented in various disciplines [8, 22, 48–56, 60, 83, 84, 121]. Nanotechnology encompasses the development of nanomaterials that range in size smaller than 100 nm in at least one dimension [5, 91]. These are highly versatile and unique in their characteristics having chemical, electrical, optical, magnetic, and biological properties including non-invasive, simple, and portable nature. The nanomaterials exhibit different advantages having high surface area to volume ratio, ease of manipulation, small size, and high sensitivity and selectivity [77, 131]. Novel nanostructures include polymeric micelles, polymeric nanoparticles, dendrimers, polymer-drug conjugates, liposomes, carbon nanomaterials, gold nanoparticles, quantum dots, magnetic nanoparticles, nanographene and metal-based nanoparticles including copper, iron, nickel, titanium, and silver etc. nanoparticles [88, 122]. The use of various nanomaterials has rigorously being employed for the development of wetland plants and soil systems. Investigators have shown effective use of metal-based nanomaterials such as copper, iron, nickel, titanium, and silver nanoparticles in regulating the plant biogeochemical processes [27, 58, 103, 151]. Subsequent studies have shown the effective use of these metal nanomaterials on the enzymatic, microbial, bioavailability, and nutrient uptake efficacy [35, 39]. Other studies have also shown the potent harmful and toxic effects of these nanomaterials in various hydrophytes [4]. Studies specifically encompassing the use of silver nanoparticles on different wetland plant species in differing soil systems have shown contrasting results (Fig. 1). This chapter focuses on the effects of silver nanoparticles on the nutrient uptake and translocation, bioaccumulation and phytoremediation, growth and development, and seed germination and fruit production in different wetland plant species. The toxic effects of these nanoparticles have also been discussed.

2 Silver Nanoparticles

Engineered silver nanoparticles enter the plant–soil systems mainly through the application of nanopesticides, or by wastewater and sludge treatments [137]. AgNPs have a size range from 1 to 100 nm and contain 20–15,000 Ag atoms. AgNPs could be synthesized by biosynthetic, chemical, and physical methods, such as the micelle method, chemical precipitation, hydrothermal, pyrolysis, tube furnace condensation, or using microbial enzymes or plants [61, 157]. The physical and biological methods

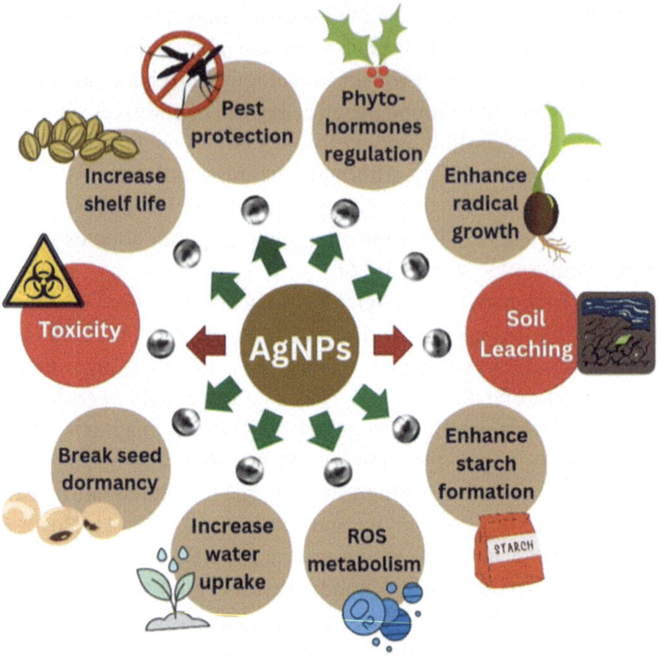

Fig. 1 Positive (green arrow) and negative (red arrow) effects of silver nanoparticles (AgNPs) on the wetland plants

do not employ any harmful chemicals, controlled speed, or the use of radiation as reducing agents, however, the drawbacks of the physical method appear to be low yield and high energy consumption as compared to the biosynthetic method having higher efficacy through the use of microbial enzymes or by the production of plant mediated AgNPs [101, 156]. Contrastingly, the chemical methods use water, organic solvents, reducing agents, metal precursors, and stabilizing agents for the preparation of AgNPs [28, 120, 157].

Many unique physicochemical properties, such as electrical and thermal conductivity, surface-enhanced Raman scattering, chemical stability, catalytic activity, and nonlinear optical behavior make silver nanoparticles unique compared to bulk Ag (Fig. 2) [37, 139]. The size of AgNPs is important for the effectiveness and phytotoxicity of AgNPs [136]. Smaller-sized AgNPs have a larger surface area to mass ratio allowing better interference with cell membrane function by directly reacting with the membrane [136, 144]. It was shown that smaller AgNPs could accumulate to higher levels in plants and be more toxic than their larger counterparts. Studies checked the impact of AgNPs with different sizes (20, 40, and 80 nm) on Arabidopsis and found that smaller AgNPs accumulated more in seeds at lower concentrations than large AgNPs (20 nm > 40 nm > 80 nm). Moreover, smaller-sized AgNPs are more susceptible to root browning [38].

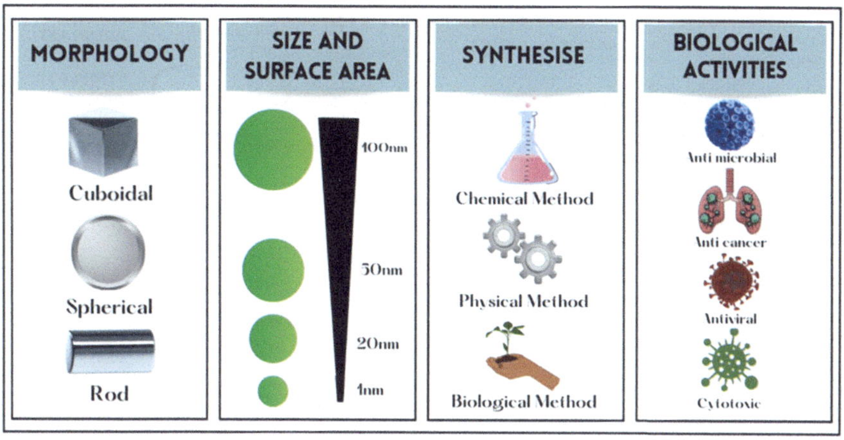

Fig. 2 Diagrammatic representation of classification, synthesis, and biological activities of silver nanoparticles (AgNPs)

Silver nanoparticles are commonly used ENPs in consumer products where they are increasingly used for their antimicrobial properties. Pure culture studies have shown that they might induce toxicity in different animal and human cells [133], fish [65], and algae species [94]. However, to date, very less studies have reported the effects of AgNPs on vascular plants, including wetland plants. These studies have shown that the effects on AgNPs can have positive and negative impacts on plants depending on concentration, exposure time, and specific plant species [59].

3 Uptake and Translocation of AgNPs in Plants

Nanoparticles primarily enter the plants through the soil and the associated food supply system [130]. In plants, AgNPs are transported through both the intercellular spaces and vascular tissues with the former being used for short transport while the latter for longer transport. In other cases, the aquaporin or ion channels in the plasma membrane may be used for the transport of AgNPs [156]. After the entry of AgNPs fundamentally in the cell walls of the plant cells following the apoplastic or symplastic pathway, they are found aggregated at plasmodesmata of the root epidermis. This occurs through a series of events that let them enter the vascular tissue, i.e., xylem of the plant and transport to the stele (Fig. 3) [21, 87]. The individual AgNP molecules could enter the cell easily as compared to the aggregated or co-exposed molecules such as AgNPs-Sb or sulfide-AgNPs [156].

The uptake of AgNPs is followed by the distribution and accumulation in different parts of the plants including cell wall, plasmodesmata, and intercellular spaces, as well as present in the cytoplasm, mitochondria, vacuoles, and chloroplast [73]. This could be manifested by the form of the nanoparticles with some being Ag ions, or

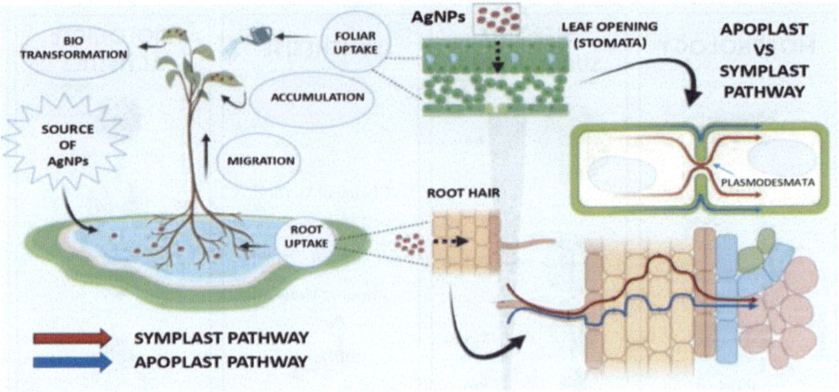

Fig. 3 Uptake, translocation, and accumulation of silver nanoparticles (AgNPs) in the wetland plants by apoplastic and symplastic pathways

zero-valent Ag (Ag^0) forms [10, 74, 117, 126, 156]. The biotransformation of the AgNP occurs in the plants to convert it into other forms such as Ag^0, Ag_2O, Ag_2S, and cysteine complex, etc. by the occurrence of redox reactions in the presence of oxidative conditions in the plants as compared to the conversion of free Ag^+ ions into Ag or nano-metallic particles in reducing conditions [23, 149, 156]. The beneficial role of iron plaques on the roots of *Oryza sativa* L. in the uptake and translocation of Ag causing oxidation of AgNPs into ionic Ag species has been demonstrated in various studies [64, 143, 150]. A study on Ag^0 and Ag_2S NPs, or $AgNO_3$ exposure to *Landoltia punctate* and *Oryza sativa* L. indicated the accumulation of similar Ag concentrations in the roots regardless of the form of Ag exposed to the plants [119, 145].

AgNPs provide the plants with protection against plant diseases and pests by foliar exposure [72]. The roots of the plants might be exposed to the AgNPs in untreated soil containing abundant AgNPs or in agricultural soil containing treated sludge or sewage wastewater [33]. The waxy hydrophobic cuticle having small polar pores (~2 nm) allows their direct entry through the cuticle into the tissues. Larger particles might be absorbed through the open pores or they might enter the cytoplasm via endocytic uptake, pore formation, or wounds. However, direct evidence is still lacking [1, 112]. AgNP uptake is regulated by stoma size, density, and cuticle thickness. Leaf stomata in abundance facilitate the quick and effective absorption of nanoparticles. Recent reports indicate the important role of stomata in the production of AgNPs in leaves [46, 81].

4 Effects of AgNPs on Ag Bioaccumulation and Phytoremediation in Wetland Plants

Certain aquatic plants play a beneficial role in the uptake and accumulation of various toxic and non-toxic metallic substances thus having an increased potential to remediate the metal-contaminated water and sediments [45]. These plants carry out phytoremediation through the phytoextraction of the contaminants from the surroundings and storing them into plant biomass [31, 45]. With the exception of freshwater submerged plant *Ceratophyllum demersum* most wetland plants are not hyper-accumulators making it easy to prevent the recycling of toxic metals on the decomposition of the plants [140]. Bao et al., in their study, indicated the important role of the plants in removing the AgNPs from the contaminated water with a significant impact on the plant biomass, root activity, peroxidase activity of leaves and biofilm biomass [11].

In a study conducted on different soil systems to assess the adsorption and uptake of wetland plants *Phragmites australis*, sand, gravel, and zeolite were used to examine the biofilm formation and activity of the plants. Sand and zeolite were found to be efficient in removing the citrate-coated AgNPs from the synthetic wastewater [7, 113]. High adsorption of AgNPs was found for a specific time which was followed by the occurrence of desorption indicating a maximum adsorption limit [113]. Silver accumulation in the belowground plant tissues in the absence of rhizosediments was found in another study on *P. australis* [36]. Yang et al. examined the role of mycorrhizal fungus *Funneliformis mosseae* on *P. australis* which indicated the significantly reduced Ag accumulation and translocation in AgNP-exposed plants versus Ag^+-exposed plants and inoculated plants having higher Ag accumulative rate and root partitioning (Ag organ distribution: root > stem > leaf) as compared to non-inoculated plants [146].

Pistia stratiotes were used to study their effective phytoremediation capabilities to treat the AgNP and Ag^+ contaminated wastewaters according to which their role as potential phytoremediators to remove the heavy metal nanoparticles and their counter ions was found [45]. Contrastingly, *Lemna gibba* was found to be more efficient than *Hydrilla verticillata* and *Pistia stratiotes* in removing the heavy metal effluents from the wastewater from the tanneries contaminated with AgNPs in another study [115]. Significant bioaccumulation of AgNPs was found in (i) roots of the *Oryza sativa* which was impacted by the surface charges of the rice root exudates [148], (ii) aquatic macrophyte *Salvinia auriculata* examined by the tolerance index, uptake index, silver removal, biomass production, bioconcentration factor, and translocation factor [98], (iii) aquatic fern Azolla filiculoides with AgNP deposited on roots and fronds subphases and accumulated in the endodermis [78] and (iv) *Iris pseudacorus* [47].

The microbial community with the interaction of the plant systems also plays a significant role in carrying out the phytoremediation processes following a series of biochemical processes: (i) compartmentalization to tolerate the potential toxins; (ii) mechanisms of tolerance against potential toxins; (iii) production of plant chelating agents; (iv) plant and microbial interactions and (v) plant uptake [31]. In the investigation of the bioavailability of Ag^0-NPs in the wetland plant *Scirpus triqueter* both in the presence and absence of sulfate-reducing bacteria (SRB), a reduced uptake of Ag ions was found in the presence of SRB [95]. The soil enzymes work as the sensors having specific biochemical and microbiological roles in the soil systems. The studies on the impact of varying concentrations of AgNPs on the rhizosphere soil exoenzymes including urease, dehydrogenase, acid phosphatase, neutral phosphatase, and alkaline phosphatase of the soil systems of wetland planta *Typha orientalis Presl, Iris wilsonii, Arundo donax,* and *Typha orientalis* indicated significantly decreased enzyme activities at high AgNP concentrations [19, 20].

5 Effects of AgNPs on Growth and Development of Wetland Plants

Studies have reported significant morphological changes in the plants after their exposure to AgNPs. Growth potential, development rates, biomass, and leaf surface area are commonly used to assess the effects of AgNPs on the growth and development of plants [6]. Stampoulis et al. reported that 100 nm AgNPs caused about 41% reduction in biomass and 57% reduction in transpiration rates in *Cucurbita pepo* (zucchini) as compared to the control group [118]. Kumari et al. found the negative impacts of AgNPs on the growth and development of *Allium cepa* root tip cells [70]. Studies conducted on the effect of Ag and AgNPs on *Lemna minor* indicated significant inhibition in the growth and development especially of the initial frond densities of the plant [41, 129]. Jiang et al. reported that AgNPs might reduce plant biomass, plant tissue nitrogen content, chlorophyll a/b levels, and chlorophyll fluorescence (Fv/Fm) of aquatic macrophytes specifically *Spirodela polyrhiza* and duckweed [59].

Yin et al. showed the difference in the effects of two AgNPs—20 nm polyvinylpyrrolidine coated silver nanoparticles (PVP-AgNPs) and 6 nm gum arabic coated silver nanoparticles (GA-AgNPs) on the growth and development of 20 wetland plants both in the presence and absence of $AgNO_3$. The inhibition level was always greater for GA-AgNPs in the presence of $AgNO_3$. Plant growth responses differed among taxa with *Lolium multiflorum* growing rapidly under both $AgNO_3$ and GA-AgNP exposures as compared to other taxa having significantly reduced growth under GA-AgNP exposure. Leaf growth of two species; *C. Scoparia* and *S. Cyperinus,* was reduced when exposed to 40 mg Ag/L GA-AgNP versus no effect on PVP-AgNP exposure. Both *C. Crinita* and *C. Vulpinoidea* showed reduced leaf growth in response to high doses of GA-AgNPs. *Phytolacca americana* and *Panicum virgatum* responded by growing significantly longer and more curved roots. The root elongation of sensitive plant species would have a dose-dependent response, although at the highest concentration of GA-AgNPs, *Panicum virgatum* root growth was significantly inhibited (Table 1) [153, 154].

Table 1 Effect of silver nanoparticles on some selected plants

Plant species	Type	Sizes (nm)	Concentration	Effects	Refs.
Cucurbita pepo (zucchini)	AgNP	100	100–500 mg L^{-1}	41 and 57% decrease in the biomass and transpiration rates	[118]
Allium cepa	AgNP	<100	N/A	Cytotoxic and genotoxic impacts on root tip cells	[70]
Lemna minor	AgNP	20–100	5 μg	Inhibit the growth of plant	[41]
Spirodela polyrhiza	AgNP	100	2.89 mg L^{-1}	Decreased plant biomass, plant tissue nitrate-nitrogen content, chlorophyll a/b, and chlorophyll fluorescence	[59]
Lolium multiflorum	GA-AgNP	6, 25	40 mg L^{-1}	Stunted leaf growth	[153]
Carex lurida	GA-AgNP	6.0 ± 1.7	40 mg L^{-1}	Increases both leaf growth and root biomass	[154]
C. scoparia	GA-AgNP	6.0 ± 1.7	40 mg L^{-1}	Reduced leaf growth and inhibited root growth	[154]
S. cyperinus	GA-AgNP	6.0 ± 1.7	40 mg L^{-1}	Reduced leaf growth	[154]
Phytolacca americana	GA-AgNP PVP-AgNP	6.0 ± 1.7 21.0 ± 17	40 mg L^{-1} 40 mg L^{-1}	Significantly longer and more curved roots	[154]
Panicum virgatum	GA-AgNP	6.0 ± 1.7	250 mg L^{-1}	Inhibit root growth	[154]
E. fistulosum	GA-AgNP	6.0 ± 1.7	40 mg L^{-1}	Inhibit root growth	[154]
Carex crinita, Carex vulpinoidea	GA-AgNP	6.0 ± 1.7	40 mg L^{-1}	Inhibit root growth	[154]

Studies have shown that 40 mg/L GA-AgNP treatment might cause cells to undergo significant changes, including vacuolization, collapse of cortical cells, damage to the epidermis, and root cap breakage. This may be due to silver-induced apoptosis involving the production of reactive oxygen species (ROS) and increased oxidative stress by both dissolved silver and AgNPs causing damage to the roots [30]. Previous studies have also shown that root tips bend away from gravity in silver treatment [154]. Contrastingly, the removal of silver is speculated to result in loss of root gravitropism due to poor auxin transport. When the plant is exposed to other stresses such as flooding, the plant will respond to toxicity through the "escape strategy" of root elongation and bending or the "silence effect" with minimal activity at high silver levels when plants are exposed to other stresses such as flooding, respectively [9].

6 Effect of AgNP on Seed Germination and Fruit Production

Inhibition or stimulation of seed germination and root elongation have been shown to depend on plant species and nanoparticle properties [110]. A study showed that treating seeds with AgNPs had no negative impact on plant growth, making them potentially useful in agriculture, and to be used effectively as compared to fungicides to protect seeds from fungi. Importantly, higher concentrations of AgNPs can affect seed germination [107]. Contrastingly, previous studies have shown that nanoparticles might have less effect on seed germination than on seed growth. This can be explained by the preservation of the seed coat which is primarily used for seed protection having anatomical features different from other plant organs or tissues [141]. Research shows that dispersed nanoparticles can increase seed germination by creating "nanopores" in the shell. Despite the slow release of Ag^+ ions, the germination rate was as high as 93.33% when *Pennisetum glaucum* seeds were treated with 50 mg/L silver nanoparticles. Studies have shown that silver nanoparticles do not affect seed growth [99].

Yin et al. have shown that seed germination of different plants is affected differently when exposed to GA-AgNP, $AgNO_3$, and PVP-AgNP. More importantly, $AgNO_3$ supported the growth of 5 out of 11 species, while 6 nm GA-AgNP at 40 mg/L concentration generally had a positive effect on *E. Fistulosum* but not *S. Cyperinus*, *J. Effusus*, and *P. Americanum*. Despite minimal effects on germination, all three silver forms significantly affected aboveground biomass [154]. Diverse studies have explored the impact of silver nanoparticles (AgNPs) on plant species. Varying dosages of AgNPs (0.2–1.6 mg/L) were observed to impede seed germination, lipase activity, and alter soluble and reducing sugar content in *Brassica nigra* germinating seeds and seedlings [3]. Notably, certain wetland plant species, including *Panicum virgatum*, *Phytolacca Americana*, *Phaseolus vulgaris*, and *Boswellia ovalifoliolata* showed good responses to seed germination after treatment [2, 111].

A study investigated seed germination of *Phaseolus vulgaris* L. by the application of AgNPs having a size range between 17 and 200 nm and synthesized from Annona leaf across varied concentrations. Germination percentages revealed a concentration-dependent trend, with optimal results at lower concentrations; 100% at 15, 30, and 60 mg L^{-1}. At higher concentrations, the germination percentage decreased—93.33% at 120 mg L^{-1}, 80% at 240 mg L^{-1}, and 73.33% at 480 mg L^{-1} [134]. Michalek et al. reported that AgNPs at low concentrations ($\leq$10 mg dm^{-3})

can increase seed germination ability, average germination time, seed germination index, seed viability index, new seedling weight, and dry weight of fenugreek and maize [85]. Studies conducted on *Oryza sativa* L. have indicated dose-dependent inhibitory effects on germination, growth, and development of the seeds with increasing negative effects with increased size and concentrations of AgNPs [96, 124].

The positive effect of low AgNP concentrations on seed germination under varying conditions has been suggested to be due to three probable causes occurring at genetic, and biosynthetic levels (Fig. 4) [80, 106]. At the genetic level, AgNPs activate genes encoding aquaporins that facilitate the transport of water and gas molecules and the efflux of ROS [82]. At the biosynthetic level, activation of gibberellin biosynthesis as well as hydrolytic enzymes, i.e., a-amylase, proteases, and lipases occurs by the action of gibberellins facilitated by AgNPs [34]. These enzymes catalyze hydrolysis of organic compounds which functions as respiratory substrates for embryo cells. Activation of a-amylase and increased water uptake by germinating rice seeds primed in AgNPs has been noted by Mahakham et al. [80]. In addition, AgNPs promote the production of ROS (hydrogen peroxide and hydroxyl radicals), which in turn activate antioxidant enzymes (catalase and superoxide dismutase) in cells thus, maintaining ROS levels in the "oxidative window" to activate the pathway that initiates seed germination [102].

Differences have also been shown in the sensitivity of plants to AgNPs and $AgNO_3$. It was observed that $AgNO_3$ generally promotes seed germination in pure cultures and when directly exposed to soil conditions, whereas AgNPs have either no effect (PVP-AgNPs) or most inhibitory effects (GA-AgNPs). For example, GA-AgNPs and $AgNO_3$ both significantly promoted seed germination of *E. Fistulosum* in pure culture, but had no effect on *E. Fistulosum* grown in soil. In contrast, *L. Multiflorum* responded negatively to GA-AgNPs and $AgNO_3$ in pure culture, but positively to both forms of Ag in soil (Table 2). Possible explanations for these changes in germination and growth have been given to be the following: (i) Decrease in effective concentration due to dosing AgNPs in a larger volume of soil versus the smaller volume in a petri dish, (ii) Decreased impact of Ag on seeds and seedlings due to surface modification and interaction of AgNPs and Ag^+ with the organic and mineral phases of soil, (iii) AgNPs may aggregate or be complexed by ligands which can cause a decrease in toxicity [13, 24].

7 Toxicity Induced by AgNP

AgNPs are known to cause toxicity in prokaryotic, eukaryotic, and aquatic organisms. The common negative impacts include disruption of cell-membrane integrity, growth inhibition, reduction in the photosynthetic activity, ROS generation, oxidative stress, protein, or DNA binding and damage, and apoptotic cell death (Fig. 5) [63, 127]. The excessively produced ROS including superoxide radicals ($O_2^{\cdot -}$) and hydrogen peroxide (H_2O_2) may bind to the DNA or proteins ultimately inducing changes in the

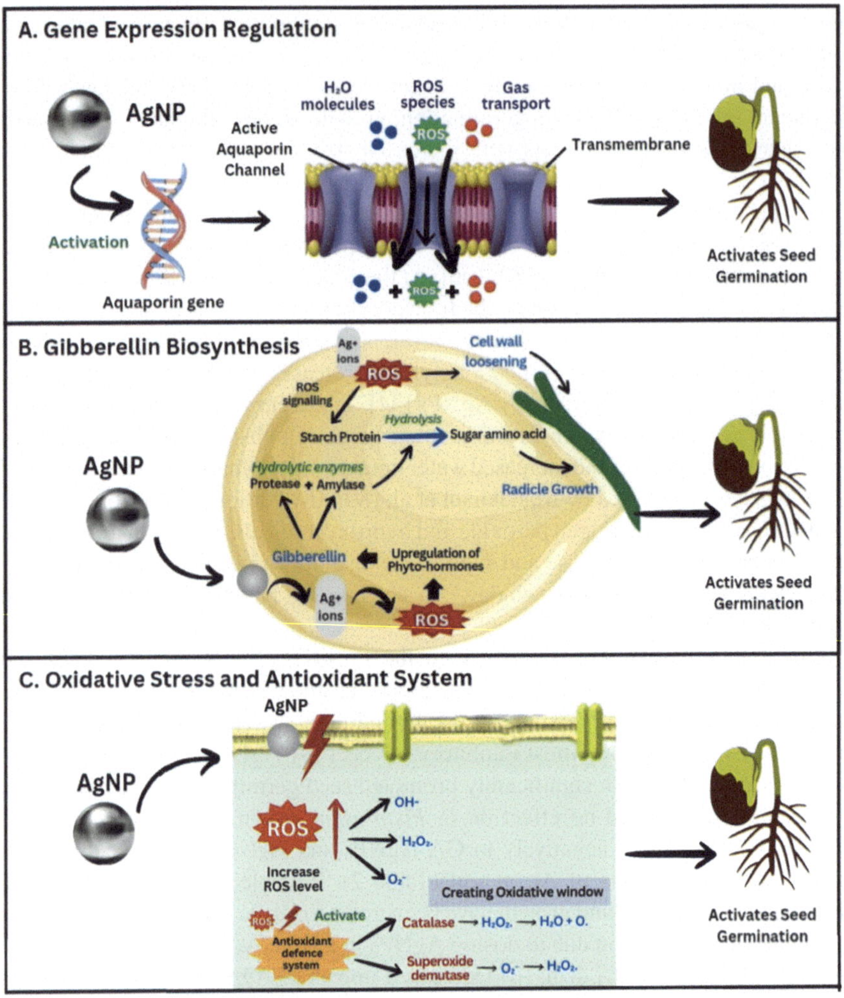

Fig. 4 Mechanisms involved in seed germination: **A** gene expression regulation for aquaporin; **B** gibberellin biosynthesis; and **C** oxidative stress and antioxidant system

cellular structures leading to apoptosis and cell death [147]. The metabolism, antioxidant activity, protein expression, chlorophyll content, superoxide dismutase (SOD) activity, level of H_2O_2, proline, carotenoids, ascorbate, and glutathione content are all altered in response to the AgNPs exposure [69, 75].

The role of AgNPs in the activation of different proteins involved in biosynthesis, folding, assembly, translocation, and degradation of proteins in many cellular processes has been demonstrated in different plants [14]. The up- and down-regulation of these proteins directly regulate the biosynthetic and metabolic pathways in the cells. AgNPs alter these pathways and directly modulate the ROS production

Table 2 Effect of silver nanoparticles on some selected monocot and dicot plants

Species	Growth form	Size and conc. of AgNPs	Seed mass (mg)	Effect on germination	Refs.
Carex lurida	Monocot	6 nm, 25 nm at 40 mg/L	2.36	Negative	[154]
Carex crinita	Monocot	6 nm, 25 nm at 40 mg/L	0.63	No effect	[154]
Carex scoparia	Monocot	6 nm, 25 nm at 40 mg/L	0.38	Positive	[154]
Carex vulpinoidea	Monocot	20–100 nm at 0.1, 1, 10, 100 and 1000 mg/L	0.35	Negative	[1, 154]
Scirpus cyperinus	Monocot	6 nm at 40 mg/L	0.01	Negative	[154]
Lolium multiflorum	Monocot	6 nm, 25 nm at 40 mg/L	0.01	Positive (in soil) Negative (pure culture)	[153, 154]
Juncus effusus	Monocot	6 nm at 40 mg/L	2.09	Negative	[154]
Panicum virgatum	Monocot	20–40 nm	1.75	Positive	[81, 154]
Eupatorium fistulosum	Dicot	6 nm at 40 mg/L	0.23	Positive (pure culture) No effect (in soil)	[154]
Pennisetum glaucum	Dicot	100 nm at 50 mg/L	N/A	Positive	[99]
Phaseolus vulgaris	Dicot	17–100 nm at 15, 30, 60, 120, 240, 480 mg/L	1.74	Positive	[134]
Boswellia ovalifoliolata	Monocot	30–40 nm at 10, 20, 30 μg/mL	N/A	Positive	[2, 111]
Lobelia cardinalis	Dicot	6 nm at 40 mg/L	0.04	No effect	[154]
Phytolacca americana	Dicot	6 nm at 40 mg/L	7.82	Negative	[2, 154]

and toxicity induction, thus deciding the fate of the cells [68]. Proteomic studies on *Oryza sativa* have shown significantly increased protein precursors inducing oxidative stress tolerance, calcium regulation and signaling, cell division, cell wall/DNA/RNA/protein direct damage, and apoptosis on encountering AgNPs [25]. In the wetland ecosystem, the plant biodiversity undergoes vast competition with plants becoming massively endangered and extinct over time due to bioaccumulation of AgNPs [89].

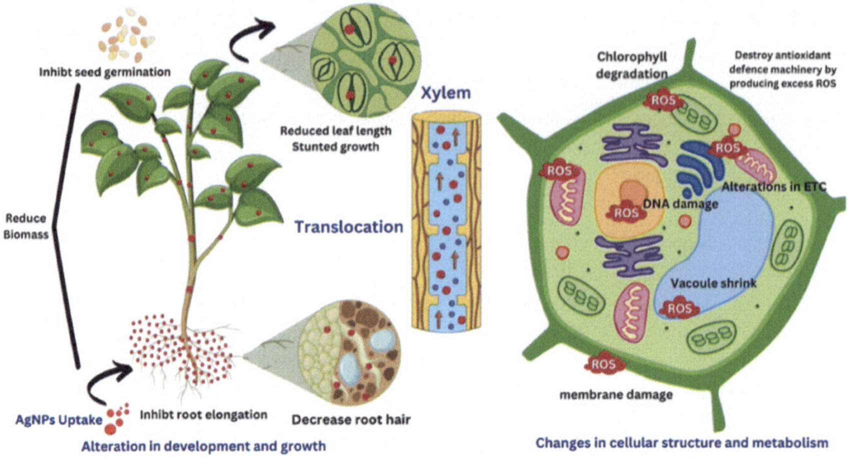

Fig. 5 Uptake, translocation, and phytotoxicity induction by AgNPs in plants. At the cellular level, AgNPs enter various organelles, causing the generation of reactive oxygen species (ROS), thereby causing cytotoxicity and genotoxicity, such as membrane damage, chlorophyll degradation, vacuole contraction, DNA damage, and chromosomal abnormalities

The amount of toxicity depends on the size and concentrations of the AgNPs thus minimizing the germination of seeds, retardation of the growth of seedlings, and affected length and mass of roots and shoots [69]. The toxicity of AgNPs in wetland plants under optimal growth conditions is only partially consistent with the results of pure culture experiments. The results suggest that increasing amounts of nanoparticles released into the environment may have an impact on aquatic plant communities [57]. One study showed agreement between AgNP toxicity observed in pure cultures and potential AgNP toxicity in the natural environment [154]. However, for some species and treatments, the benefits of parasites from plants in soil are reduced compared to pure cultures, or sometimes their effects are different (e.g., promoting rather than inhibiting growth) [104].

A study on the aquatic plant *Lemna gibba* demonstrated the agglomerations of AgNP leading to the production of ROS, ultimately inducing oxidative stress within the plant cells with the increasing concentration of AgNP exposure in the medium [97]. In another study conducted by Minogiannis et al., on *L. gibba* indicated a 40% decrease in the total number of fronds at AgNP concentrations as low as 5 $\mu g\ L^{-1}$, and by 91% at 40 $\mu g\ L^{-1}$ concentrations with the existence of only one frond leaf and color change from green to yellow or white [86]. The study on *Lemna minor* conducted to assess the impact of varying sizes, concentrations, and accumulation of AgNPs indicated that after a month of continuous examination, AgNPs-sized 30 nm were found to be most toxic causing negative growth in all the concentrations of AgNP solution given, with the maximum mortality rate of 60% in the worst case. Highest toxicity was induced at 50 $\mu g\ mL^{-1}$ concentration causing negative growth in the given period [117].

The dose-dependent AgNP accumulation with increasing negative impacts on the N, P, K, and Mg levels in the plant tissue ultimately increasing the inhibition in the plant growth and photosynthesis in *Lemna minor* and *Wolffia globose* has been demonstrated [40, 159]. Triggered tolerance mechanisms were found to be corroborated by chlorophyll a/b, carotenoid content, and soluble proteins found in both the plant [67, 158]. Increased generation of ROS and lipid peroxidation were found with active detoxification enzymes to counteract AgNP-induced oxidative stress and balance the intracellular redox status [40]. AgNP and fumonism B1 (FB1) induced toxicity analysis in *Lemna minor* L., revealed that individually both caused an increase in the antioxidant enzyme activity. However, in combination, AgNP influenced the uptake of FB1 into plant cells with increased cellular accumulation of Ag, higher toxic effects, and disturbed cell redox status and nutrient levels [105]. Interestingly, Ding et al. revealed the effective role of humic acid in decreasing ROS production and phytotoxity in *Lemna minor* [29].

The morphological and ultrastructural analysis to investigate the potential toxicity in *Landoltia punctate* revealed significant damage to the ultrastructure with decreased content of photosynthetic pigments and starch grains, and higher percentage of phenolic compounds as well as physiological changes, primarily the loss in color of the plant [71]. AgNPs were found to induce lipid membrane peroxidation and oxidative damage in *Spirodela punctate*, inhibiting photosynthesis rate and root activity causing atrophy and cell death [152]. In a research conducted by Nair et al., a significant increase in H_2O_2 production, increased lipid peroxidation in the shoots and roots, increased foliar proline accumulation, decreased sugar contents, dose-dependent increase in ROS generation and induction of cytotoxicity were observed in *Oryza sativa* L. [93]. In another study, enhanced activity of antioxidant enzymes including catalase, ascorbate peroxidase, superoxide dismutase, glutathione peroxidase, and glutathione reductase was found in AgNP-treated *Oryza sativa* seeds [125].

Submerged macrophytes are the most important inorganic carbon (Ci) utilizing organisms. The negative impact on the utilization and processing of Ci led the researchers to investigate the role of AgNP as an emerging contaminant which indicated its severely negative role in the carbon cycle. Inhibition of external carbonic anhydrase and Rubisco activity, reduction in HCO_3^- usage capacity of the plants, aggregation of AgNPs in the cell wall and chloroplast, blocking CO_2 fixation, and disturbed C4 and metabolic activities have been found in *Ottelia alismoides* [138]. A dose-dependent increase in oxidative stress and non-specific ROS production was found to be present in the fern *Azolla caroliana* with the increased deposition of callose in the cell wall [42]. The phytotoxicity analysis of AgNPs corresponding to physiologic and enzymatic responses in aquatic plants *Egeria densa* and *Juncus effuses* indicated increased superoxide dismutase (SOD) and peroxidase (POD) activities in the plants to effectively tolerate the AgNP toxicity by scavenging free radicals. However, higher malondialdehyde content in submerged *E. densa* than in emergent *J. effuses* indicated increased membrane damage in the former [155].

8 Conclusion

The increasing utilization of nanoparticles with emerging technological developments has both positive and negative impacts. Diverse research on different organisms has revealed their important role in important physiological and biochemical processes. AgNPs have merged as one of the important metallic nanoparticles having varying effects on different organisms based on their size, concentration, time, and, morphology. Studies on wetland plants have demonstrated both the useful and hazardous impacts of AgNPs on different species of plants along with the microbial community in the soil systems. AgNPs have been found to have an important role in the growth and development, seed germination, and phytoremediation in the plant–soil systems. However, the effects are mostly dose and size-dependent. With varying conditions including concentration, size, and time of AgNPs exposure, severe toxicity has been found to be induced in the seeds, roots, leaves, fronds, and overall growth and development of the plants. The metabolic enzymes present in the soil systems of the associated wetland plants have also been found to be severely affected by the subsequent use of AgNPs. As compared to the studies conducted on the role of AgNPs on terrestrial plants, a limited number of studies have been found on their role on wetland plants for which rigorous research needs to be conducted for a better understanding of the roles of AgNPs in the wetland plant–soil systems.

References

1. Ahmad A, Hashmi SS, Palma JM, Corpas FJ (2022) Influence of metallic, metallic oxide, and organic nanoparticles on plant physiology. Chemosphere 290:133329
2. Almutairi ZM, Alharbi A (2015) Effect of silver nanoparticles on seed germination of crop plants. Int J Nucl Quantum Eng 9(6):689–693
3. Amooaghaie R, Tabatabaei F, Ahadi A-M (2015) Role of hematin and sodium nitroprusside in regulating Brassica nigra seed germination under nanosilver and silver nitrate stresses. Ecotoxicol Environ Saf 113:259–270
4. Anjum NA, Adam V, Kizek R, Duarte AC, Pereira E, Iqbal M et al (2015) Nanoscale copper in the soil–plant system-toxicity and underlying potential mechanisms. Environ Res 138:306–325
5. Anjum S, Ishaque S, Fatima H, Farooq W, Hano C, Abbasi BH, Anjum I (2021) Emerging applications of nanotechnology in healthcare systems: grand challenges and perspectives. Pharmaceuticals 14(8):707
6. Aslani F, Bagheri S, Muhd Julkapli N, Juraimi AS, Hashemi FSG, Baghdadi A (2014) Effects of engineered nanomaterials on plants growth: an overview. Sci World J 2014
7. Auvinen H, Sepúlveda VV, Rousseau DP, Du Laing G (2016) Substrate- and plant-mediated removal of citrate-coated silver nanoparticles in constructed wetlands. Environ Sci Pollut Res 23:21920–21926
8. Bachheti RK, Bachheti A, Husen A (2024) Metal and metal-oxide based nanomaterials. In: Synthesis, agricultural, biomedical and environmental interventions. Springer Nature Singapore Pte Ltd., Singapore
9. Bailey-Serres J, Voesenek LA (2010) Life in the balance: a signaling network controlling survival of flooding. Curr Opin Plant Biol 13(5):489–494

10. Bao D, Oh ZG, Chen Z (2016) Characterization of silver nanoparticles internalized by Arabidopsis plants using single particle ICP-MS analysis. Front Plant Sci 7:32
11. Bao S, Liang L, Huang J, Liu X, Tang W, Yi J, Fang T (2019) Removal and fate of silver nanoparticles in lab-scale vertical flow constructed wetland. Chemosphere 214:203–209
12. Batty LC (2003) Wetland plants-more than just a pretty face? Land Contam Reclam 11(2):173–180
13. Bernhardt ES, Colman BP, Hochella MF, Cardinale BJ, Nisbet RM, Richardson CJ, Yin L (2010) An ecological perspective on nanomaterial impacts in the environment. J Environ Qual 39(6):1954–1965
14. Biba R, Štefanić PP, Cvjetko P, Tkalec M, Balen B (2021) Silver nanoparticles phytotoxicity mechanisms. In: Silver nanomaterials for agri-food applications. Elsevier, pp 317–356
15. Boutin C, Keddy PA (1993) A functional classification of wetland plants. J Veg Sci 4(5):591–600
16. Butt MA, Zafar M, Ahmed M, Shaheen S, Sultana S, Butt MA et al (2021) Types of wetland and wetland plants. In: Wetland plants: a source of nutrition and ethno-medicines, pp 35–54
17. Butt MA, Zafar M, Ahmed M, Shaheen S, Sultana S, Butt MA et al (2021) Wetland and wetland plants. In: Wetland plants: a source of nutrition and ethno-medicines, pp 1–15
18. Cahoon DR, McKee KL, Morris JT (2021) How plants influence resilience of salt marsh and mangrove wetlands to sea-level rise. Estuaries Coasts 44(4):883–898
19. Cao C, Huang J, Cai W-S, Yan C-N, Liu J-L, Jiang Y-D (2017) Effects of silver nanoparticles on soil enzyme activity of different wetland plant soil systems. Soil Sediment Contam: Int J 26(5):558–567
20. Cao C, Huang J, Cai W, Yan C, Jiang Y, Peng C (2016) Effect of silver nanoparticles on the rhizosphere exoenzyme of *T. orientalis* Presl. Paper presented at the proceedings of 2016 world congress on advances in civil, environmental, and materials research
21. Carpita NC, Gibeaut DM (1993) Structural models of primary cell walls in flowering plants: consistency of molecular structure with the physical properties of the walls during growth. Plant J 3(1):1–30
22. Chawla S, Singh S, Husen A (2023) Smart nanomaterials targeting pathological hypoxia. Springer Nature Singapore Pte Ltd., Singapore. ISBN: 978-981-99-1717-4. https://doi.org/10.1007/978-981-99-1718-1
23. Chen Z, Sheng X, Wang J, Wen Y (2018) Silver nanoparticles or free silver ions work? An enantioselective phytotoxicity study with a chiral tool. Sci Total Environ 610:77–83
24. Colman BP, Wang S-Y, Auffan M, Wiesner MR, Bernhardt ES (2012) Antimicrobial effects of commercial silver nanoparticles are attenuated in natural streamwater and sediment. Ecotoxicology 21:1867–1877
25. Cox A, Venkatachalam P, Sahi S, Sharma N (2016) Silver and titanium dioxide nanoparticle toxicity in plants: a review of current research. Plant Physiol Biochem 107:147–163
26. Cronk JK, Fennessy MS (2016) Wetland plants: biology and ecology. CRC Press
27. Cyrusová T, Petrová Š, Vaněk T, Podlipná R (2017) Responses of wetland plant Carex vulpina to copper and iron nanoparticles. Water Air Soil Pollut 228:1–8
28. Deepak V, Umamaheshwaran PS, Guhan K, Nanthini RA, Krithiga B, Jaithoon NMH, Gurunathan S (2011) Synthesis of gold and silver nanoparticles using purified URAK. Colloids Surf, B 86(2):353–358
29. Ding Y, Bai X, Ye Z, Gong D, Cao J, Hua Z (2019) Humic acid regulation of the environmental behavior and phytotoxicity of silver nanoparticles to Lemna minor. Environ Sci Nano 6(12):3712–3722
30. Dinneny JR, Long TA, Wang JY, Jung JW, Mace D, Pointer S et al (2008) Cell identity mediates the response of Arabidopsis roots to abiotic stress. Science 320(5878):942–945
31. Ebrahimbabaie P, Meeinkuirt W, Pichtel J (2020) Phytoremediation of engineered nanoparticles using aquatic plants: mechanisms and practical feasibility. J Environ Sci 93:151–163
32. Edwards KR, Čižková H, Zemanová K, Šantrůčková H (2006) Plant growth and microbial processes in a constructed wetland planted with Phalaris arundinacea. Ecol Eng 27(2):153–165

33. Fageria N, Filho MB, Moreira A, Guimarães C (2009) Foliar fertilization of crop plants. J Plant Nutr 32(6):1044–1064
34. Feregrino-Perez AA, Magaña-López E, Guzmán C, Esquivel K (2018) A general overview of the benefits and possible negative effects of the nanotechnology in horticulture. Sci Hortic 238:126–137
35. Fernandes JP, Almeida CMR, Andreotti F, Barros L, Almeida T, Mucha AP (2017) Response of microbial communities colonizing salt marsh plants rhizosphere to copper oxide nanoparticles contamination and its implications for phytoremediation processes. Sci Total Environ 581:801–810
36. Fernandes JP, Mucha AP, Francisco T, Gomes CR, Almeida CMR (2017) Silver nanoparticles uptake by salt marsh plants—implications for phytoremediation processes and effects in microbial community dynamics. Mar Pollut Bull 119(1):176–183
37. Frattini A, Pellegri N, Nicastro D, De Sanctis O (2005) Effect of amine groups in the synthesis of Ag nanoparticles using aminosilanes. Mater Chem Phys 94(1):148–152
38. Geisler-Lee J, Wang Q, Yao Y, Zhang W, Geisler M, Li K et al (2013) Phytotoxicity, accumulation and transport of silver nanoparticles by Arabidopsis thaliana. Nanotoxicology 7(3):323–337
39. Geitner NK, Cooper JL, Avellan A, Castellon BT, Perrotta BG, Bossa N et al (2018) Size-based differential transport, uptake, and mass distribution of ceria (CeO_2) nanoparticles in wetland mesocosms. Environ Sci Technol 52(17):9768–9776. https://doi.org/10.1021/acs.est.8b02040
40. Glavaš Ljubimir K, Domijan A-M, Radić Brkanac S (2023) Phytotoxic action of silver nanoparticles on lemna minor: multi-parameter analysis of different physiological processes. Plants 12(2):343
41. Gubbins EJ, Batty LC, Lead JR (2011) Phytotoxicity of silver nanoparticles to Lemna minor L. Environ Pollut 159(6):1551–1559
42. Gunn S (2018) Induction of stress response, cell wall damage, and cell death in determination of silver nanoparticle toxicity threshold of the heavy-metal accumulating Fern azolla caroliniana. Southern Illinois University at Carbondale
43. Guntenspergen GR, Stearns F, Kadlec J (2020) Wetland vegetation. In: Constructed wetlands for wastewater treatment. CRC Press, pp 73–88
44. Gutknecht JL, Goodman RM, Balser TC (2006) Linking soil process and microbial ecology in freshwater wetland ecosystems. Plant Soil 289:17–34
45. Hanks NA, Caruso JA, Zhang P (2015) Assessing Pistia stratiotes for phytoremediation of silver nanoparticles and Ag (I) contaminated waters. J Environ Manage 164:41–45
46. Huang D, Dang F, Huang Y, Chen N, Zhou D (2022) Uptake, translocation, and transformation of silver nanoparticles in plants. Environ Sci Nano 9(1):12–39
47. Huang J, Yan C-N, Cao C, Peng C, Liu J-L, Guan W-Z (2018) Performance evaluation of Iris pseudacorus constructed wetland for advanced wastewater treatment under long-term exposure to nanosilver. Ecol Eng 116:188–195
48. Husen A (2019) Natural product-based fabrication of zinc oxide nanoparticles and their application. In: Husen A, Iqbal M (eds) N*anomaterials* and plant potential. Springer International Publishing AG, Cham, pp 193–291. https://doi.org/10.1007/978-3-030-05569-1_7
49. Husen A (2020) Introduction and techniques in nanomaterials formulation. In: Husen A, Jawaid M (eds) Nanomaterials for agriculture and forestry applications. Elsevier Inc., Cambridge, USA, pp 1–14. https://doi.org/10.1016/B978-0-12-817852-2.00001-9
50. Husen A (2022) Engineered nanomaterials for sustainable agricultural production, soil improvement and stress management. Elsevier Inc., Cambridge, USA. https://doi.org/10.1016/C2021-0-00054-7
51. Husen A (2023) Secondary metabolites based green synthesis of nanomaterials and their applications. Springer Nature Singapore Pte Ltd., Singapore. https://doi.org/10.1007/978-981-99-0927-8
52. Husen A (2023) Nanomaterials from agricultural and horticultural products. Springer Nature Singapore Pte Ltd., Singapore. https://link.springer.com/book/9789819934348

53. Husen A (2023) Nanomaterials and nanocomposites exposures to plants. In: Response, interaction, phytotoxicity and defense mechanisms. Springer Nature Singapore Pte Ltd., Singapore. https://doi.org/10.1007/978-981-99-2419-6
54. Husen A, Bachheti RK, Bachheti A (2023) Current trends in green nano-emulsions. In: Food, agriculture and biomedical sectors. Springer Nature Singapore Pte Ltd., Singapore
55. Husen A, Iqbal M (2019) Nanomaterials and plant potential. Springer International Publishing AG, Cham, Switzerland. https://doi.org/10.1007/978-3-030-05569-1
56. Husen A, Siddiqi KS (2023) Advances in smart nanomaterials and their applications. Elsevier Inc., Cambridge, USA. https://doi.org/10.1016/C2021-0-02202-1
57. Ihtisham M, Noori A, Yadav S, Sarraf M, Kumari P, Brestic M et al (2021) Silver nanoparticle's toxicological effects and phytoremediation. Nanomaterials 11(9):2164
58. Jacob DL, Borchardt JD, Navaratnam L, Otte ML, Bezbaruah AN (2013) Uptake and translocation of Ti from nanoparticles in crops and wetland plants. Int J Phytorem 15(2):142–153
59. Jiang HS, Li M, Chang FY, Li W, Yin LY (2012) Physiological analysis of silver nanoparticles and $AgNO_3$ toxicity to Spirodela polyrhiza. Environ Toxicol Chem 31(8):1880–1886
60. Jin-Chul K, Madhusudhan A, Husen A (2021) Smart nanomaterials in biomedical applications. Springer Nature Switzerland AG, Cham, Switzerland. https://doi.org/10.1007/978-3-030-84262-8
61. Kalimuthu K, Babu RS, Venkataraman D, Bilal M, Gurunathan S (2008) Biosynthesis of silver nanocrystals by Bacillus licheniformis. Colloids Surf, B 65(1):150–153
62. Kellogg CH, Bridgham SD, Leicht SA (2003) Effects of water level, shade and time on germination and growth of freshwater marsh plants along a simulated successional gradient. J Ecol 91(2):274–282
63. Khan M, Khan MSA, Borah KK, Goswami Y, Hakeem KR, Chakrabartty I (2021) The potential exposure and hazards of metal-based nanoparticles on plants and environment, with special emphasis on ZnO NPs, TiO_2 NPs, and AgNPs: a review. Environ Adv 6:100128
64. Khan N, Seshadri B, Bolan N, Saint C, Kirkham M, Chowdhury S et al (2016) Root iron plaque on wetland plants as a dynamic pool of nutrients and contaminants. Adv Agron 138:1–96
65. Kim J, Kim S, Lee S (2011) Differentiation of the toxicities of silver nanoparticles and silver ions to the Japanese medaka (Oryzias latipes) and the cladoceran Daphnia magna. Nanotoxicology 5(2):208–214
66. Kim K-J, Kim J-S, Kim L-H, Yang K-C (2012) Characteristics of nutrient uptake by aquatic plant in constructed wetlands for treating livestock wastewater. J Wetl Res 14(1):121–130
67. Kocer ZI, Ayişiği M, Haseki S, Aktaş L (2021) Phyto-synthesized silver nanoparticle toxicity effect on aquatic plant Lemna minor L. Avrupa Bilim ve Teknoloji Dergisi (27):1087–1094
68. Kosová K, Vítámvás P, Prášil IT, Renaut J (2011) Plant proteome changes under abiotic stress—contribution of proteomics studies to understanding plant stress response. J Proteomics 74(8):1301–1322
69. Kumar S, Masurkar P, Sravani B, Bag D, Sharma KR, Singh P et al (2023) A review on phytotoxicity and defense mechanism of silver nanoparticles (AgNPs) on plants. J Nanopart Res 25(4):54
70. Kumari M, Mukherjee A, Chandrasekaran N (2009) Genotoxicity of silver nanoparticles in Allium cepa. Sci Total Environ 407(19):5243–5246
71. Lalau CM, Simioni C, Vicentini DS, Ouriques LC, Mohedano RA, Puerari RC, Matias WG (2020) Toxicological effects of AgNPs on duckweed (Landoltia punctata). Sci Total Environ 710:136318
72. Lamsal K, Kim SW, Jung JH, Kim YS, Kim KS, Lee YS (2011) Application of silver nanoparticles for the control of Colletotrichum species in vitro and pepper anthracnose disease in field. Mycobiology 39(3):194
73. Leonardo T, Farhi E, Pouget S, Motellier S, Boisson A-M, Banerjee D et al (2016) Silver accumulation in the green microalga coccomyxa actinabiotis: toxicity, in situ speciation, and localization investigated using synchrotron XAS, XRD, and TEM. Environ Sci Technol 50(1):359–367

74. Li C-C, Dang F, Li M, Zhu M, Zhong H, Hintelmann H, Zhou D-M (2017) Effects of exposure pathways on the accumulation and phytotoxicity of silver nanoparticles in soybean and rice. Nanotoxicology 11(5):699–709
75. Li F, Li R, Lu F, Xu L, Gan L, Chu W et al (2023) Adverse effects of silver nanoparticles on aquatic plants and zooplankton: a review. Chemosphere 139459
76. Li J, Chen Q, Li Q, Zhao C, Feng Y (2021) Influence of plants and environmental variables on the diversity of soil microbial communities in the Yellow River Delta Wetland, China. Chemosphere 274:129967
77. Lines M (2008) Nanomaterials for practical functional uses. J Alloy Compd 449(1–2):242–245
78. López-Herrera A, Avalos-Borja M, García-Nava J, Trejo-Téllez L, Alarcón A, Patrón-Soberano A et al (2021) Interaction of silver nanoparticles with the aquatic fern Azolla filiculoides: root structure, particle distribution, and silver accumulation. J Nanopart Res 23(1):13
79. Magee TK, Kentula ME (2005) Response of wetland plant species to hydrologic conditions. Wetl Ecol Manag 13:163–181
80. Mahakham W, Sarmah AK, Maensiri S, Theerakulpisut P (2017) Nanopriming technology for enhancing germination and starch metabolism of aged rice seeds using phytosynthesized silver nanoparticles. Sci Rep 7(1):8263
81. Mason C, Vivekanandhan S, Misra M, Mohanty AK (2012) Switchgrass (Panicum virgatum) extract mediated green synthesis of silver nanoparticles. World J Nano Sci Eng 2(2):47
82. Maurel C, Boursiac Y, Luu D-T, Santoni V, Shahzad Z, Verdoucq L (2015) Aquaporins in plants. Physiol Rev 95(4):1321–1358
83. Madhusudhan A, Purohit SD, Prasad R, Husen A (2024) Functional smart nanomaterials and their theranostics approaches. Springer Nature Singapore Pte Ltd., Singapore
84. Mandal AK, Ghorai S, Husen A (2023) Functionalized smart nanomaterials for point-of-care testing. Springer Nature Singapore Pte Ltd., Singapore
85. Michalek S, Swiecilo A, Molas J (2018) Effect of silver nanoparticles and ions on seeds epiphytic microorganisms activity and early stages of sweetcorn development. Przem Chem 97(10):1654–1658
86. Minogiannis P, Valenti M, Kati V, Kalantzi O-I, Biskos G (2019) Toxicity of pure silver nanoparticles produced by spark ablation on the aquatic plant Lemna minor. J Aerosol Sci 128:17–21
87. Miralles P, Church TL, Harris AT (2012) Toxicity, uptake, and translocation of engineered nanomaterials in vascular plants. Environ Sci Technol 46(17):9224–9239
88. Mitragotri S, Anderson DG, Chen X, Chow EK, Ho D, Kabanov AV et al (2015) Accelerating the translation of nanomaterials in biomedicine. ACS Nano 9(7):6644–6654
89. Monica RC, Cremonini R (2009) Nanoparticles and higher plants. Caryologia 62(2):161–165
90. Moor H, Rydin H, Hylander K, Nilsson MB, Lindborg R, Norberg J (2017) Towards a trait-based ecology of wetland vegetation. J Ecol 105(6):1623–1635
91. Mughal SS (2022) Diagnosis and treatment of diseases by using metallic nanoparticles—a review. Authorea Preprints
92. Musyimi D, Netondo G, Owuor B, Gichimu B (2010) Growth and nutrient uptake among three wetland plant species occurring in Lake Victoria basin in Kenya
93. Nair PMG, Chung IM (2014) Physiological and molecular level effects of silver nanoparticles exposure in rice (Oryza sativa L.) seedlings. Chemosphere 112:105–113
94. Navarro E, Piccapietra F, Wagner B, Marconi F, Kaegi R, Odzak N et al (2008) Toxicity of silver nanoparticles to Chlamydomonas reinhardtii. Environ Sci Technol 42(23):8959–8964
95. Niu Z-S, Yang Y, Tou F-Y, Guo X-P, Huang R, Xu J et al (2020) Sulfate-reducing bacteria (SRB) can enhance the uptake of silver-containing nanoparticles by a wetland plant. Environ Sci Nano 7(3):912–925
96. Ottoni C, Neto ML, Léo P, Ortolan B, Barbieri E, De Souza AO (2020) Environmental impact of biogenic silver nanoparticles in soil and aquatic organisms. Chemosphere 239:124698

97. Oukarroum A, Barhoumi L, Pirastru L, Dewez D (2013) Silver nanoparticle toxicity effect on growth and cellular viability of the aquatic plant Lemna gibba. Environ Toxicol Chem 32(4):902–907
98. Palácio SM, Nogueira DA, Espinoza-Quiñones FR, de Campos ÉA, Veit MT (2020) Silver nanoparticles bioaccumulation by aquatic macrophyte Salvinia auriculata. Water Air Soil Pollut 231:1–13
99. Parveen A, Rao S (2015) Effect of nanosilver on seed germination and seedling growth in Pennisetum glaucum. J Cluster Sci 26:693–701
100. Pérez-Sirvent C, Hernández-Pérez C, Martínez-Sánchez MJ, García-Lorenzo ML, Bech J (2017) Metal uptake by wetland plants: implications for phytoremediation and restoration. J Soils Sediments 17:1384–1393
101. Pluym T, Powell Q, Gurav A, Ward T, Kodas T, Wang L, Glicksman H (1993) Solid silver particle production by spray pyrolysis. J Aerosol Sci 24(3):383–392
102. Prażak R, Święciło A, Krzepiłko A, Michałek S, Arczewska M (2020) Impact of Ag nanoparticles on seed germination and seedling growth of green beans in normal and chill temperatures. Agriculture 10(8):312
103. Qu H, Ma C, Xing W, Xue L, Liu H, White JC et al (2022) Effects of copper oxide nanoparticles on Salix growth, soil enzyme activity and microbial community composition in a wetland mesocosm. J Hazard Mater 424:127676
104. Quah B, Musante C, White JC, Ma X (2015) Phytotoxicity, uptake, and accumulation of silver with different particle sizes and chemical forms. J Nanopart Res 17:1–13
105. Radić S, Domijan A-M, Ljubimir KG, Maldini K, Ivešić M, Štefanić PP, Krivohlavek A (2019) Toxicity of nanosilver and fumonisin B1 and their interactions on duckweed (Lemna minor L.). Chemosphere 229:86–93
106. Rajjou L, Duval M, Gallardo K, Catusse J, Bally J, Job C, Job D (2012) Seed germination and vigor. Annu Rev Plant Biol 63:507–533
107. Razak NQA, Yusoff MHM, Aziz WNAA, Kamal ML, Hasan S, Uyup NH et al (2023) Effects of silver nanoparticles on seed germination and seedling growth: a review. J Indian Chem Soc 100(1):100866
108. Rejmankova E (2011) The role of macrophytes in wetland ecosystems. J Ecol Environ 34(4):333–345
109. Salvato M, Borin M, Doni S, Macci C, Ceccanti B, Marinari S, Masciandaro G (2012) Wetland plants, micro-organisms and enzymatic activities interrelations in treating N polluted water. Ecol Eng 47:36–43
110. Santás-Miguel V, Arias-Estévez M, Rodríguez-Seijo A, Arenas-Lago D (2023) Use of metal nanoparticles in agriculture. A review on the effects on plant germination. Environ Pollut 122222
111. Savithramma N, Ankanna S, Bhumi G (2012) Effect of nanoparticles on seed germination and seedling growth of Boswellia ovalifoliolata an endemic and endangered medicinal tree taxon. Nano Vision 2(1):2
112. Schwab F, Zhai G, Kern M, Turner A, Schnoor JL, Wiesner MR (2016) Barriers, pathways and processes for uptake, translocation and accumulation of nanomaterials in plants—critical review. Nanotoxicology 10(3):257–278
113. Sepúlveda VV (2014) Fate of silver nanoparticles in treatment wetlands. In: Master of Science in Environmental Sanitation Dissertation, Faculty of Bioscience Engineering, Gent University
114. Shen R, Lan Z, Huang X, Chen Y, Hu Q, Fang C et al (2020) Soil and plant characteristics during two hydrologically contrasting years at the lakeshore wetland of Poyang Lake, China. J Soils Sediments 20:3368–3379
115. Shravan R (2018) A comparitive study on the efficacy of floating aquatic plants, marine macro algae and their silver nanoparticles in the treatment of the tannery effluent
116. Simonin M, Colman BP, Anderson SM, King RS, Ruis MT, Avellan A et al (2018) Engineered nanoparticles interact with nutrients to intensify eutrophication in a wetland ecosystem experiment. Ecol Appl 28(6):1435–1449

117. Souza LR, Corrêa TZ, Bruni AT, da Veiga MA (2021) The effects of solubility of silver nanoparticles, accumulation, and toxicity to the aquatic plant Lemna minor. Environ Sci Pollut Res 28:16720–16733
118. Stampoulis D, Sinha SK, White JC (2009) Assay-dependent phytotoxicity of nanoparticles to plants. Environ Sci Technol 43(24):9473–9479
119. Stegemeier JP, Colman BP, Schwab F, Wiesner MR, Lowry GV (2017) Uptake and distribution of silver in the aquatic plant Landoltia punctata (duckweed) exposed to silver and silver sulfide nanoparticles. Environ Sci Technol 51(9):4936–4943
120. Tao A, Sinsermsuksakul P, Yang P (2006) Polyhedral silver nanocrystals with distinct scattering signatures. Angew Chem Int Ed 45(28):4597–4601
121. Taghiyari HR, Morrell JF, Husen A (2023) Emerging nanomaterials. In: Opportunities and challenges in forestry sectors. Springer Nature Switzerland AG, Cham, Switzerland. https://doi.org/10.1007/978-3-031-17378-3
122. Tariq Z, Qadeer MI, Anjum I, Hano C, Anjum S (2023) Thalassemia and nanotheragnostics: advanced approaches for diagnosis and treatment. Biosensors 13(4):450
123. Thies JE, Grossman JM (2023) The soil habitat and soil ecology. In: Biological approaches to regenerative soil systems, pp 69–84
124. Thuesombat P, Hannongbua S, Akasit S, Chadchawan S (2014) Effect of silver nanoparticles on rice (Oryza sativa L. cv. KDML 105) seed germination and seedling growth. Ecotoxicol Environ Saf 104:302–309
125. Thuesombat P, Hannongbua S, Ekgasit S, Chadchawan S (2016) Effects of silver nanoparticles on hydrogen peroxide generation and antioxidant enzyme responses in rice. J Nanosci Nanotechnol 16(8):8030–8043
126. Thwala M, Klaine SJ, Musee N (2016) Interactions of metal-based engineered nanoparticles with aquatic higher plants: a review of the state of current knowledge. Environ Toxicol Chem 35(7):1677–1694
127. Tkalec M, Štefanić PP, Balen B (2019) Phytotoxicity of silver nanoparticles and defence mechanisms. Compr Anal Chem 84:145–198
128. Tong X, Wang X, He X, Sui Y, Shen J, Feng J (2019) Effects of antibiotics on nitrogen uptake of four wetland plant species grown under hydroponic culture. Environ Sci Pollut Res 26:10621–10630
129. Tran IT, Heiman JA, Lydy VR, Kissoon LT (2023) Silver inhibits Lemna minor growth at high initial frond densities. Plants 12(5):1104
130. Tripathi DK, Tripathi A, Shweta, Singh S, Singh Y, Vishwakarma K et al (2017) Uptake, accumulation and toxicity of silver nanoparticle in autotrophic plants, and heterotrophic microbes: a concentric review. Front Microbiol 8:07
131. Tundisi LL, Ataide JA, Costa JSR, de Freitas Coêlho D, Liszbinski RB, Lopes AM et al (2023) Nanotechnology as a tool to overcome macromolecules delivery issues. Colloids Surf, B 222:113043
132. van der Valk A (2012) The biology of freshwater wetlands. Oxford University Press
133. Van der Zande M, Vandebriel RJ, Van Doren E, Kramer E, Herrera Rivera Z, Serrano-Rojero CS et al (2012) Distribution, elimination, and toxicity of silver nanoparticles and silver ions in rats after 28 day oral exposure. ACS Nano 6(8):7427–7442
134. Verma DK, Patel S, Kushwah KS (2020) Green biosynthesis of silver nanoparticles and impact on growth, chlorophyll, yield and phytotoxicity of Phaseolus vulgaris L. Vegetos 33:648–657
135. Vymazal J, Kröpfelová L (2008) Wastewater treatment in constructed wetlands with horizontal sub-surface flow, vol 14. Springer Science & Business Media
136. Wang P, Lombi E, Zhao F-J, Kopittke PM (2016) Nanotechnology: a new opportunity in plant sciences. Trends Plant Sci 21(8):699–712
137. Wang P, Menzies NW, Dennis PG, Guo J, Forstner C, Sekine R et al (2016) Silver nanoparticles entering soils via the wastewater–sludge–soil pathway pose low risk to plants but elevated Cl concentrations increase Ag bioavailability. Environ Sci Technol 50(15):8274–8281
138. Wang W, Yuan L, Zhou J, Zhu X, Liao Z, Yin L et al (2023) Inorganic carbon utilization: a target of silver nanoparticle toxicity on a submerged macrophyte. Environ Pollut 318:120906

139. Warheit DB, Borm PJ, Hennes C, Lademann J (2007) Testing strategies to establish the safety of nanomaterials: conclusions of an ECETOC workshop. Inhalation Toxicol 19(8):631–643
140. Weis JS, Weis P (2004) Metal uptake, transport and release by wetland plants: implications for phytoremediation and restoration. Environ Int 30(5):685–700
141. Wierzbicka M, Obidzińska J (1998) The effect of lead on seed imbibition and germination in different plant species. Plant Sci 137(2):155–171
142. Wu FY, Chung AKC, Tam NFY, Wong MH (2012) Root exudates of wetland plants influenced by nutrient status and types of plant cultivation. Int J Phytorem 14(6):543–553
143. Wu Y, Yang L, Gong H, Dang F, Zhou D-M (2020) Contrasting effects of iron plaque on the bioavailability of metallic and sulfidized silver nanoparticles to rice. Environ Pollut 260:113969
144. Yan A, Chen Z (2019) Impacts of silver nanoparticles on plants: a focus on the phytotoxicity and underlying mechanism. Int J Mol Sci 20(5):1003
145. Yan X, Pan Z, Chen S, Shi N, Bai T, Dong L et al (2022) Rice exposure to silver nanoparticles in a life cycle study: effect of dose responses on grain metabolomic profile, yield, and soil bacteria. Environ Sci Nano 9(6):2195–2206
146. Yang D, Wang L, Ma F, Wang G, You Y (2023) Effects of Ag nanoparticles on plant growth, Ag bioaccumulation, and antioxidant enzyme activities in Phragmites australis as influenced by an arbuscular mycorrhizal fungus. Environ Sci Pollut Res 30(2):4669–4679
147. Yang J, Cao W, Rui Y (2017) Interactions between nanoparticles and plants: phytotoxicity and defense mechanisms. J Plant Interact 12(1):158–169
148. Yang J, Duan H, Wang X, Zhang H, Zhang Z (2022) Effects of rice root exudates on aggregation, dissolution and bioaccumulation of differently-charged Ag nanoparticles. RSC Adv 12(15):9435–9444
149. Yang Q, Shan W, Hu L, Zhao Y, Hou Y, Yin Y et al (2018) Uptake and transformation of silver nanoparticles and ions by rice plants revealed by dual stable isotope tracing. Environ Sci Technol 53(2):625–633
150. Yang Q, Xu W, Liu G, Song M, Tan Z, Mao Y et al (2020) Transformation and uptake of silver nanoparticles and silver ions in rice plant (Oryza sativa L.): the effect of iron plaque and dissolved iron. Environ Sci: Nano 7(2):599–609
151. Yang X, Chen Y, Liu X, Guo F, Su X, He Q (2018) Influence of titanium dioxide nanoparticles on functionalities of constructed wetlands for wastewater treatment. Chem Eng J 352:655–663
152. Yang X, He Q, Guo F, Liu X, Chen Y (2021) Translocation and biotoxicity of metal (oxide) nanoparticles in the wetland-plant system. Front Environ Sci Eng 15:1–13
153. Yin L, Cheng Y, Espinasse B, Colman BP, Auffan M, Wiesner M et al (2011) More than the ions: the effects of silver nanoparticles on Lolium multiflorum. Environ Sci Technol 45(6):2360–2367
154. Yin L, Colman BP, McGill BM, Wright JP, Bernhardt ES (2012) Effects of silver nanoparticle exposure on germination and early growth of eleven wetland plants
155. Yuan L, Richardson CJ, Ho M, Willis CW, Colman BP, Wiesner MR (2018) Stress responses of aquatic plants to silver nanoparticles. Environ Sci Technol 52(5):2558–2565
156. Zhang N, Sun J, Yin L, Liu J, Chen C (2023) Silver nanoparticles: From in vitro green synthesis to in vivo biological effects in plants. Adv Agrochem
157. Zhang X-F, Liu Z-G, Shen W, Gurunathan S (2016) Silver nanoparticles: synthesis, characterization, properties, applications, and therapeutic approaches. Int J Mol Sci 17(9):1534
158. Zou X, Li P, Huang Q, Zhang H (2016) The different response mechanisms of Wolffia globosa: light-induced silver nanoparticle toxicity. Aquat Toxicol 176:97–105
159. Zou X, Li P, Lou J, Zhang H (2017) Surface coating-modulated toxic responses to silver nanoparticles in Wolffia globosa. Aquat Toxicol 189:150–158

Impact of Silver Nanoparticles in Aquatic Plants

Noureddine Chaachouay, Lahcen Zidane, and Azamal Husen

Abstract Nanotechnology is a burgeoning field that involves manipulating materials at the nanoscale. Nanoparticles are utilized in various fields, including biomedical sciences, catalysis, chemical industries, cosmetics, drug-gene delivery, electronics, energy science, environmental health, food and feed, healthcare, light emitters, mechanics, optics, optoelectronics, space industries, single electron transistors, nonlinear optical devices, and photo-electrochemical applications. Due to the significant advancements in nanotechnology, silver nanoparticles (Ag-NPs) have been extensively used across several industries and have emerged as one of the most prevalent nanomaterials. Consequently, releasing Ag-NPs into aquatic environments is unavoidable, occurring via numerous routes. Gaining a comprehensive understanding of the impact of Ag-NPs on aquatic plants is crucial. These plants are prevalent and exhibit various variations, making them significant contributors to the aquatic ecosystem. This chapter examines the production of Ag-NPs using several techniques, including physical, chemical, and biological processes. In addition, we explore the characteristics of Ag-NPs and the methodologies and detection techniques used. Similarly, we thoroughly examine these substances' movement and ultimate destiny in aquatic ecosystems. Furthermore, we analyze these factors' immediate and indirect effects on underwater vegetation. Lastly, we will examine the regulatory aspects and potential developments regarding Ag-NPs.

Keywords Aquatic plants · Ecosystem · Nanotechnology · Ag-NPs

N. Chaachouay (✉)
ENS of Berrechid, Agri-Food and Health Laboratory (AFHL), Hassan First University, PO Box 382, 26000 Settat, Morocco
e-mail: nour.chay@gmail.com

L. Zidane
Department of Biology, Faculty of Sciences, Plant, Animal Productions and Agro-Industry Laboratory, Ibn Tofail University, B.P. 133, 14000 Kenitra, Morocco

A. Husen
Wolaita Sodo University, PO Bx 138, Wolaita, Ethiopia

Department of Biotechnology, Graphic Era (Deemed to be University), Dehradun 248002, Uttarakhand, India

A. Husen (ed.), *Plant Response to Silver Nanoparticles*, Smart Nanomaterials Technology, https://doi.org/10.1007/978-981-97-7352-7_14

List of Abbreviations

AAS	Atomic absorption spectroscopy
AFM	Atomic force microscopy
Ag-NPs	Silver nanoparticles
DLS	Dynamic light scattering
DNA	Deoxyribonucleic acid
ELISA	Enzyme-linked immunosorbent assay
EXAFS	Extended X-ray absorption fine structure
ICP-MS	Inductively coupled plasma mass spectrometry
SEM	Scanning electron microscopy
SERS	Surface-enhanced Raman spectroscopy
TEM	Transmission electron microscopy
XANES	X-ray absorption near-edge structure
XAS	X-ray absorption spectroscopy

1 Introduction

Silver nanoparticles (Ag-NPs) are minuscule silver particles with distinct physical, chemical, and biological characteristics. They are extensively utilized in diverse applications, including medicine, cosmetics, textiles, electronics, and environmental cleanup [3, 27, 69]. Nevertheless, the escalating manufacturing and utilization of Ag-NPs likewise give rise to apprehensions regarding their prospective ramifications on the environment and human well-being [60]. Wastewater discharge is a significant pathway for introducing Ag-NPs into the environment, leading to contamination of aquatic ecosystems and subsequent impacts on aquatic life [9]. Aquatic plants play a crucial role in aquatic environments by supplying oxygen, nutrients, shelter, and sustenance for other creatures [64]. Hence, comprehending the impacts of Ag-NPs on aquatic plants is crucial in evaluating the ecological hazards associated with Ag-NPs.

The influence of Ag-NPs on aquatic plants can be variable and contingent upon multiple aspects, including the dimensions, configuration, concentration, coating, and dissolution of Ag-NPs, as well as the species, developmental stage, and ambient circumstances of the plants [37]. Ag-NPs may interact with aquatic plants via many processes, such as adsorption, absorption, translocation, accumulation, and biotransformation [35]. Ag-NPs can cause various physiological, biochemical, and molecular changes in aquatic plants [32]. These include oxidative stress, damage to cell membranes, decreased levels of chlorophyll, inadequate nutrition, hormonal imbalances, enzyme inhibition, altered gene expression, and DNA damage [28]. These modifications may affect the growth, development, reproduction, and survival of aquatic plants and their interactions with other organisms. Ag-NPs may also alter aquatic plant communities' functioning and composition, influencing their productivity, biomass, stability, and diversity [22].

Several issues need to be improved in knowledge, attention, and resolution. These factors encompass the lack of established methods to assess and quantify the presence of Ag-NPs, limited information on how Ag-NPs behave and persist in water environments, uncertainty regarding the harmful effects and buildup of Ag-NPs in aquatic plants, and a lack of comprehension regarding the routes and mechanisms by which Ag-NPs impact aquatic plants [17]. This chapter aims to investigate the intricate interactions between aquatic plants and Ag-NPs, offering insights into the present knowledge and the broader ecological implications of their presence in aquatic environments.

2 Silver Nanoparticles and Their Properties

Ag-NPs are minuscule silver particles with dimensions ranging from 1 to 100 nm [3]. They exhibit distinct physical, chemical, and biological characteristics compared to their larger counterparts, including a large surface area, increased reactivity, and unique optical qualities. Ag-NPs may be produced by several techniques, including physical, chemical, and biological processes, which are chosen based on the desired characteristics of the particles, such as form, size, and dispersion [46, 55]. Chemical reduction, thermal decomposition, irradiation, and green synthesis using plant extracts or microorganisms are among the prevalent techniques for synthesis [51]. Ag-NPs have many uses across several domains, including nanomedicine, cosmetics, renewable energy, environmental remediation, and biomedicine [32]. They have been used as antibacterial agents, wound dressings, methods for delivering drugs, sensors for biological substances, substances that facilitate chemical reactions, devices that convert sunlight into electricity, and other applications [15]. Their extensive use is primarily attributed to their capacity to eradicate a wide range of bacteria, fungi, and viruses while avoiding the development of resistance or harm to human cells. Ag-NPs may also demonstrate anticancer, anti-inflammatory, and wound-healing characteristics via their interaction with biological components and ability to induce apoptosis, inflammation, or tissue regeneration [19, 40].

Nevertheless, Ag-NPs have inherent dangers to human health and the environment due to their tendency to accumulate, persist, and exhibit toxicity [33]. Upon introduction into aquatic environments, Ag-NPs may adversely affect aquatic creatures, including algae, fish, and microbes, by perturbing their physiological processes, inducing oxidative stress, DNA impairment, or cellular demise. Ag-NPs may enter the human body by inhalation, ingestion, or skin contact and can harm the respiratory, gastrointestinal, or immunological systems [18].

3 Transport and Fate in Aquatic Environments

Ag-NPs are extensively used in diverse goods and applications due to their antimicrobial characteristics. Ag-NPs may be introduced into aquatic habitats via many pathways, including industrial discharges, runoff from treated surfaces, and releasing items containing nanomaterials [2]. Ag-NPs may change while they are in the water, which can be affected by pH, temperature, and organic matter [38]. These changes may impact substances' movement, availability, and harmfulness to aquatic creatures, such as plants, algae, bacteria, and fish. Ag-NPs in aquatic settings often change, such as oxidation, reduction, aggregation, and complexation [25]. The process of oxidizing Ag-NPs might result in the release of Ag^+ ions, which possess a higher toxicity level than Ag-NPs. Sunlight, pH, and dissolved oxygen may influence the oxidation rate [53]. The reduction of Ag^+ ions may result in the regeneration of Ag-NPs, which may exhibit varying sizes and shapes compared to their initial forms [70]. Dissolved organic matter, pH, and redox potential may affect the reduction rate. The aggregation of Ag-NPs may decrease their surface area and reactivity while increasing their sedimentation rate. The collection level may vary based on ionic strength, pH, and divalent cations [1]. The process of complexation using Ag-NPs or Ag^+ ions may give rise to many silver species, including silver sulfide, silver chloride, and silver thiosulfate. The type and concentration of ligands may influence the stability and solubility of these complexes [25].

4 Impact on Aquatic Plants

Aquatic plants are essential to the health of marine environments because they provide air, food, and a place for other creatures to live. Ag-NPs can affect water plants in two ways: directly and indirectly.

4.1 Direct Effects

Ag-NPs on aquatic plants might differ based on nanoparticle size, concentration, exposure length, and the plant species' unique properties. Ag-NPs possess antibacterial characteristics and are often included in diverse consumer goods, resulting in their subsequent dispersion into the environment [61]. The following are probable immediate impacts of Ag-NPs on aquatic plants (Fig. 1):

- **Bioaccumulation**: Aquatic plants can absorb and collect Ag-NPs from nearby water. Bioaccumulation may manifest in several plant organs, such as roots, stems, and leaves. Ag-NPs in plant tissues can interfere with regular physiological processes [50].

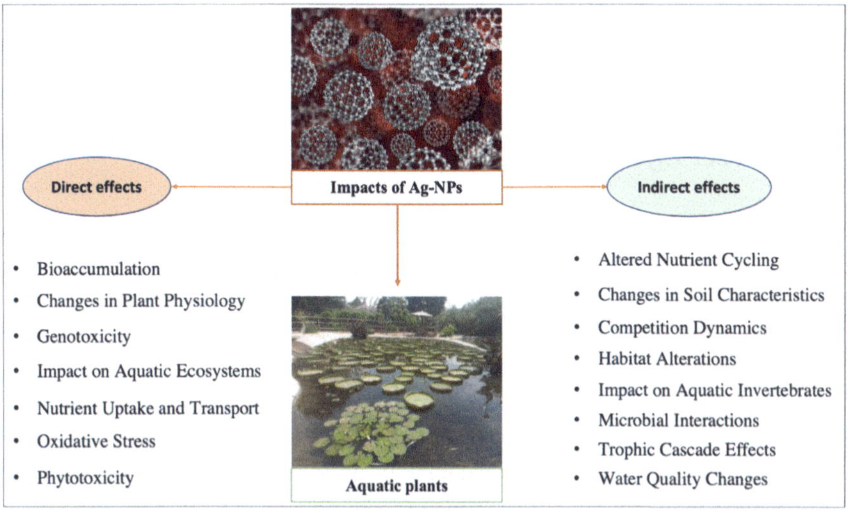

Fig. 1 Direct and indirect impacts of Ag-NPs on aquatic plants

- **Changes in Plant Physiology**: Aquatic plants may experience physiological changes due to exposure to Ag-NPs. This may include alterations in chlorophyll levels, photosynthetic functionality, and other vital biochemical processes necessary for plant development [41].
- **Genotoxicity**: Data indicates that Ag-NPs may have genotoxic effects on aquatic plants. This implies that they have the potential to cause harm to the genetic material of the plant, which might result in mutations or other undesirable impacts on the plant's overall well-being and ability to reproduce in the future [16, 26].
- **Impact on Aquatic Ecosystems**: The effect of Ag-NPs on aquatic plants may have far-reaching consequences for whole ecosystems. Aquatic plants are essential for creating habitat, impacting water quality, and maintaining the general well-being of aquatic habitats [10]. Disturbances to the well-being of aquatic plants might, therefore, lead to a series of consequences that impact the overall ecosystem.
- **Nutrient Uptake and Transport**: Ag-NPs may disrupt the absorption and movement of vital nutrients in aquatic plants. This interference can potentially result in nutritional imbalances inside the plants, impacting their general health and development [5].
- **Oxidative Stress**: Aquatic plants may experience oxidative stress due to exposure to Ag-NPs. This phenomenon arises when the generation of reactive oxygen species (ROS) surpasses the plant's ability to eliminate them [52, 62]. Oxidative stress may result in the impairment of cellular constituents, including proteins, lipids, and DNA.
- **Phytotoxicity**: Aquatic plants may be adversely affected by Ag-NPs, which can display phytotoxic properties. They may disrupt plant growth and development by impacting processes including seed germination, root elongation, and

total biomass output [17]. Nanoparticles may infiltrate plant tissues, potentially affecting cellular structures and functioning [8].

4.2 *Indirect Effects*

The indirect impacts of Ag-NPs on aquatic plants often include alterations in the surrounding ecosystem or interactions with other animals [59]. The consequences may be intricate and may arise from changes in the aquatic habitat's physical, chemical, or biological attributes (Fig. 1). The following are probable secondary impacts of Ag-NPs on aquatic vegetation:

- **Altered Nutrient Cycling**: Ag-NPs can impact the process of nutrient cycling in aquatic environments. For instance, the presence of AgNP may change microbial populations, thus impacting the availability of nutrients. The modified nutrient cycle may indirectly affect the growth and development of aquatic plants that depend on precise nutrient concentrations [54].
- **Changes in Soil Characteristics**: Aquatic plants interact with the sediment or substrate in which they thrive. Sediment-accumulated Ag-NPs may impact soil properties, including nutrient accessibility and pH levels. These modifications may subsequently affect the well-being and development of underwater vegetation [10].
- **Competition Dynamics**: Exposure to Ag-NPs may cause alterations in the development and health of aquatic plants, which can affect competition dynamics with other plant species [68]. This phenomenon may result in alterations in the makeup of plant communities within water habitats.
- **Habitat Alterations**: Ag-NPs in aquatic settings might result in habitat changes, impacting the accessibility of favorable places for plant development [4]. Alterations in the environment's physical characteristics may affect the spatial arrangement and population size of underwater plant species.
- **Impact on Aquatic Invertebrates**: Aquatic invertebrates are essential for the functioning of aquatic ecosystems, and their interactions with aquatic plants may impact the plants' overall health [59]. Ag-NPs may exert direct toxicity on invertebrates, resulting in alterations in population dynamics or community composition [48]. These modifications might indirectly impact aquatic vegetation by modifying herbivory or other ecological interactions.
- **Microbial Interactions**: Ag-NPs can impact microbial populations in aquatic habitats. Given that several plants depend on symbiotic associations with microorganisms for nutrient absorption, any changes in microbial communities might indirectly affect the nutrient levels and general well-being of aquatic plants [21].
- **Trophic Cascade Effects**: Disruptions at a particular level of the food chain might result in consequences propagating across the whole food web [14, 39]. For instance, if Ag-NPs influence the quantity or behavior of herbivores that

devour aquatic plants, it might indirectly influence plant populations and modify the structure of the whole ecosystem.

- **Water Quality Changes**: Ag-NPs may influence water quality metrics like pH, conductivity, and dissolved oxygen levels [67]. Alterations in water quality may impact the general well-being and development of aquatic vegetation, especially those susceptible to specific environmental circumstances.

5 Ecological Implications

The ecological impacts of Ag-NPs on aquatic plants are beyond the direct effects on individual species [20]. These nanoparticles can disturb vital ecological processes, such as nutrient cycling, which is essential for maintaining aquatic ecosystems [31]. Nutrient cycling encompasses transferring and reusing crucial components like nitrogen and phosphorus, which are required for the proliferation and maturation of aquatic vegetation. When Ag-NPs are introduced into marine habitats, they may disrupt these processes, affecting the availability and distribution of nutrients [12]. This disruption can have cascading effects on the entire ecosystem, influencing the composition of plant communities and altering the delicate balance of the aquatic environment.

Moreover, the deposition of Ag-NPs in aquatic species via biological processes raises additional concerns about their ecological consequences. As these nanoparticles gather in the tissues of different aquatic species, they can enter the food chain and maybe reach higher levels, such as fish that are devoured by humans [36]. The entire extent of the long-term effects of bioaccumulation remains uncertain. Still, it prompts concerns over the possible transmission of Ag-NPs throughout the food chain and the resulting hazards to human health. Hence, it is crucial to thoroughly comprehend the ecological consequences of Ag-NPs on aquatic plants to make well-informed decisions and establish sustainable measures to alleviate any detrimental impacts on marine ecosystems and human populations.

6 Methodologies and Detection Techniques

Investigating the effects of Ag-NPs on aquatic plants requires using many methodologies and detection techniques to provide a thorough understanding of the possible ecological outcomes [20, 33]. Initially, scientists often do control laboratory studies to subject aquatic plants to varying levels of Ag-NPs. These tests evaluate the dose–response relationship, investigate the absorption processes, and observe any physiological or biochemical alterations in the plants. Typical approaches include using hydroponic systems or placing organisms in aquatic settings that replicate natural circumstances. The influence is often quantified by measuring parameters such as plant growth, photosynthetic activity, and biochemical markers.

Detection methods are vital in evaluating the movement and behavior of Ag-NPs in aquatic environments [47, 66]. Coupled Plasma Mass Spectrometry (ICP-MS) and Atomic Absorption Spectroscopy (AAS) are used to measure the amount of silver in plant tissues and the water around them. Furthermore, sophisticated imaging methods such as Transmission Electron Microscopy (TEM) and Scanning Electron Microscopy (SEM) assist in visualizing the internalization of nanoparticles inside plant cells. These approaches aid in comprehending the spatial arrangement of Ag-NPs and their possible interactions with biological components (Table 1).

Field studies are conducted in natural aquatic environments to assess the comprehensive ecological effect [1, 20, 58]. These studies include observing variations in the variety of underwater plants, changes in the arrangement of communities, and the possible transmission of Ag-NPs down the food chain. Scientists use environmental monitoring instruments, such as water quality evaluations and genetic analysis of aquatic animals, to measure the enduring impacts. Integrating laboratory experiments, detection methodologies, and field investigations offers a complete strategy to fully comprehend the influence of Ag-NPs on aquatic vegetation and their interconnected ecosystems.

7 Regulatory Considerations

Examine existing regulations and guidelines related to releasing nanoparticles into aquatic environments. Discuss the adequacy of current regulatory frameworks in addressing the potential risks to aquatic plants. The REACH rule, which encompasses the safe use of all chemicals and combinations in the European Union, including nanoforms of substances, is one of the current laws about the discharge of nanoparticles into aquatic habitats [6]. The REACH rule mandates the evaluation of the dangerous characteristics of nano forms of chemicals and guarantees their secure use [57]. The Water Framework Directive is an additional legislation that seeks to safeguard and enhance the quality of all European Union water bodies, including surface waters and groundwater [11]. The Water Framework Directive establishes environmental goals and criteria for water quality and mandates the identification and surveillance of priority chemicals and emerging contaminants, which may include nanoparticles [29].

The efficacy of existing regulatory frameworks in mitigating the potential hazards to aquatic flora is a contention among scientists and policymakers [11]. Some contend that the current laws adequately address and adapt to the difficulties of nanoparticles since they are grounded on the precautionary principle and the risk assessment technique. Some argue that the current restrictions need to be revised because they fail to include nanoparticles' distinct characteristics and actions, such as their dimensions, configuration, surface composition, clumping, dissolution, and accumulation in living organisms. Furthermore, it is said that the existing restrictions need more dependable techniques and information for identifying, describing, and measuring nanoparticles in aquatic ecosystems. Hence, conducting more studies and fostering

Table 1 Methodologies and techniques for detecting the effects of Ag-NPs on aquatic plants

Technique	Characteristics	Refs.
Atomic force microscopy (AFM)	Utilize AFM to investigate the surface characteristics of aquatic plants and identify the presence of Ag-NPs. AFM offers precise visualization with exceptional detail at the nanometer level	[45]
Chemical extraction and analysis	Create chemical extraction protocols to separate Ag-NPs from plant tissues. Employ analytical methods such as atomic absorption spectroscopy (AAS) or inductively coupled plasma mass spectrometry (ICP-MS) to examine the collected samples	[23, 56]
Dynamic light scattering (DLS)	Utilize DLS to examine water samples containing Ag-NPs to ascertain the hydrodynamic size distribution of the nanoparticles	[24]
Environmental monitoring	Employ field-based monitoring techniques, such as passive samplers, sediment analysis, and water quality evaluations, to identify and measure the existence of Ag-NPs in aquatic environments	[42]
Enzyme-linked immunosorbent assay (ELISA)	Apply ELISA methods to identify and measure Ag-NPs by using targeted antibodies. The use of this immunological method may provide very accurate detection	[30]
Fluorescence microscopy	Employ fluorescent dyes or probes designed explicitly for silver to see and measure the presence of Ag-NPs in samples of aquatic plants. Fluorescence microscopy enables direct visual examination	
Genotoxicity assays	Assess the harmful effects of Ag-NPs on the DNA of aquatic plants by using methods such as comet assays or molecular markers to detect DNA damage	[43, 58]
Inductively coupled plasma mass spectrometry (ICP-MS)	Conduct plant sample digestion and analyze the resulting solution using ICP-MS to accurately quantify the amount of silver. This approach has extraordinary sensitivity and can detect minuscule quantities of Ag-NPs	[34, 63]
Phytotoxicity assays	Evaluate aquatic plants' physiological and biochemical reactions when exposed to Ag-NPs. This may include doing growth inhibition experiments, measuring chlorophyll content, and analyzing other biomarkers	[65]
Scanning electron microscopy (SEM)	Employ SEM to analyze the physical characteristics of aquatic plants and identify the presence of Ag-NPs on the surfaces of these plants. Combining SEM with energy-dispersive X-ray spectroscopy allows for investigating elemental composition	[13]
Surface-enhanced Raman spectroscopy (SERS)	Employ SERS for improved detection of Ag-NPs. This method integrates Raman spectroscopy with nanostructured surfaces to provide sensitive and specific detection	[7]

(continued)

Table 1 (continued)

Technique	Characteristics	Refs.
Transmission electron microscopy (TEM)	Produce skinny slices of aquatic plant tissues using TEM to observe the structure, dimensions, and arrangement of Ag-NPs at the nanoscale	[49]
X-ray absorption spectroscopy (XAS)	Use XAS methods, like extended X-ray absorption fine structure (EXAFS) and X-ray absorption near-edge structure (XANES), to find what kind of silver is in samples of water plants	[44]

cooperation among relevant parties is essential to enhance and refine the regulatory frameworks governing nanoparticles in marine ecosystems, aiming to safeguard their well-being and long-term viability.

8 Future Perspectives

Further investigation into the utilization of Ag-NPs in aquatic plants should prioritize the acquisition of a thorough comprehension of the enduring ecological ramifications and possible hazards linked to their implementation. Studying the bioaccumulation and translocation of Ag-NPs in various aquatic plant species is essential because it may provide valuable information on how Ag-NPs might enter food chains and affect higher levels of the food web [39]. Furthermore, it is essential to study the ecotoxicological impacts of Ag-NPs on aquatic plant communities, considering variables such as alterations in species composition, nutrient cycling, and overall ecosystem functioning [12]. Evaluating the combined effects of Ag-NPs and other environmental stressors, such as pollution or climate change, is crucial for developing a complete risk assessment methodology. To reduce the adverse effects of Ag-NPs on aquatic plants, it is necessary to devise techniques for controlling their use and minimizing their presence in the environment. By implementing sustainable nanotechnology principles and integrating life-cycle evaluations, the responsible development and implementation of Ag-NPs in different sectors may be effectively guided.

Furthermore, it is essential to investigate the creation of ecologically sustainable substitutes for Ag-NPs that may be used in specific contexts. One might also explore engineering methods, such as altering the surface characteristics of Ag-NPs to decrease their harmfulness or improving their ability to be eliminated from aquatic environments. Effective laws and best practices that ensure the safe use of Ag-NPs and protect aquatic ecosystems need collaboration among researchers, regulators, and industry.

9 Conclusion

Ag-NPs are extensively used in many sectors and products owing to their potent antibacterial and anticancer characteristics. Nevertheless, the growing discharge of Ag-NPs into the environment presents significant hazards to aquatic ecosystems, particularly to plants, which serve as the leading producers and foundation of the food chain. The impact of Ag-NPs on aquatic plants is contingent upon various factors, including the dimensions, configuration, concentration, coating, and metamorphosis of Ag-NPs, as well as the physicochemical circumstances of the water, the species and physiological characteristics of the plants, and the existence of other contaminants or organisms. Exposing aquatic plants to Ag-NPs may result in a range of negative consequences, including diminished growth, biomass, photosynthesis, nutrient absorption, transpiration, and hormone levels. Additionally, it can lead to heightened oxidative stress, membrane impairment, cell demise, and genotoxicity. Ag-NPs can concentrate and move through plant tissues, causing cellular structure and function changes.

Furthermore, Ag-NPs can modify the microbial communities connected to aquatic plants, impacting their symbiotic associations and the process of nutrient cycling. Certain aquatic plants can absorb and convert Ag-NPs, altering their toxicity and behavior in the environment. The presence of Ag-NPs in aquatic plants has profound consequences for aquatic ecosystems' ecological activities and services, including water quality, biodiversity, carbon sequestration, and food security. Hence, it is crucial to oversee and control the emission and disposal of Ag-NPs while also evaluating their ecological and health hazards. Further investigation is required to comprehend the processes and interactions between Ag-NPs and aquatic plants and to devise approaches to alleviate their adverse impacts and harness their potential advantages.

References

1. Afshinnia K (2018) Effect of environmental factors on the aggregation behavior of silver nanoparticles in the aquatic environment. PhD thesis, University of South Carolina. https://search.proquest.com/openview/2bb48f8ee24af8dd5ed8e064fd63a44c/1?pq-origsite=gscholar&cbl=18750
2. Aguilar-Pérez KM, Avilés-Castrillo JI, Ruiz-Pulido G, Medina DI, Parra-Saldivar R, Iqbal HM (2021) Nanoadsorbents in focus for the remediation of environmentally-related contaminants with rising toxicity concerns. Sci Total Environ 779:146465
3. Ahmad N, Sharma S, Alam MK, Singh VN, Shamsi SF, Mehta BR, Fatma A (2010) Rapid synthesis of silver nanoparticles using dried medicinal plant of basil. Colloids Surf, B 81(1):81–86
4. Alcívar MA, Sendra M, Silva DC, González-Ortegón E, Blasco J, Moreno-Garrido I, Araújo CV (2021) Could Contamination avoidance be an endpoint that protects the environment? An overview on how species respond to copper, glyphosate, and silver nanoparticles. Toxics 9(11):301
5. Ali S, Mehmood A, Khan N (2021) Uptake, translocation, and consequences of nanomaterials on plant growth and stress adaptation. J Nanomater 2021:1–17

6. Allan J, Belz S, Hoeveler A, Hugas M, Okuda H, Patri A, Rauscher H, Silva P, Slikker W, Sokull-Kluettgen B, Tong W, Anklam E (2021) Regulatory landscape of nanotechnology and nanoplastics from a global perspective. Regul Toxicol Pharmacol 122:104885. https://doi.org/10.1016/j.yrtph.2021.104885
7. Alyami A, Quinn AJ, Iacopino D (2019) Flexible and transparent surface enhanced Raman scattering (SERS)-active Ag NPs/PDMS composites for in-situ detection of food contaminants. Talanta 201:58–64. https://doi.org/10.1016/j.talanta.2019.03.115
8. Avellan A, Yun J, Morais BP, Clement ET, Rodrigues SM, Lowry GV (2021) Critical review: role of inorganic nanoparticle properties on their foliar uptake and in Planta translocation. Environ Sci Technol 55(20):13417–13431. https://doi.org/10.1021/acs.est.1c00178
9. Azimzada A, Tufenkji N, Wilkinson KJ (2017) Transformations of silver nanoparticles in wastewater effluents: links to Ag bioavailability. Environ Sci Nano 4(6):1339–1349
10. Bakshi MS (2020) Impact of nanomaterials on ecosystems: mechanistic aspects in vivo. Environ Res 182:109099
11. Brack W, Dulio V, Ågerstrand M, Allan I, Altenburger R, Brinkmann M, Bunke D, Burgess RM, Cousins I, Escher BI, Hernández FJ, Hewitt LM, Hilscherová K, Hollender J, Hollert H, Kase R, Klauer B, Lindim C, Herráez DL et al (2017) Towards the review of the European Union water framework directive: recommendations for more efficient assessment and management of chemical contamination in European surface water resources. Sci Total Environ 576:720–737. https://doi.org/10.1016/j.scitotenv.2016.10.104
12. Chaachouay N, Azeroual A, Benkhnigue O, Zidane L (2023) Alkaloids: a suitable precursor for nanomaterials synthesis, and their various applications. In: Husen A (eds) secondary metabolites based green synthesis of nanomaterials and their applications. Smart Nanomater Technol. Springer, Singapore. https://doi.org/10.1007/978-981-99-0927-8_2
13. Chaachouay N, Zidane L (2024) Role of carbon nanotubes for herbicide detection and remediation. In: bachheti A., Bachheti RK, Husen A (eds) carbon-based nanomaterials. Smart Nanomater Technol. Springer, Singapore. https://doi.org/10.1007/978-981-97-0240-4_8
14. Chaachouay N, Azeroual A, Bencherki B, Douira A, Zidane L (2023) Various metabolites and or bioactive compounds from vegetables, and their use nanoparticles synthesis, and applications. In: Husen A (eds) nanomaterials from agricultural and horticultural products. Smart Nanomater Technol. Springer, Singapore. https://doi.org/10.1007/978-981-99-3435-5_10
15. Dawadi S, Katuwal S, Gupta A, Lamichhane U, Thapa R, Jaisi S, Lamichhane G, Bhattarai DP, Parajuli N (2021) Current research on silver nanoparticles: synthesis, characterization, and applications. J Nanomater 2021:1–23
16. De Lima R, Seabra AB, Durán N (2012) Silver nanoparticles: a brief review of cytotoxicity and genotoxicity of chemically and biogenically synthesized nanoparticles. J Appl Toxicol 32(11):867–879. https://doi.org/10.1002/jat.2780
17. Domingo G, Bracale M, Vannini C (2019) Phytotoxicity of silver nanoparticles to aquatic plants, algae, and microorganisms. In: Nanomaterials in plants, algae and microorganisms. Elsevier, pp 143–168. https://www.sciencedirect.com/science/article/pii/B9780128114889000081
18. Dubey P, Matai I, Kumar SU, Sachdev A, Bhushan B, Gopinath P (2015) Perturbation of cellular mechanistic system by silver nanoparticle toxicity: cytotoxic, genotoxic and epigenetic potentials. Adv Coll Interface Sci 221:4–21
19. Dutt Y, Pandey RP, Dutt M, Gupta A, Vibhuti A, Raj VS, Chang C-M, Priyadarshini A (2023) Silver nanoparticles phytofabricated through Azadirachta indica: anticancer, apoptotic, and wound-healing properties. Antibiotics 12(1):121
20. Fabrega J, Luoma SN, Tyler CR, Galloway TS, Lead JR (2011) Silver nanoparticles: behaviour and effects in the aquatic environment. Environ Int 37(2):517–531
21. Fabrega J, Zhang R, Renshaw JC, Liu W-T, Lead JR (2011) Impact of silver nanoparticles on natural marine biofilm bacteria. Chemosphere 85(6):961–966
22. Fernandes JP, Mucha AP, Francisco T, Gomes CR, Almeida CMR (2017) Silver nanoparticles uptake by salt marsh plants—implications for phytoremediation processes and effects in microbial community dynamics. Mar Pollut Bull 119(1):176–183

23. Haq F, Ahmad H, Alam M (2011) Traditional uses of medicinal plants of Nandiar Khuwarr catchment (District Battagram), Pakistan. J Med Plants Res 5(1):39–48
24. Hassanien AS, Khatoon UT (2019) Synthesis and characterization of stable silver nanoparticles, Ag-NPs: discussion on the applications of Ag-NPs as antimicrobial agents. Physica B 554:21–30. https://doi.org/10.1016/j.physb.2018.11.004
25. He D, Garg S, Wang Z, Li L, Rong H, Ma X, Li G, An T, Waite TD (2019) Silver sulfide nanoparticles in aqueous environments: formation, transformation and toxicity. Environ Sci Nano 6(6):1674–1687
26. Heikal YM, Şuţan NA, Rizwan M, Elsayed A (2020) Green synthesized silver nanoparticles induced cytogenotoxic and genotoxic changes in Allium cepa L. varies with nanoparticles doses and duration of exposure. Chemosphere 243:125430
27. Husen A, Siddiqi KS (2023) Advances in smart nanomaterials and their applications. Elsevier Inc., Cambridge, USA. https://doi.org/10.1016/C2021-0-02202-1
28. Jiang HS, Yin L, Ren NN, Xian L, Zhao S, Li W, Gontero B (2017) The effect of chronic silver nanoparticles on aquatic system in microcosms. Environ Pollut 223:395–402
29. Kern K (2014) New standards for the chemical quality of water in Europe under the new directive 2013/39/EU. J Eur Environ Plan Law 11(1):31–48. https://doi.org/10.1163/18760104-01101002
30. Khoris IM, Takemura K, Lee J, Hara T, Abe F, Suzuki T, Park EY (2019) Enhanced colorimetric detection of norovirus using in-situ growth of Ag shell on Au NPs. Biosens Bioelectron 126:425–432. https://doi.org/10.1016/j.bios.2018.10.067
31. Kumar P, Burman U, Kaul RK (2018) Ecological risks of nanoparticles: effect on soil microorganisms. In: Nanomaterials in plants, algae, and microorganisms. Elsevier, pp 429–452. https://www.sciencedirect.com/science/article/pii/B9780128114872000190
32. Kumar V, Yadav SK (2009) Plant-mediated synthesis of silver and gold nanoparticles and their applications. J Chem Technol Biotechnol: Int Res Process, Environ Clean Technol 84(2):151–157
33. Levard C, Hotze EM, Lowry GV, Brown GE (2012) Environmental transformations of silver nanoparticles: impact on stability and toxicity. Environ Sci Technol 46(13):6900–6914. https://doi.org/10.1021/es2037405
34. Loula M, Kaňa A, Koplík R, Hanuš J, Vosmanská M, Mestek O (2019) Analysis of silver nanoparticles using single-particle inductively coupled plasma—mass spectrometry (ICP-MS): parameters affecting the quality of results. Anal Lett 52(2):288–307. https://doi.org/10.1080/00032719.2018.1459657
35. Lv J, Christie P, Zhang S (2019) Uptake, translocation, and transformation of metal-based nanoparticles in plants: recent advances and methodological challenges. Environ Sci Nano 6(1):41–59
36. Maharramov AM, Hasanova UA, Suleymanova IA, Osmanova GE, Hajiyeva NE (2019) The engineered nanoparticles in food chain: potential toxicity and effects. SN Appl Sci 1(11):1362. https://doi.org/10.1007/s42452-019-1412-5
37. Mahmoudi M, Landry MP, Moore A, Coreas R (2023) The protein corona from nanomedicine to environmental science. Nat Rev Mater 1–17
38. Mat Lazim Z, Salmiati S, Marpongahtun M, Arman NZ, Mohd Haniffah MR, Azman S, Yong EL, Salim MR (2023) Distribution of silver (Ag) and silver nanoparticles (AgNPs) in aquatic environment. Water 15(7):1349
39. Montoya JoséM, Woodward G, Emmerson MC, Solé RV (2009) Press perturbations and indirect effects in real food webs. Ecology 90(9):2426–2433. https://doi.org/10.1890/08-0657.1
40. Naseer F, Ahmed M, Majid A, Kamal W, Phull AR (2022) Green nanoparticles as multifunctional nanomedicines: insights into anti-inflammatory effects, growth signaling and apoptosis mechanism in cancer. Semin Cancer Biol. https://www.sciencedirect.com/science/article/pii/S1044579X22001584
41. Obaid W (2016) The impact of silver nanoparticles on plant physiology. PhD thesis, University of Essex. https://repository.essex.ac.uk/16747/

42. O'Connor Šraj L, Almeida MIGS, Sharp SM, McKelvie ID, Morrison R, Kolev SD (2020) Monitoring of ammonia in marine waters using a passive sampler with biofouling resistance and neural network-based calibration. Environ Pollut 267:115457. https://doi.org/10.1016/j.envpol.2020.115457
43. Patlolla AK, Tchounwou PB (2020) Genotoxicity of silver nanoparticles (Ag-NPs) in in vitro and in vivo models. In: Rai M (ed) Nanotechnology in skin, soft tissue, and bone infections. Springer International Publishing, pp 269–281. https://doi.org/10.1007/978-3-030-35147-2_15
44. Pelli Cresi JS, Principi E, Spurio E, Catone D, O'Keeffe P, Turchini S, Benedetti S, Vikatakavi A, D'Addato S, Mincigrucci R, Foglia L, Kurdi G, Nikolov IP, De Ninno G, Masciovecchio C, Nannarone S, Kopula Kesavan J, Boscherini F, Luches P (2021) Ultrafast dynamics of plasmon-mediated charge transfer in Ag@CeO_2 studied by free electron laser time-resolved X-ray absorption spectroscopy. Nano Lett 21(4):1729–1734. https://doi.org/10.1021/acs.nanolett.0c04547
45. Pletikapić G, Žutić V, Vinković Vrček I, Svetličić V (2012) Atomic force microscopy characterization of silver nanoparticles interactions with marine diatom cells and extracellular polymeric substance. J Mol Recognit 25(5):309–317. https://doi.org/10.1002/jmr.2177
46. Pryshchepa O, Pomastowski P, Buszewski B (2020) Silver nanoparticles: synthesis, investigation techniques, and properties. Adv Coll Interface Sci 284:102246
47. Pu H, Xie X, Sun D-W, Wei Q, Jiang Y (2019) Double strand DNA functionalized Au@AgNps for ultrasensitive detection of 17β-estradiol using surface-enhanced raman spectroscopy. Talanta 195:419–425
48. Ramskov T, Forbes VE, Gilliland D, Selck H (2015) Accumulation and effects of sediment-associated silver nanoparticles to sediment-dwelling invertebrates. Aquat Toxicol 166:96–105
49. Reymond-Laruinaz S, Saviot L, Potin V, Marco de Lucas MdC (2016) Protein–nanoparticle interaction in bioconjugated silver nanoparticles: a transmission electron microscopy and surface enhanced Raman spectroscopy study. Appl Surf Sci 389:17–24. https://doi.org/10.1016/j.apsusc.2016.07.082
50. Rico CM, Majumdar S, Duarte-Gardea M, Peralta-Videa JR, Gardea-Torresdey JL (2011) Interaction of nanoparticles with edible plants and their possible implications in the food chain. J Agric Food Chem 59(8):3485–3498. https://doi.org/10.1021/jf104517j
51. Salem SS, Fouda A (2021) Green synthesis of metallic nanoparticles and their prospective biotechnological applications: an overview. Biol Trace Elem Res 199(1):344–370. https://doi.org/10.1007/s12011-020-02138-3
52. Sharma S, Singh VK, Kumar A, Mallubhotla S (2019) Effect of nanoparticles on oxidative damage and antioxidant defense system in plants. In: Roychoudhury A, Tripathi D (eds) Molecular plant abiotic stress, 1st edn. Wiley, pp 315–333. https://doi.org/10.1002/9781119463665.ch17
53. Sharma VK, Siskova KM, Zboril R, Gardea-Torresdey JL (2014) Organic-coated silver nanoparticles in biological and environmental conditions: fate, stability and toxicity. Adv Coll Interface Sci 204:15–34
54. Sibiya A, Jeyavani J, Santhanam P, Preetham E, Freitas R, Vaseeharan B (2022) Comparative evaluation on the toxic effect of silver (Ag) and zinc oxide (ZnO) nanoparticles on different trophic levels in aquatic ecosystems: a review. J Appl Toxicol 42(12):1890–1900. https://doi.org/10.1002/jat.4310
55. Siddiqi KS, Husen A, Rao RAK (2018) A review on biosynthesis of silver nanoparticles and their biocidal properties. J Nanobiotechnol 16(14):1–28
56. Suárez-Cerda J, Alonso-Nuñez G, Espinoza-Gómez H, Flores-López LZ (2015) Synthesis, kinetics and photocatalytic study of "ultra-small" Ag-NPs obtained by a green chemistry method using an extract of Rosa 'Andeli' double delight petals. J Colloid Interface Sci 458:169–177
57. Subhan MA, Subhan T (2022) Chapter 5—safety and global regulations for application of nanomaterials. In: Rai M, Nguyen TA (eds) Nanomaterials recycling. Elsevier, pp 83–107. https://doi.org/10.1016/B978-0-323-90982-2.00005-6

58. Tavares P, Balbinot F, de Oliveira HM, Fagundes GE, Venâncio M, Ronconi JVV, Merlini A, Streck EL, da Silva Paula MM, de Andrade VM (2012) Evaluation of genotoxic effect of silver nanoparticles (Ag-Nps) in vitro and in vivo. J Nanopart Res 14(4):791. https://doi.org/10.1007/s11051-012-0791-y
59. Tortella GR, Rubilar O, Durán N, Diez MC, Martínez M, Parada J, Seabra AB (2020) Silver nanoparticles: toxicity in model organisms as an overview of its hazard for human health and the environment. J Hazard Mater 390:121974
60. Upadhyay SK, Kumar S, Rout C, Vashistha G, Aggarwal D (2022) Risks and concerns of use of nanoparticles in agriculture. In: Rajput VD, Verma KK, Sharma N, Minkina T (eds) The role of nanoparticles in plant nutrition under soil pollution. Springer International Publishing, pp 371–394. https://doi.org/10.1007/978-3-030-97389-6_16
61. Verma P, Maheshwari SK (2019) Applications of silver nanoparticles in diverse sectors. Int J Nano Dimens 10(1):18–36
62. Von Moos N, Slaveykova VI (2014) Oxidative stress induced by inorganic nanoparticles in bacteria and aquatic microalgae—state of the art and knowledge gaps. Nanotoxicology 8(6):605–630. https://doi.org/10.3109/17435390.2013.809810
63. Wei W-J, Yang Y, Li X-Y, Huang P, Wang Q, Yang P-J (2022) Cloud point extraction (CPE) combined with single particle-inductively coupled plasma-mass spectrometry (SP-ICP-MS) to analyze and characterize nano-silver sulfide in water environment. Talanta 239:123117. https://doi.org/10.1016/j.talanta.2021.123117
64. Wersal RM, Madsen JD (2012) Aquatic plants their uses and risks. In: A review of the global status of aquatic plants. FAO, Rome. https://www.gri.msstate.edu/publications/docs/2012/03/10109FAO_IPPC_Aquatic_Plants_2012x.pdf
65. Zhang H, Chen S, Jia X, Huang Y, Ji R, Zhao L (2021) Comparation of the phytotoxicity between chemically and green synthesized silver nanoparticles. Sci Total Environ 752:142264. https://doi.org/10.1016/j.scitotenv.2020.142264
66. Zhang K, Liu X, Man B, Yang C, Zhang C, Liu M, Zhang Y, Liu L, Chen C (2018) Label-free and stable serum analysis based on Ag-NPs/PSi surface-enhanced Raman scattering for noninvasive lung cancer detection. Biomed Opt Express 9(9):4345–4358
67. Zhang L, Li J, Yang K, Liu J, Lin D (2016) Physicochemical transformation and algal toxicity of engineered nanoparticles in surface water samples. Environ Pollut 211:132–140
68. Zhang W, Xiao B, Fang T (2018) Chemical transformation of silver nanoparticles in aquatic environments: mechanism, morphology and toxicity. Chemosphere 191:324–334
69. Zhang X-F, Liu Z-G, Shen W, Gurunathan S (2016) Silver nanoparticles: synthesis, characterization, properties, applications, and therapeutic approaches. Int J Mol Sci 17(9):1534
70. Zou X, Shi J, Zhang H (2014) Coexistence of silver and titanium dioxide nanoparticles: enhancing or reducing environmental risks? Aquat Toxicol 154:168–175

Use of Silver Nanoparticles as a Nano-weapon in Plant Disease Management

Murugan Karuvelan, Salai S. Sumukhi, Greeshma Odukkathil, Ramachandran Chelliah, and Deog Hwan Oh

Abstract The increasing global population, coupled with rising demands from the agricultural sector amid limited resources, has spurred extensive research aimed at enhancing both the quality and yield of crops. One of the primary challenges in plant management is the presence of phytopathogens, leading to crop losses, reduced yield quality and quantity, and ultimately elevated production costs with broader economic implications. Traditional methods of biocontrol in plant disease management often involve the use of chemical substances, which, unfortunately, bring about undesirable consequences such as non-targeted effects, residue accumulation, soil and water contamination, ecological imbalances, heightened phytotoxicity, and other long-term impacts. As a response to these challenges, there has been a surge in exploring the applications of engineered nanomaterials in agriculture and plant disease management, particularly focusing on silver nanoparticles (AgNPs) due to their unique physicochemical properties and antimicrobial activity. The efficacy of AgNP in combating plant diseases is supported by various studies, employing characterization techniques like TEM, SEM, UV-Vis spectrophotometry, which unveil their mode of action and cellular interactions. Additionally, extensive field research provides insights into the influence of AgNPs on plant growth, health, and disease management. The findings collectively suggest that AgNPs represent an innovative tool in precision plant disease management. While concerns exist about the phytotoxic effects of conventionally synthesized chemically derived AgNPs, the green-synthesized AgNPs offer sustainable approaches with minimized ecological

M. Karuvelan · S. S. Sumukhi
Department of Microbiology, Vels Institute of Science, Technology and Advanced Studies (VISTAS), Pallavaram, Chennai, Tamil Nadu 603117, India

G. Odukkathil
Centre for Clean Environment, Vellore Institute of Technology, Vellore, India

R. Chelliah (✉) · D. H. Oh
Department of Food Science and Biotechnology, College of Agriculture and Life Sciences, Kangwon National University, Chuncheon 24341, South Korea
e-mail: ramachandran865@gmail.com

R. Chelliah
Saveetha School of Engineering, (SIMATS) University, Sriperumbudur, Chennai 600124, India

A. Husen (ed.), *Plant Response to Silver Nanoparticles*, Smart Nanomaterials Technology, https://doi.org/10.1007/978-981-97-7352-7_15

impacts. Thus, they emerge as a promising, potential, and sustainable nano-weapon in the realm of plant disease management.

Keywords Nanoparticles · Plant · Disease · Silver · Nano-weapon · Environment

1 Introduction

Plant diseases have profound economic repercussions, impacting not just the agricultural sector but also exerting influence on global trade, food security, and the well-being of individuals and communities. It is imperative to implement effective disease management strategies and research initiatives to alleviate these economic consequences and establish a sustainable and resilient agricultural industry. The effective management of plant diseases is a pivotal component of agricultural practices, directly influencing both crop productivity and food security. In recent years, there has been growing interest in utilizing silver nanoparticles as a nano-weapon for plant disease management, as these nanoparticles possess notable antimicrobial properties. The incorporation of silver nanoparticles in plant disease management presents a promising and environmentally friendly solution for combating plant pathogens. Nanoparticles, characterized by their size ranging from 1 to 100 nm, play vital roles in diverse fields, including medicine, pharmacy, biotechnology, and environmental science. Nanotechnology operates on a scale of factors of 10^{-9}, with nanoparticles serving as carriers for drugs and genes in medical applications, particularly in the treatment of tumors and cancer [2]. Silver nanoparticles (AgNPs) have been extensively researched and applied in various sectors owing to their exceptional physicochemical properties. They are highly valued in commercial industries due to their remarkable chemical stability, conductivity, catalytic activity, and antimicrobial capabilities [1, 12].

2 Synthesis of Silver Nanoparticles

The production of silver nanoparticles can be accomplished through diverse techniques, encompassing chemical, physical, and biological methods. Notably, biological synthesis has attracted considerable attention owing to its environmentally friendly and non-toxic characteristics. Biological synthesis involves employing various biological agents, such as plants, bacteria, fungi, and algae. Plants, particularly, have emerged as noteworthy candidates for silver nanoparticle synthesis due to their cost-effectiveness, abundance, and non-toxic properties. In this process, plant extracts or different plant parts, including leaves, stems, or roots, act as both reducing and stabilizing agents for silver ions, facilitating the creation of silver nanoparticles. The utilization of silver nanoparticles in plant disease management has gained prominence recently, with plants proving to be effective and environmentally sustainable sources for their synthesis. In contrast, chemical synthesis commonly employs substances like sodium borohydride, polyvinyl pyrrolidone, or even graphene oxide as major components in the synthesis of silver nanoparticles. Physical methods, such as laser ablation, electron beam irradiation, and chemical reduction by photochemicals, are also employed in the preparation of silver nanoparticles. However, these physical and chemical approaches face drawbacks, including the generation of toxic waste, high energy consumption, and environmental and safety concerns, limiting their encouragement. Addressing the need for alternative synthesis methods, green synthesis techniques have emerged, relying on plant extracts, bacteria, yeast, etc., to efficiently synthesize silver nanoparticles (Fig. 1).

3 Characterization of Silver Nanoparticles

The characterization of silver nanoparticles is crucial for comprehending their physical and chemical attributes, directly influencing their potential applications in managing plant diseases. Various methods, such as reduction reactions and sonoelectrochemical deposition, have been employed for both the synthesis and characterization of silver nanoparticles. Characterization techniques include transmission electron microscopy (TEM), UV-Vis spectroscopy, Fourier transform infrared spectroscopy (FTIR), atomic force microscopy (AFM), X-ray diffractometry (XRD), and scanning probe microscopy (SPM). AFM and SPM are utilized to explore the surface morphology and topography of the nanoparticles, while XRD analysis unveils the crystal structure. These techniques play a pivotal role in understanding the properties and behavior of silver nanoparticles, essential for their diverse applications. To study the formation of AgNPs, UV-Vis spectroscopy provides information on size, structure, stabilization, and aggregation of nanoparticles. Furthermore, XRD analysis offers insights into the crystalline structure, while TEM analysis enables visualization. Dynamic light scattering analysis is also employed to determine size distribution and stability. AFM provides information on size, form, physical structure, and morphology of the nanoparticles. Techniques like FTIR, XPSM, and Zeta potential are employed to study the chemical composition, surface functionalization, and stability of silver nanoparticles, including emission, absorption, photoconductivity, characteristic bonding, and surface charge stability.

4 Antimicrobial Effects of Silver Nanoparticles

The antimicrobial properties of silver nanoparticles have been extensively documented in numerous studies. Silver nanoparticles, synthesized using plant extracts, have demonstrated antibacterial efficacy against *Escherichia coli*, *Staphylococcus*

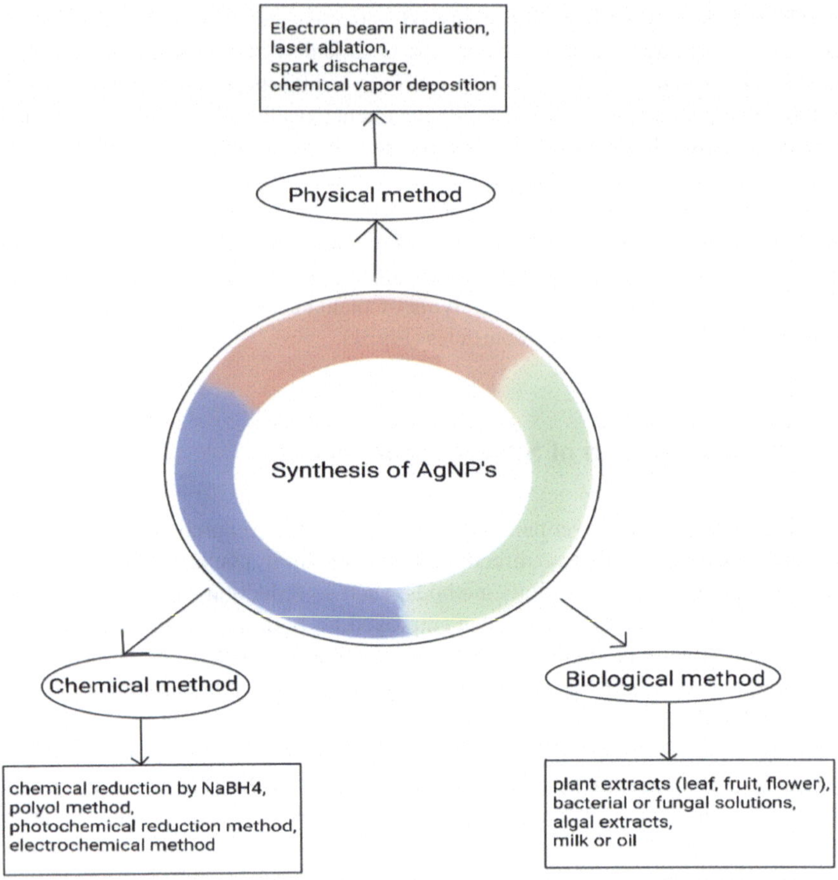

Fig. 1 Provides a comprehensive visual guide to the synthesis of silver nanoparticles through three distinct methodologies: physical, chemical, and biological. This figure intricately maps out each synthesis route, highlighting the unique processes and conditions required for each method. In the physical synthesis section, the figure illustrates techniques like evaporation–condensation and laser ablation, focusing on how they utilize physical forces to generate nanoparticles. The chemical synthesis part delves into methods such as chemical reduction and sol–gel processes, detailing the chemical reactions and agents involved, like reducing agents and stabilizers. Lastly, the biological synthesis section showcases eco-friendly and sustainable methods using biological entities like plants, fungi, and bacteria. This part of the figure explains the bioreduction mechanisms where these organisms convert silver ions into nanoparticles. Additionally, the figure compares the advantages and limitations of each method, such as purity, size control, energy consumption, and environmental impact. Overall, this figure serves as an educational tool for scientists and students alike, providing a clear and detailed overview of the various pathways for synthesizing silver nanoparticles, each with its own set of applications and efficiencies

aureus, *Pseudomonas aeruginosa*, *Azotobacter chroococcum*, and *Bacillus licheniformis*. Similarly, silver nanoparticles produced through chemical reduction of silver nitrate exhibited antibacterial activity against *Staphylococcus aureus* and *Escherichia coli*, with the extent of bacterial inhibition depending on particle aggregation. Silver nanoparticles have emerged as a novel solution to combat bacterial multidrug resistance, inhibiting bacterial growth and interfering with bacterial physiology, quorum sensing, and biofilm development. The mechanism of action involves direct contact with the bacterial cell wall, hindering cellular adhesion and attachment without penetrating the cell. The antibacterial activity of silver nanoparticles is influenced by physicochemical properties, environmental conditions, bacterial strain, and exposure time. Additionally, the synthesis of silver nanoparticles using plant extracts has been shown to confer antibacterial activity against various bacteria, highlighting their potential as effective antibacterial agents. Silver nanoparticles also exhibit antifungal properties against various fungal species. For instance, AgNPs synthesized by Aspergillus flavus inhibit the growth of *Aspergillus ochraceus*, *Penicillium expansum*, and *Fusarium oxysporum*. AgNPs from *Juglans regia* and *Camellia sinensis* leaves demonstrate strong antifungal activity against *Aspergillus ochraceus* and *Candida albicans*, respectively. Moreover, AgNPs from pomegranate peel extract and neem leaf extract exhibit antifungal activity against *Aspergillus flavus* and *Alternaria solani*, respectively. These findings suggest the potential of AgNPs as effective antifungal agents, with applications in areas such as food packaging, plant protection, and disease management. Silver nanoparticles (AgNPs) have demonstrated potent antiviral properties against various viruses, including alfalfa mosaic virus (AMV), influenza A virus, and tobacco mosaic virus (TMV). AgNPs inhibit viral replication and transmission, presenting effectiveness against viral infections. For instance, AgNPs from *Punica granatum* peel extract alleviates the negative effects of TMV infection in tomato plants by reducing virus accumulation and enhancing the expression of pathogenesis-related genes. Additionally, AgNP–TPU materials, incorporating AgNPs, exhibit significant antiviral activity against the Spring viraemia of carp virus (SVCV) and the severe acute respiratory syndrome coronavirus 2 (SARS-CoV-2). These findings underscore the potential of AgNPs as effective antiviral agents against various viral infections. Moreover, silver nanoparticles have shown significant antihelminthic activity in several studies. The assessment involves in vitro experiments with different nanoparticle concentrations and helminth parasites, evaluating parasite viability through the trypan blue exclusion test. Silver nanoparticles synthesized from sources like *Momordica charantia* fruit extract and biological silver nanoparticles from *Hibiscus rosasinensis* extract exhibit strong anthelmintic effects against various worms. Notably, the use of silver nanoparticles as a potential scolicidal agent for cystic hydatid disease has displayed promising results. Overall, these findings suggest that silver nanoparticles hold promise as effective antihelminthic agents for treating helminth infections (Table 1).

Table 1 An extensive compilation that methodically presents the antimicrobial effects of silver nanoparticles against a wide array of microorganisms. This table systematically categorizes different pathogens, including bacteria, fungi, and viruses, and illustrates the efficacy of silver nanoparticles in inhibiting or destroying these microbes. For each pathogen, the table lists specific studies or experiments that demonstrate the antimicrobial action, noting the concentrations of nanoparticles used and the observed outcomes. It includes details on the mechanisms by which silver nanoparticles exert their antimicrobial effects, such as disrupting cell membranes, interfering with cellular processes, and inducing oxidative stress in pathogens. The table also provides comparative insights into the effectiveness of silver nanoparticles against various strains of pathogens, including antibiotic-resistant ones, highlighting their potential as alternative or adjunctive treatments in combating antimicrobial resistance. Additionally, it addresses the influence of nanoparticle size, shape, and surface modification on their antimicrobial potency. This table also serves as a comprehensive resource for researchers and healthcare professionals, offering a detailed overview of the promising role of silver nanoparticles in the field of antimicrobial therapy

Sl. No.	Antimicrobial activity	Example
1	Antibacterial activity	*Escherichia coli*, *Staphylococcus aureus*, *Pseudomonas aeruginosa*, *Azotobacter chroococcum*, and *Bacillus licheniformis* etc
2	Antifungal activity	*Aspergillus ochraceus*, *Penicillium expansum*, *Fusarium oxysporum*
3	Antiviral activity	*Alfalfa mosaic* virus (AMV), Influenza A virus, Tobacco mosaic virus (TMV)
4	Antihelminthic activity	Studies have shown that silver nanoparticles synthesized from different sources, such as *Momordica charantia* fruit extract, biological silver nanoparticles from *Hibiscus rosasinensis* extract have strong anthelmintic effects against various types of worms
5	Biofilm destruction	The mechanism of action of silver nanoparticles involves direct contact with the bacterial cell wall, inhibiting cellular adhesion and attachment, without the need to penetrate into the cell

5 Comparison of Silver Nanoparticles with Conventional Methods in Plant Disease Management

Silver nanoparticles have proven effective in controlling numerous plant diseases caused by fungi, bacteria, and viruses. Their diminutive size and expansive surface area enable interaction with pathogens at the nanoscale, resulting in their antibacterial properties. AgNPs possess the capability to disrupt cell membranes, impede enzymatic activity, and induce oxidative stress in pathogens, ultimately leading to their demise. In contrast, conventional pesticides and insecticides are also successful in disease control. However, their efficacy may diminish over time as target organisms develop resistance. Additionally, certain chemical pesticides exert broad-spectrum effects, impacting unintended species and disrupting ecosystems. Environmental concerns arise due to the potential accumulation of traditional pesticides in soil and water, leading to soil degradation, water contamination, and harm to non-target organisms such as beneficial insects and wildlife. Furthermore, the ingestion of pesticide residues in food products poses potential risks to human health. In contrast, silver nanoparticles offer a more sustainable alternative. AgNPs are biodegradable and are less likely to persist in the environment. Green chemistry methods that minimize the use of hazardous chemicals can be employed in their production. However, the environmental impact of AgNPs may vary based on their application and manufacturing processes.

6 Modes of Action of Silver Nanoparticles and Factors Affecting

Silver nanoparticles (AgNPs) exhibit diverse mechanisms of action, with a common mode involving the absorption of free silver ions that disrupt ATP molecules and hinder DNA replication. AgNPs can also generate reactive oxygen species or directly damage cell membranes through the release of silver ions (Ag^+). In gram-negative organisms, AgNPs create pits in the cell wall, enhancing permeability and leading to cell death. Overall, AgNPs cause denaturation and oxidation of the cell wall, resulting in organelle rupture and cell lysis. Additionally, the presence of AgNPs alters the phosphotyrosine profile of peptides, disrupting signal transduction processes in organisms and inhibiting their multiplication. The primary antibacterial effect of AgNPs is attributed to the release of Ag^+ ions, with smaller AgNPs (<10 nm particle size) exhibiting more pronounced ion discharge compared to larger AgNPs. AgNPs can hinder the binding and replication of viruses by interacting with the outer proteins of viruses, although the exact antiviral mechanism is not fully understood, leaving room for further research in this area. The action of AgNPs against Candida albicans may involve the disruption of cell membrane integrity, leading to the inhibition of cell growth. The disintegration status of silver nanoparticles in exposed media significantly influences their antibacterial impact. Disintegration proficiency depends on manufacturing and processing factors, such as the type of media in which the nanoparticles are suspended. The Ostwald-Freundlich condition theoretically describes the impact of particle size and shape on the release of silver particles. Capping agents are utilized to modify the surfaces of silver nanoparticles, altering their dissolution behavior. Besides intrinsic properties, the surrounding media can affect silver ion release. Smaller silver nanoparticles with spherical or quasi-spherical formats are more prone to silver release, elucidating the lower release of silver in aggregated nanoparticles compared to isolated ones. Organic or inorganic components in media can also influence silver nanoparticle dissolution. Gram-negative bacteria are more susceptible to silver nanoparticles due to their narrower cell walls compared to gram-positive strains. Researchers have demonstrated that silver nanoparticles release ions more rapidly in acidic solutions than in neutral ones. It is widely recognized that silver nanoparticles smaller than 10 nm can directly alter cell permeability, enter bacterial cells, and cause cell damage. The distinct antibacterial effects of silver nanoparticles on gram-negative and gram-positive bacteria underscore the importance of nanoparticle uptake in their antibacterial efficacy (Fig. 2).

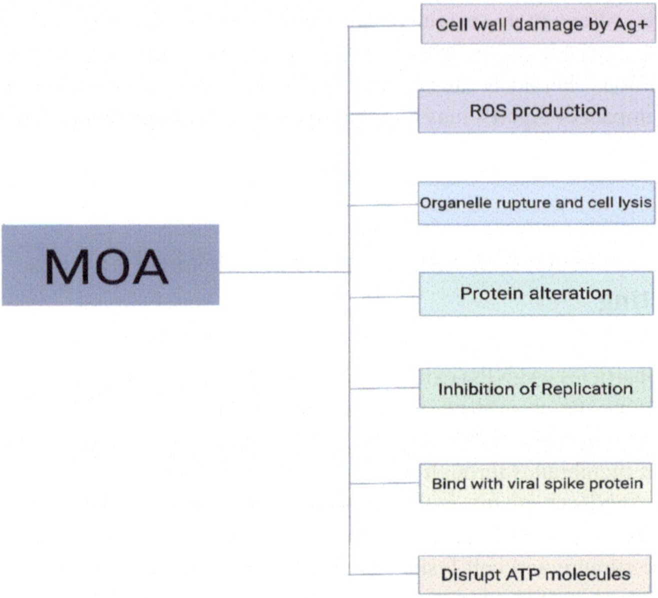

Fig. 2 Offers a detailed and insightful visualization of the diverse modes of action of silver nanoparticles, alongside the various factors that influence these mechanisms. This figure meticulously illustrates how silver nanoparticles interact with biological systems, highlighting their antimicrobial, anti-inflammatory, and catalytic activities. For the antimicrobial action, it details the process by which these nanoparticles disrupt microbial cell membranes, interfere with cellular metabolism, and impede DNA replication. In terms of anti-inflammatory effects, the figure depicts the interaction of silver nanoparticles with immune cells and the subsequent modulation of inflammatory pathways. Additionally, the catalytic activities of these nanoparticles in various chemical reactions are visually represented. The latter part of the figure focuses on the factors that affect the efficacy and behavior of silver nanoparticles, such as particle size, shape, surface coating, concentration, and environmental conditions. By doing so, it provides a comprehensive understanding of the dynamic and multifaceted nature of silver nanoparticles, making it an invaluable resource for researchers and practitioners in nanotechnology, medicine, and environmental science

7 Advantages of Silver Nanoparticles Over Conventional Methods

The utilization of silver nanoparticles (AgNPs) in plant disease management offers several advantages over traditional methods. In comparison to chemical pesticides, AgNPs often leave a smaller environmental footprint, exhibiting rapid degradation and reduced likelihood of accumulation in soil and water. This characteristic minimizes harm to non-target organisms, aligning with principles of sustainable agriculture. Moreover, AgNPs generally exhibit lower toxicity toward humans and animals than some chemical pesticides, thereby decreasing the risk of harm to consumers, wildlife, and farm workers. Unlike chemical pesticides, AgNPs' unique mode of action reduces the likelihood of pathogens developing resistance over time, potentially leading to more persistent disease control. AgNPs have demonstrated effectiveness against a diverse range of plant diseases caused by bacteria, fungi, and viruses. Their broad-spectrum activity makes them versatile tools for addressing various diseases in different crops. Integration of AgNPs into Integrated Pest Management (IPM) plans, which emphasize a comprehensive and ecologically balanced approach to pest and disease management, can enhance their efficacy. Combining AgNPs with other IPM techniques like crop rotation, biological control, and cultural practices can make AgNPs more effective and reduce reliance on chemical pesticides. The application of AgNPs to harvested crops results in minimal or no hazardous residues, lowering the potential for pesticide residues in food products, a critical consideration for crops intended for human consumption. Studies suggest that AgNPs may stimulate plant growth and development, leading to more robust and resilient crops capable of resisting disease pressure. Additionally, AgNPs can be engineered to encapsulate or deliver other active substances, such as nutrients or biocontrol agents, through nano-enhanced delivery, enhancing the overall effectiveness of disease management techniques. Compared to chemical pesticides, AgNPs often require fewer applications for disease control, leading to less disruption and cost savings for agricultural enterprises. Furthermore, AgNPs may encounter fewer regulatory restrictions than certain chemical pesticides, particularly if classified as low-risk chemicals, potentially expediting their adoption in agriculture.

8 Impacts of Silver Nanoparticles on Agricultural Environment (Organisms, Soil and Plant)

In the expanding realm of nanotechnology, the investigation of crucial subjects, including the potential interactions of nanoparticles, has garnered significant attention in both past and present studies. This exploration extends across diverse components of the ecosystem, encompassing soil, humans, plants, and macro-microorganisms. Notably, global reports indicate an annual production of 500 tonnes of AgNPs. Consequently, AgNPs are released into the environment through various pathways, including landfills, waste incineration plants, air, water sewage sludge, and consumer goods containing AgNPs. Focusing on the significance of agroecosystems, this discussion centers on three primary interactions involving AgNPs, with specific emphasis on their effects on soil, microbial populations, and, most importantly, plants. Once introduced into the soil, AgNPs have the potential to contaminate the environment. The fate, transport, bioavailability, and toxicity of AgNPs in the soil are strongly influenced by physico-chemical soil characteristics such as texture, pH, cation exchange capacity, and organic matter [11]. Understanding these interactions forms a baseline for assessing the environmental advantages and risks associated with AgNPs. Soil factors, including pH, cation exchange capacity, and organic matter content, play a crucial role in the toxicity behavior of AgNPs, influencing their bioavailability and interactions within the soil environment [4]. Aggregated nanoparticles, influenced by particle size, shape, and surface charge, may exhibit altered toxicity behavior as they interact with the soil matrix [14]. Factors like increased ionic strength and divalent cations impact AgNP aggregation and retention in soil, highlighting the importance of stabilizing agents [13]. Despite the widely acknowledged broad-spectrum antibacterial ability of AgNPs, their impact on soil biota remains less understood. Recent research emphasizes the significant interaction of AgNPs with soil microbial communities, affecting denitrifying bacteria and microbial diversity in activated sludge systems [9, 15, 19, 20]. The observed effects on microbial functions and diversity underscore the need for a comprehensive understanding of the ecotoxicity of AgNPs in agroecosystems. The phytotoxicity of AgNPs to higher plants is an increasingly significant aspect, considering plants' crucial role in the environment. Previous studies have produced varied results, indicating that factors such as seed type, species, growth stage, AgNP characteristics (size, concentration, method of application), and experimental conditions influence phytotoxicity outcomes [3, 18] [3]. The concentration and size of AgNPs largely determine their phytotoxic effects on plant biomass [18]. The growth medium, whether soil or agar, also influences the bioavailability of AgNPs for plant uptake. In summary, the multifaceted interactions of AgNPs within agroecosystems necessitate comprehensive research to ascertain their potential environmental impacts and benefits. Investigations evaluating soil conditions supporting AgNP aggregation and their reduced toxicity are particularly crucial for their sustainable utilization in agriculture. Additionally, the interactions of AgNPs with soil biota and higher plants remain key areas requiring further exploration and understanding.

9 Positive Effects of Silver Nanoparticles on Plants

In the realm of nanotechnology, studies have noted the positive impacts of silver nanoparticles (AgNPs) on various aspects of plant growth. Salama [6] observed favourable effects of AgNPs (10–30 nm) on root/shoot length and leaf surface area. The seed germination and seedling development of corn, watermelon, and zucchini plants were collectively influenced by AgNPs, with higher germination rates across all tested concentrations compared to untreated plants. Watermelon and zucchini exhibited improved germination rates despite an increased mean germination time for zucchini. Another study by Sharma et al. [10] reported a dose-dependent positive effect of AgNPs (29, 25 and 50 ppm) on the vigor index and plant growth of *B. juncea* when using MS medium as a growth medium. Reactive oxygen species levels were reduced due to the induction of antioxidant enzymes. PVP-coated AgNPs (10 nm) stimulated root elongation at a concentration of 10 mg Ag/L but had no significant effect on shoot elongation. Changes in proteins associated with redox regulation and sulfur metabolism indicated the maintenance of cell homeostasis. These findings underscore the potential agricultural applications and benefits of optimal AgNP use, emphasizing the need to optimize specific nanoparticle concentrations. Examining the impact of silver nanoparticles (20 nm width) on Fenugreek seeds, Sathishkumar et al. conducted experiments using various AgNP concentrations (0, 10, 20, 30, and 40 $\mu g\ mL^{-1}$). The results revealed that the highest seed germination (76.11%), speed of germination (4.102), root length (76.94 mm), root fresh weight (2.783), and root dry weight (1.204) occurred at a concentration of 10 $\mu g\ mL^{-1}$. Similarly, Pallavi investigated the effects of AgNPs on Brassica and cowpea, demonstrating significant enhancements in seed germination potential, mean germination time, seed germination index, seed vigor index, and seedling fresh weight and dry weight in cowpea. Plants treated with silver nanoparticles exhibited tolerance to abiotic stress, overcoming nutrient shortages, enhancing enzymatic processes, and facilitating the adhesion of plant growth-promoting bacteria to roots. These collective findings suggest that the application of silver nanoparticles has the potential to positively influence plant growth and development, with specific concentrations requiring optimization for optimal results.

10 Challenges in Using Silver Nanoparticles for Plant Disease Management

Silver nanoparticles and silver ions have both been confirmed to exhibit phytotoxic properties. The short-term (3 days) and long-term responses of Arabidopsis thaliana to exposure to silver nanoparticles (AgNPs) and Ag^+ particles ($AgNO_3$) in separate hydroponic and soil environments were investigated. In hydroponics, AgNPs (1.0 and 2.5 mg/L) displayed more potent inhibitory effects on plant growth and photosynthesis than Ag^+ ions at the same concentration over a three-day-period. The presence of AgNPs in plants induced alterations in gene and protein expression, suggesting a molecular response to AgNPs. These findings raise concerns about the uncontrolled release and increasing use of AgNPs in aquatic and terrestrial settings, potentially causing detrimental effects on the environment and living organisms, particularly plants. Exposure to 10 mg L^{-1} AgNPs negatively impacted the growth of germinating wheat seedlings, leading to morphological changes in root tip cells, primarily due to the release of Ag ions from the AgNPs. Further analysis of genetic and protein changes induced by AgNPs in wheat seedlings revealed alterations in the expression of various proteins related to primary metabolism and cell defense [16]. The presence of citrate-coated silver nanoparticles (AgNP) and silver ions ($AgNO_3$) was found to impede the respiration and photosynthesis of periphyton, along with the functioning of extracellular enzymes. This suggests potential disruptions in the metabolic functions and enzymatic resource acquisition of stream periphyton. The toxicity of AgNP suspensions was primarily attributed to the dissolved Ag(I) from the particles. However, distinct effects caused by the nanoparticles themselves, independent of the dissolved Ag(I), were observed, particularly in the inhibition of a specific extracellular enzyme.

11 Overcoming Challenges with an Eco-friendly Approach: Green Synthesized Silver Nanoparticles

As evident from previous investigations, a variety of physical and chemical methods, such as high-energy ball milling, laser pyrolysis or ablation, electrochemical and chemical vapor deposition, micro-emulsion sol–gel, and reverse precipitation, have been employed for the synthesis of metallic nanoparticles [7, 8]. It is crucial to note that these methods pose environmental threats due to the use of numerous toxic chemicals and are both costly and unstable. Consequently, in recent times, the biosynthesis of nanoparticles using plant extracts and microorganisms has gained prominence. The biomolecules involved in the biosynthetic process are non-toxic, cost-effective, and, importantly, environmentally friendly. Additionally, the nanoparticles produced are more biocompatible. In this context, researchers have turned to biological agents for AgNP synthesis, recognizing the promising applications of AgNPs in plant disease management. To create more environmentally friendly and biocompatible nanoparticles suitable for agroecosystems, microbes crucial to agriculture may play a role in AgNP synthesis. Recently, the well-known plant growth-promoting bacterium Serratia sp. has been employed to produce AgNPs with antifungal activity against the wheat spot blotch pathogen *B. sorokiniana.* Additionally, beneficial microbes like Bacillus sp., Trichoderma sp., and Spirulina platensis have been utilized in the biosynthesis of AgNPs with antimicrobial properties against a wide variety of phytopathogens. Beyond microorganisms, plant sources such as Acalypha indica, Piper nigrum, and pine cones have been used to synthesize AgNPs, demonstrating significant efficacy in inhibiting phytopathogens [5, 17]. Despite this, the biological synthesis of AgNPs, especially for the management of plant diseases, has received limited attention to date. Therefore, incorporating beneficial microorganisms and plant extracts in AgNP synthesis could mitigate biosafety concerns associated with their application in agroecosystems. To assess phytotoxicity effects, chemical silver nanoparticles (CH-AgNPs) and green-synthesized silver nanoparticles (Bio-AgNPs) were compared. CH-AgNPs exhibited a significant toxic effect on plant germination and growth in studies involving wheat seed germination, Bacillus subtilis, and water spinach and mustard greens seeds. Conversely, Bio-AgNPs did not inhibit seed germination and actually had positive effects on plant growth parameters such as biomass, shoot, and root growth. However, the phytotoxicity of Bio-AgNPs was found to be dose-dependent, with higher concentrations negatively affecting insect reproductive traits. In conclusion, Bio-AgNPs exhibited lower phytotoxicity compared to CH-AgNPs, making them a more sustainable and environmentally friendly option for various applications, including agriculture.

12 Conclusion

First of all, AgNPs have shown remarkable antimicrobial properties against a wide variety of plant pathogens, including viruses, fungi, and bacteria. Their exceptional physicochemical properties, like size, shape, and surface charge, impact their viability in restraining pathogenic development and expansion. Because of their adaptability, AgNPs can be used to combat a variety of plant diseases. Furthermore, AgNPs' potential in plant disease management has been demonstrated in both the laboratory and the field. When applied as foliar sprays, seed treatments, or soil amendments, AgNPs have been shown to significantly reduce disease incidence and severity in numerous studies. These encouraging results suggest that AgNPs could be a useful component of integrated strategies for disease management. Nevertheless, it is essential to acknowledge that although AgNPs have tremendous potential, their application in agriculture also raises concerns. Comprehensive risk assessments and regulatory considerations are required due to the potential ecological and human health effects of AgNP release into the environment. In addition, a crucial area for future research is the development of AgNPs synthesis techniques that are both environmentally friendly and cost-effective. In conclusion, the use of silver nanoparticles as a nano-weapon in the management of plant diseases is a promising area that provides novel approaches to combating plant pathogens. Proceeded with innovative work endeavors is crucial for saddling the maximum capacity of AgNPs while guaranteeing their security and maintainability in agrarian practices. Eventually, AgNPs can possibly contribute fundamentally to worldwide food security by upgrading crop well-being and yield.

References

1. Husen A, Siddiqi KS (2023) Advances in smart nanomaterials and their applications. Elsevier Inc., Cambridge, USA. https://doi.org/10.1016/C2021-0-02202-1
2. Jin-Chul K, Madhusudhan A, Husen A (2021) Smart nanomaterials in biomedical applications. Springer Nature Switzerland AG, Cham, Switzerland. https://doi.org/10.1007/978-3-030-84262-8
3. Navarro E, Baun A, Behra R, Hartmann NB, Filser J, Miao AJ, Quigg A, Santschi PH, Sigg L (2008) Environmental behavior and ecotoxicity of engineered nanoparticles to algae, plants, and fungi. Ecotoxicology 17:372–386
4. Oromieh AG (2011) Evaluating solubility, aggregation and sorption of nanosilver particles and silver ions in soils. Master's thesis in environmental science, Swedish University of Agricultural Sciences, Department of Soil and Environment, Sweden
5. Paulkumar K, Gnanajobitha G, Vanaja M, Rajeshkumar S, Malarkodi C, Pandian K, Annadurai G (2014) Piper nigrum leaf and stem assisted green synthesis of silver nanoparticles and evaluation of its antibacterial activity against agricultural plant pathogens. Sci World J 2014(1):829894
6. Salama HMH (2012) Effects of silver nanoparticles in some crop plants, common bean (*Phaseolus vulgaris* L.) and corn (*Zea mays* L.). Int Res J Biotechnol 3:190–197
7. Sastry M, Ahmad A, Khan MI, Kumar R (2003) Biosynthesis of metal nanoparticles using fungi and actinomycete. Curr Sci 85:162–170

8. Sastry M, Ahmad A, Khan MI, Kumar R (2004) Microbial nanoparticle production. In: Niemeyer CM, Mirkin CA (eds) Nanobiotechnology. Wiley, Weinheim, pp 126–135
9. Shah V, Collins D, Walker VK, Shah S (2014) The impact of engineered cobalt, iron, nickel and silver nanoparticles on soil bacterial diversity under field conditions. Environ Res Lett 9:024001
10. Sharma P, Bhatt D, Zaidi MGH, Saradhi PP, Khanna PK, Arora S (2012) Silver nanoparticle-mediated enhancement in growth and antioxidant status of Brassica juncea. Appl Biochem Biotechnol 167:2225–2233
11. Shoults-Wilson WA, Reinsch BC, Tsyusko OV, Bertsch PM, Lowry GV, Unrine JM (2011) Role of particle size and soil type in toxicity of silver nanoparticles to earthworms. Soils Sci Soc Am J 75:365–377
12. Siddiqi KS, Husen A, Rao RAK (2018) A review on biosynthesis of silver nanoparticles and their biocidal properties. J Nanobiotechnol 16(14):1–28
13. Thio BJR, Montes MO, Mahmoud MA, Lee DW, Zhou D, Keller AA (2012) Mobility of capped silver nanoparticles under environmentally relevant conditions. Environ Sci Technol 46:6985–6991
14. Tourinho PS, Van Gestel CA, Lofts S, Svendsen C, Soares AM, Loureiro S (2012) Metal-based nanoparticles in soil: fate, behavior, and effects on soil invertebrates. Environ Toxicol Chem 31:1679–1692
15. VandeVoort AR, Arai Y (2012) Effect of silver nanoparticles on soil denitrification kinetics. Ind Biotechnol 8:358–364
16. Vannini C, Domingoa G, Onellib E, Mattiac FD, Bruni I, Marsonia M, Bracale M (2014) Phytotoxic and genotoxic effects of silver nanoparticles exposure on germinating wheat seedlings. J Plant Physiol 171:1142–1148
17. Velmurugan P, Lee SM, Iydroose M, Lee KJ, Oh BT (2013) Pine cone-mediated green synthesis of silver nanoparticles and their antibacterial activity against agricultural pathogens. Appl Microbiol Biotechnol 97:361–368
18. Wang J, Koo Y, Alexander A, Yang Y, Westerhof S, Zhang Q, Schnoor JL, Colvin VL, Braam J, Alvarez PJ (2013) Phytostimulation of poplars and Arabidopsis exposed to silver nanoparticles and Ag^{+} at sublethal concentrations. Environ Sci Technol 47:5442–5449
19. Yang Y, Quensen J, Mathieu J, Wang Q, Wang J, Lia M, Tiedje JM, Alvarez PJJ (2014) Pyrosequencing reveals higher impact of silver nanoparticles than Ag^{+} on the microbial community structure of activated sludge. Water Res 48:317–325
20. Zhang C, Liang Z, Hu Z (2014) Bacterial response to a continuous long-term exposure of silver nanoparticles at sub-ppm silver concentrations in a membrane bioreactor activated sludge system. Water Res 50:350–358

Comparative Evolution of Phytotoxicity Between Chemically and Green Synthesized Silver Nanoparticles

Murugan Karuvelan, Kaliyamoorthy Suganya, Sudharsan Kasirajan, Kathireshan Alagapuram Kaliyaperumal, Ramachandran Chelliah, Ghazala Sultan, Momna Rubab, and Deog Hwan Oh

Abstract Plant extracts can be used to manufacture nanoparticles inexpensively and without harming the environment. The primary determinants of the effects of nanoparticles made from green materials are their dosage and the specific characteristics of both the nanoparticles and the plant species involved. The potential impact of chemically synthesized nanoparticles on plants is a subject of considerable scientific interest. It has been observed that these nanoparticles may exert profoundly adverse impacts on plant physiology and growth. Silver nanoparticle exposure resulted in growth and reproductive toxicity in plants because of the overproduction of reactive oxygen species (ROS) triggered by silver nanoparticles or Ag^+ released from silver nanoparticles. Silver nanoparticle exposure would affect crop productivity and quality over the whole growing cycle, which could pose a danger to food security. In the context of phytotoxicity, it was observed that the chemically synthesized nanoparticles exhibited an ability to enhance the production of ROS within plants. This, in turn, triggered the activation of the plants antioxidant defence systems. Consequently, there was an observed increase in the levels of malondialdehyde and zinc content, while the antioxidant capacity, carotenoid levels, globulin levels, and molybdenum content experienced a decline. In contrast, it was observed that the utilization of

M. Karuvelan · K. Suganya · S. Kasirajan · K. A. Kaliyaperumal
Department of Microbiology, Vels Institute of Science, Technology and Advanced Studies (VISTAS), Pallavaram, Chennai, Tamil Nadu 603117, India

R. Chelliah (✉)
Department of Food Science and Biotechnology, College of Agriculture and Life Sciences, Kangwon National University, Chuncheon 24341, South Korea
e-mail: ramachandran865@gmail.com

R. Chelliah · D. H. Oh
Saveetha School of Engineering, (SIMATS) University, Sriperumbudur, Chennai 600124, India

G. Sultan
Department of Computer Science, Faculty of Science, Aligarh Muslim University, Aligarh, Uttar Pradesh 202002, India

M. Rubab
Department of Food Engineering and Biotechnology, University of Engineering and Technology, New Campus, Lahore 54000, Pakistan

A. Husen (ed.), *Plant Response to Silver Nanoparticles*, Smart Nanomaterials Technology, https://doi.org/10.1007/978-981-97-7352-7_16

green synthesis techniques for the production of silver nanoparticles resulted in an augmentation of protein contents. Furthermore, the application of green synthesized nanoparticles was found to induce an increase in Mn levels and a decrease in Al levels, thereby eliciting a series of favourable outcomes. Green synthesized nanoparticles, conversely, have exhibited some favourable rather than detrimental effects on plants. The comparative analysis revealed that nanoparticles synthesized through green methods exhibited enhanced biocompatibility and environmental friendliness towards plants when compared to their chemically synthesized counterparts. This observation may be attributed to the potential interaction between amino acids, sugars, organic acids, or fatty acids and the nanoparticle surface, resulting in a more favourable outcome. Overall, green synthesized silver nanoparticles showed long-term antimicrobial capabilities and reduced plant toxicity compared to chemically synthesized nanoparticles, making them ideal nano pesticides or nanoscale growth regulators in agriculture.

Keywords Silver nanoparticles · Green synthesis · Phytotoxicity · ROS · Eco-friendly

1 Introduction

The rapid advancement of nanotechnology has provided novel insights into the transformative potential of products from agriculture [1–8]. Silver nanoparticles (AgNPs) have a broad antibacterial spectrum, making them excellent pesticides. Chemically synthesized silver nanoparticles (chem-AgNPs), while widely used, pose significant drawbacks due to their high cost and potential adverse effects on both the environment and public health [9]. Consequently, the synthesis of green silver nanoparticles (g-AgNPs) has garnered increasing interest owing to their environmentally benign characteristics [10]. A multitude of investigations have been carried out to explore the methodologies of biologically mediated synthesis for the purpose of producing silver nanoparticles (AgNPs). These studies have focused on utilizing a diverse range of microorganisms including bacteria, fungi, actinomycetes, yeasts, viruses, and plants. The studies conducted by Singh et al. [10] and Tarannum et al. [11] have investigated the topic at hand. Plant-mediated green synthesis of g-AgNPs has been widely studied because of the presence of many plant species that include abundant functional metabolites, such as reducing and capping agents, which serve as reactants for the synthesis process [12].

The utilization of plant-mediated g-AgNPs holds significant potential in several applications, primarily attributed to their notable antioxidant, anticancer, and antibacterial properties [13]. According to studies, plant-mediated g-AgNPs can be employed as antioxidants because of their reducing power and ROS scavenging capabilities [14]. Moreover, previous studies have demonstrated the inhibitory effects of plant-mediated g-AgNPs on the growth of various cancer cell types. For instance,

Sarkar and Kotteeswaran [15] reported the inhibitory effects of these nanoparticles on human cervical cancer cells. Similarly, Cyril et al. [16] specifically investigated the inhibitory effects on lung cancer cells, specifically the A549 cell line. Hamelian et al. [17] conducted a study on human cervix carcinoma (hela NCBI-C115) cells and observed inhibitory effects. Additionally, Hajebi et al. [18] explored the impact of plant-mediated g-AgNPs on breast cancer carcinogenesis. Mortazavi-Derazkola et al. [19] investigated the inhibitory effects on human gastric cancer cells, while Samari et al. [20] and Verma et al. (2017) separately examined the impact on human colon cancer cells. More crucially, plant-mediated g-AgNPs have been demonstrated to have significant growth-inhibiting effects on a variety of microorganisms, including bacteria (*Pseudomonas nitroreducens* [21], *Escherichia coli* [22–24], *Staphylococcus aureus* [22, 25, 26], *Enterococcus*, *Acinetobacter baumannii*, *Proteus mirabilis*, *Pseudomonas*, *Klebsiella* [22], *Bacillus cereus* [27], *Bacillus subtilis* [28] and *Pseudomonas aeruginosa* [29]) and fungi (*Aspergillus unguis* [21], *Aspergillus flavus*, *Aspergillus parasiticus* [25], *Aspergillus japonicas* [24], *Aspergillus nidulans* and *Penicillium chrysogenum* [29]). Plant-mediated g-Ag nanoparticles (NPs) exhibit potent pathogen suppression, making them highly suitable for utilization as nano-pesticides in agricultural applications.

One crucial measure for ensuring the safe and sustainable utilization of silver nanoparticles (AgNPs) is the implementation of an environmental impact assessment. Prior research has demonstrated that plants have pronounced adverse effects when exposed to chemically produced nanoparticles (NPs). Exposure to silver nanoparticles (AgNPs) has been found to induce growth [30] and reproductive [31] toxicity in plants, primarily attributed to the excessive generation of reactive oxygen species produced by the presence of AgNPs (Zhang et al. 2018) or the release of Ag^+ ions from AgNPs [32]. The possible impact of AgNPs on crop productivity and quality over the entire growing period raises concerns regarding food security [33]. In contrast, it has been observed that green-produced nanoparticles (NPs) exhibit favourable impacts on plants as opposed to exhibiting hazardous consequences. In a study conducted by Acharya et al. [34], it was shown that the utilization of green-generated nanoparticles (NPs) resulted in notable improvements in the growth of seedlings as well as the production and quality of onion crops. The proposition was made that green-manufactured nanoparticles (NPs) exhibit greater biocompatibility and environmental friendliness towards plants in comparison to chemically synthesized NPs. This observation may be attributed to the presence of amino acids, sugars, organic acids, or fatty acids that form a protective layer on the surface of the NPs [35]. Nevertheless, there has been limited scholarly focus on investigating the phytotoxicity associated with nanoparticles (NPs) generated using green methods. However, limited research has been dedicated to investigating the phytotoxicity associated with the synthesis of nanoparticles using green methods.

The present chapter aims to investigate the antibacterial activity and phytotoxicity of silver nanoparticles (AgNPs) that are generated using green synthesis methods as well as those synthesized chemically. The outcomes of this investigation will not solely contribute to the existing knowledge regarding the responses of plants and bacteria towards green and chemically synthesized silver nanoparticles (AgNPs), but will also yield significant insights into the environmental implications associated with the utilization of green nanoparticles. The aforementioned insights are anticipated to make a significant contribution towards the advancement and endorsement of sustainable practices within the domains of nanotechnology and nanomaterials.

2 History of Nanomaterials

The human cognitive ability to engage in dreaming and imagination has been observed to often act as a catalyst for the emergence of novel scientific advancements and technological innovations. The field of nanotechnology, which has gained prominence in the twenty-first century, can be traced back to its inception driven by ambitious goals. Nanotechnology is a field of study that involves the thorough understanding and control of matter at the scale of 1 to 100 nm. At this scale, unique phenomena arise, which can be harnessed for the development of novel applications. The historical record reveals a notable occurrence of human exposure to nanoparticles, which has persisted over the course of human civilization. Notably, a substantial increase in such exposure was observed during the era commonly referred to as the industrial revolution. The exploration and analysis of nanoparticles is a field of study that has been ongoing for a considerable period of time. Richard Zsigmondy, a distinguished scientist who was awarded the prestigious Nobel Prize in Chemistry in 1925, is credited with the introduction of the term "nanometer." The person under consideration is acknowledged for their contribution in coining the term "nanometer," with a specific focus on its application in delineating particle sizes. In addition, the researchers in question were at the forefront of employing a microscope as a means to quantify the dimensions of various particles, such as gold colloids [36].

The inception of modern nanotechnology can be attributed to the visionary contributions of Richard Feynman, a distinguished physicist who was honoured with the Nobel Prize in Physics in 1965. At the 1959 American Physical Society meeting held at Caltech, a lecture titled "There's Plenty of Room at the Bottom" was delivered by the esteemed individual. This lecture served as a platform for the introduction of a groundbreaking concept, namely the manipulation of matter at the atomic scale. The novel concept put forth by Feynman exhibited a paradigm shift in cognitive frameworks, and subsequent empirical investigations have substantiated the veracity of his hypotheses. Due to these aforementioned factors, he is widely acknowledged as the progenitor of contemporary nanotechnology.

Approximately 15 years subsequent to Feynman's speech, a Japanese scientist named Norio Taniguchi became the pioneer in employing the term "nanotechnology" to delineate semiconductor procedures transpiring at the nanoscale scale. The individual expressed the viewpoint that nanotechnology encompasses the manipulation of materials at the atomic or molecular level, involving processes such as processing, separation, consolidation, and deformation. The inception of the nanotechnology era can be traced back to the 1980s, during which significant advancements were made. Notably, the discovery of fullerenes by Kroto, Smalley, and Curl, as well as the utilization of concepts from Feynman's seminal work "There is Plenty of Room at the Bottom" and Taniguchi's term "nanotechnology" by Eric Drexler of the Massachusetts Institute of Technology (MIT) in his 1986 book titled "Engines of Creation: The Coming Era of Nanotechnology," played pivotal roles in shaping this transformative field. The concept of a nanoscale "assembler" was introduced by Drexler, suggesting its capability to construct replicas of itself and other objects with varying levels of intricacy. The concept of nanotechnology as envisioned by Drexler is commonly referred to as "molecular nanotechnology". The field of nanotechnology has witnessed notable advancements following the pioneering work of Iijima, a distinguished Japanese scientist, who introduced carbon nanotubes.

The onset of the twenty-first century witnessed a surge in attention towards the nascent domains of nanoscience and nanotechnology. Feynman's significance and his theoretical framework for the manipulation of atomic matter had a profound impact on the establishment of scientific goals within the United States. During a lecture delivered at the California Institute of Technology on January 21, 2000, President Bill Clinton expressed his support for the allocation of financial resources towards the advancement of research in this nascent field of technology. Three years after its initial proposal, President George W. Bush enacted the 21st Century Nanotechnology Research and Development Act. The enactment of the Act established nanotechnology research as a paramount concern at the national level, leading to the establishment of the National Technology Initiative (NNI). Currently, the National Nanotechnology Initiative (NNI) is overseen by a hierarchical structure that includes the National Science and Technology Council (NSTC), led by the President's Cabinet, and its Committee on Technology. The Subcommittee on Nanoscale Science, Engineering, and Technology (NSET) of the Committee is tasked with the strategic planning, financial allocation, execution, and evaluation of the National Nanotechnology Initiative (NNI). This subcommittee is composed of delegates from 20 governmental departments as well as independent agencies and commissions inside the United States.

3 Different Approaches to Nanoparticle Synthesis

In general, the synthesis of nanoparticles (NPs) is conducted by 3 distinct processes, which are outlined as follows.

3.1 Physical Synthesis Methods

The synthesis of AgNPs employs many physical techniques, including evaporation–condensation, spark discharge, and transformation. Nanoparticles (NPs) composed of metals can also be manufactured using the thermal breaking approach, resulting in an average size falling within the range of 9.5 nm [37]. Jung et al. found that the best way to make metal nanoparticles is to use a ceramic ware heater with a small heating area to help evaporate the materials that go into making the nanoparticles. The findings of the study revealed a positive correlation between the rise in ambient temperature and the increase in the concentration of nanoparticles (NPs) [38]. As soon as the surface heater ceases to generate any flux, the aggregation of nanoparticles will progressively intensify. The determination of spherical-shaped nanoparticles (NPs) without any agglomeration is possible even at high concentrations and elevated temperatures [39]. A different way to make nanoparticles with an arc discharge is to make silver nanoparticles (AgNPs) in double-deionized water without using any extra surfactants. The physical synthesis method employed for the production of AgNPs typically encompasses the application of diverse forms of energy, including an arc discharge, an arc power, heating energy, and other related sources. The physical formulation methodology is a commonly employed methodology in the synthesis of silver nanoparticles (AgNPs) on a large scale. Nevertheless, it is crucial to acknowledge that this particular approach is associated with a comparatively elevated upfront investment in equipment, substantial energy usage, potential solvent contaminants, and an absence of an even distribution.

3.2 *Methods for Chemical Synthesis*

The implementation of chemical methodologies, as outlined in Table 1, typically involves the utilization of metal-based precursors, agents for reducing, and restoring stability or sealing substances. The reduction of silver salts entails a series of steps, commencing with the formation of nucleation sites, which are subsequently succeeded by the growth of nanoparticles. The chemo-reduction process is widely acknowledged as the predominant technique utilized in the synthesis of stable nanoparticles. The chemicals utilized in the synthesis of AgNPs, including elemental hydrogen, ascorbate, citrate, borohydride thio-glycerol, and 2-mercaptoethanol, exhibit inherent properties that are associated with potential hazards.

Additionally, the synthetic nanoparticles may not achieve the expected level of purity due to the possibility of chemical deposition on their surfaces. The preparation of AgNPs of the desired size can pose a challenge, necessitating an additional step to mitigate particle aggregation. Furthermore, it is worth noting that a significant number of toxic and hazardous byproducts are removed throughout the synthesis process. Previous studies have demonstrated that the utilization of solid reductants, such as borohydride, induces the creation of minute particles that exhibit a certain degree of monodispersity. Nevertheless, the production of bigger particles can be achieved by employing low reductants, such as citrate, which exhibit a reduced rate of reduction. Therefore, the range of sizes can exhibit significant variation [48]. Various chemical techniques are utilized in scientific research, including cytochemical synthesis, laser ablation, lithography, electrochemical reduction, laser irradiation, sonolytic breakdown, thermal decomposition, and chemical reduction. Furthermore, metal nanoparticles (NPs) that are disseminated and exhibit instability can be effectively stabilized

Table 1 Overview of Silver Nanoparticle Synthesis Parameters and the implementation of chemical methodologies, typically involves the utilization of metal-based precursors restoring stability or sealing

S.no	Silver salt	Weakening agent	Agent for Capping	Size of Ag (nm)	The source
1.	Silver nitrate	Sodium borohydride	Tri-sodium citrate	5	Rahimi et al. [40]
2.	Silver nitrate	Ethylene glycol	Poly vinyl pyrrolidone	50–175	Diaz-Alvarez et al. [41]
3.	Silver nitrate	Sodium hydroxide	Polyanionic Na + poly (γ-glutamic acid)	17.2 ± 3.4–37.3 ± 5.5	Liu et al. [42]
4.	Silver nitrate	Poly (vinyl pyrrolidone) and gelatin	Glucose, fructose, lactose, and sucrose	35	Yu et al. [43]
5.	Silver nitrate	Aniline	Ethyltrimethylammonium bromide	10–30	Murgunde et al. [44]
6.	Silver nitrate	D-Glucose	Carboxy methyl cellulose, Sodium hydroxide	5–15	Proessl et al. [45]
7.	Silver nitrate	Sodium hydroxide	Alkali lignin (low sulfonate)	5–100	Kuntyi et al. [46]
8.	Silver nitrate	Hydrazine hydrate and sodium citrate	Sodium dodecyl sulfate	10–20	Massironi etal. [47]

through the utilization of particular protective agents. The aforementioned technique is commonly employed to provide protection to nanoparticles (NPs) by the utilization of protective agents that can either bind to or be absorbed onto the surface of NPs, thereby impeding their aggregation [49].

In this case, silver nanoparticles (AgNPs) can be made smaller by using sodium borohydride and dodecanethiol, which act as stabilizers to keep the nanoparticles from sticking together in certain solvents [50]. The aforementioned observation suggests that by making minor modifications to the established parameters, notable enhancements can be achieved in various characteristics of the nanomaterial, including its thickness, distribution of size, average dimension, stability, and shape. In the realm of gold nanoparticle production, it is customary to employ stabilizers, including but not limited to polymethyl methacrylate, polyvinyl pyrrolidone, polymethacrylic acid, polyethylene glycol (PEG), and various other polymers [51].

The utilization of a branching polymer, specifically methylene-bis-acrylamide aminoethyl piperazine with dimethylamine groups at the termini, denoted as HPAMAM-$N(CH_3)_2$, has been documented in the synthesis of silver and gold colloids. The predominant factors contributing to the branching architecture of HPAMAM-$N(CH_3)_2$ can be ascribed to the inclusion of the piperazine ring, tertiary amine moiety, and amide functionality. The structural elements in question are of utmost importance in augmenting the stabilizing and reducing characteristics exhibited by the nanoparticles (NPs). The distinctive characteristics exhibited by metal nanoparticles (NPs) can be achieved by manipulating the molar ratio concentration of gold and silver [52]. This manipulation leads to the manifestation of low and narrow distribution rates in the NPs. The production of silver nanoparticles (AgNPs) can be achieved through a biphasic methodology, which entails the utilization of two separate phases, specifically aqueous and organic. The efficacy of this particular approach is contingent upon the successful segregation of the reactants within solvents that exhibit negligible intermolecular interactions. The kinetics of the reaction between the reducing agent and the precursor to metal are affected by the interaction between the outermost barrier and the exchange of substances between the organic and aqueous phases. The response in question is facilitated by a compound known as a quaternary alkylammonium salt [53]. This methodology provides a precise and analogous means of manufacturing nanoparticles (NPs) by manipulating the choice of solvent. The process employed for the production of AgNPs may incur significant costs, and the resulting nanoparticles cannot be separated from the mixture in which they are formed [54]. The particles made by this process have several benefits, such as the ability to spread quickly in natural environments, to form colloids in liquids that are not water, and to speed up chemical reactions in non-polar solvents. The practical applications of nanoparticles have garnered significant attention, leading to efforts to assemble them in various chemo-physical settings. Nevertheless, the occurrence of a solution produced in a non-polar solvent is infrequent [55]. Hence, the preferred method for reducing Ag and gold is the utilization of photoreduction. The photoreduction of metal ions in DNA complexes takes place in aquatic environments, leading to the formation of solid metallic nanoparticles. The approach employed in organizing the AgNPs has been found to yield enhanced characteristics when subjected to UV irradiation [56].

3.3 Green or Biological Processes for AgNP Synthesis

Biologically green synthetic approaches present a highly convenient option for addressing the limitations inherent in synthetic chemical methods. The utilization of a greener approach presents a straightforward and cost-efficient method for synthesizing AgNPs. This methodology is also known for its reliability and environmentally friendly nature. Additionally, an advantageous aspect of this strategy is the ease with which the appropriate solvent medium can be selected, and depicted. Significant attention has been devoted to the production of silver nanoparticles (AgNPs) characterized by accurate dimensions and geometries. Numerous biological systems, encompassing plant extracts and microorganisms such as fungi, bacteria, and minute biomolecules like vitamins and amino acids, have been utilized as alternative sources for the specific objective of diminishing and capping agents. The utilization of starch as a coating mediator in the manufacture of AgNPs has been observed to have potential as a reducing agent, specifically β-D-glucose, in a mildly heated setup [57].

In order to mitigate the reliance on organic solvents, a solvent alternative for the synthesis of silver nanoparticles (AgNPs) involves the utilization of an aqueous solution of starch. Several bacterial species, namely *Pseudomonas stutzeri*, different strains of *Lactobacillus*, *Bacillus licheniformis*, *Escherichia coli*, and *Brevibacterium casei*, have been utilized in conjunction with fungi such as *Fusarium oxysporum* and *Ganoderma neo-japonicum Imazeki*, as well as plant extracts derived from *Allophylus cobbe*, *Artemisia princeps*, and *Typha angustifolia*, for the purpose of synthesizing silver nanoparticles (AgNPs) [58].

In a study published in a scientific journal, it was found that a considerable amount of silver nanoparticles (AgNPs) can be produced by utilizing an aqueous solution comprising the ions of silver (Ag^+) and employing *Bacillus licheniformis* as the stabilizing agent. The utilization of a biogenic strategy including microorganisms for the production of stable AgNPs has several advantages compared to alternative procedures. One notable advantage is that the microorganisms employed in this process are nonpathogenic [59]. Various biomolecules, such as biopolymers, starch, fibrinolytic enzymes, and amino acids, are utilized in the production of AgNPs (Table 2).

The green synthesis method employs plant extracts as a means of obtaining capping and reducing agents for the manufacture of silver nanoparticles (AgNPs). It is hypothesized that the biomolecules included in these extracts, particularly secondary metabolites, possess the ability to effectively decrease Ag ions. The composition of these entities includes vitamins, polysaccharides, amino acids, proteins, enzymes,

Table 2 This table focuses on the biological processes used in the production of silver nanoparticles (AgNPs), specifically highlighting the use of plant extracts as reducing agents. It details various plant extracts that have been effectively used, their specific roles and mechanisms in the reduction process, and the resulting characteristics of the AgNPs produced. The table also includes information on the concentration and conditions of the reaction, as well as the observed benefits and potential applications of using plant extracts over chemical methods. This comprehensive overview serves as an invaluable resource for understanding the eco-friendly and sustainable methods of producing AgNPs through biological means

Silver salt	Plant source	Shape/form	Size of Ag (nm)	Key refs.
Silver nitrate	*Coffee arabica*	Round and ellipsoidal	20–30	[60]
Silver nitrate	*Artocarpus heterophyllus lam*	Irregular	10.78	[61]
Silver nitrate	*Adenium obesum*	Round	10–30	[62]
Silver nitrate	*Aloe barbadensis*	Round	15.2 ± 4.2	[63]
Silver nitrate	*Nyctanthes arbor-tristis* L.	Round	50–80	[64]
Silver nitrate	*Vigna radiata*	Round and oval	5–30	[65]
Silver nitrate	*Tribulus terrestris* L.	Round	16–28	[66]
Silver nitrate	*Bunium persicum*	Round	20–50	[67]
Silver nitrate	*Prunus yedoensis*	Round and oval	20–70	[68]
Silver nitrate	*Cinnamomum camphora*	Round	55–80	[69]

polyphenolics, flavonoids, and various other secondary metabolites [70, 71]. Similarly, the proteins included in the plant extracts have a crucial function in reducing Ag^+ ions and regulating the size of AgNPs during their creation. The presence of functional groups, specifically the carboxylic group (COOH) in glutamine and aspartic residues, as well as the hydroxyl group (OH) in tyrosine residues, has been identified as the primary factors contributing to the stabilization of silver ions and the formation of small nanoplates of silver with minimal variation in size [72]. Various plant extracts, including alfalfa, lemongrass soup, geranium leaves, capsicum annum, and aqueous geranium extract, have been employed as bio-reductants for the synthesis of nanoparticles [73–76]. The proteins present in mushrooms have the ability to function as a stabilizing agent, leading to the reduction of Ag ions and the subsequent formation of nanoparticles (NPs) with distinct shapes. These NPs have an average size of roughly 30.6 nm [73]. The utilization of glutathione as a stabilizing agent for the preparation of water-soluble AgNPs has been extensively studied, with numerous medical applications being explored [77]. The reduction of Ag ions is attributed to the tryptophan residues located at the C-terminus of the structured oligopeptides [77]. The synthesis of AgNPs can be achieved through the utilization of vitamin E as a reducing agent, as proposed by the Langmuir Blodgett technique [78–80]. Oleic acid has been employed in the synthesis of environmentally friendly and biodegradable AgNPs [81]. Different types of microorganisms are utilized for the extracellular or intracellular expansion of AgNPs [82, 83]. One example involves the utilization of Pseudomonas stutzeri bacteria to synthesize nanocrystals with modified compositions, specifically those containing Ag [84]. Fungal species, including *Fusarium oxysporum*, *Chrysosporium phanerochaete*, and *Trichoderma viride*, have been observed to participate in the reduction of Ag^{1+} ions during the creation of silver nanoparticles [85]. In addition, the reduction of Ag ions can also be achieved by Klebsiella pneumonia, Enterobacter cloacae, and Escherichia coli, as stated in the reference [86]. Biologically produced silver nanoparticles (AgNPs) exhibit distinct attributes, including hydrophilicity, a substantial surface area, and enhanced stability.

The synthesis approach possesses numerous advantages, including its cost-effectiveness, simplicity, regenerability, and low energy requirement in comparison to particles created using chemical methods. The macrophages containing AgNPs have been successfully acquired, and their physical characteristics indicate a lack of stability, as evidenced by the presence of both rod-shaped and spherical nanoparticles ranging in size from 5 to 40 nm [87, 88]. The investigation focused on the analysis of nanoscale particles that were generated by biosynthesis. These particles were examined in the periplasmic region of the bacterial cell, which is situated between the inner and outer cellular membranes.

Lactobacillus species have also been employed as potential sources of capping and reducing agents. The formulation of AgNPs by lactic acid bacteria is contingent upon the specific lactobacillus species, as both the distribution and positioning of the particles within the cell are influenced by this factor. Similarly, the thickness of the biosynthesized silver nanoparticles (AgNPs) generated using this method exhibited variations depending on the specific strains of lactobacillus utilized. Under standard temperature and pressure settings, the rate of reduction of metal ions using the biological process is significantly higher compared to alternative procedures [89].

Similarly, Zuorro et al. were able to effectively produce silver nanoparticles (AgNPs) using waste materials from bilberry and red currant. These waste materials served as both reducing and stabilizing agents [90]. In a cost-effective and environmentally sustainable approach, AgNPs were synthesized using phenolic extracts derived from agricultural and industrial byproducts, including bilberry and wasted coffee grounds, as reducing agents [91]. The effectiveness of AgNPs in biological systems is contingent upon their morphology and structure, which are regulated by the size and form of the particles. Despite the synthesis of AgNPs with distinct form and size ranges in various research studies, these nanoparticles nevertheless possess significant limitations. The synthesis of uniform-sized silver colloids was conducted using a highly effective reducing agent, namely sodium borohydride ($NaBH_4$) [89, 92]. While the green synthesis methodology offers advantages compared to chemical synthesis, there may be a limited number of AgNPs produced utilizing this method. The optimization of the form, size, and distribution of the generated nanoparticles can be achieved by controlling synthesis procedures, reduction and stabilization factors, as well as parameters including temperature and pH [93].

3.3.1 Green Synthesis of Nanoparticles

The process of synthesizing nanoparticles from biological sources is commonly referred to as green synthesis or biosynthesis of nanoparticles. Plants, including their root, stem, leaf, fruit, and seeds [94], as well as microorganisms such as bacteria, fungi, and algae [95], together with biopolymers including starch, alginate, cellulose, chitosan, and proteins [96], have been employed for this particular objective. The application of green synthesis methodologies in nanoparticle production is widely recognized as a favourable and environmentally friendly approach (Fig. 1). When juxtaposed with conventional physical and chemical methodologies, the utilization of this specific technology presents a multitude of distinct benefits. However, it is important to note that the green synthesis of nanoparticles may be subject to certain limitations. These limitations include the potential ecological imbalance that may arise from the exploitation of natural biological sources, as well as the seasonal variation in the availability and composition of phytochemicals [96].

The process of green synthesis of nanoparticles necessitates the inclusion of three fundamental constituents, specifically reducing agents, solvents, and capping agents. Biomolecules, including polyphenols, flavonoids, terpenoids, tannins, alkaloids, polysaccharides, proteins, amino acids, and vitamins, can serve as reducing agents in the creation of nanoparticles, either through their natural occurrence or through production by plants and microbes [97]. These substances facilitate the reduction of metal ions to a valence state of zero. Subsequently, the functional groups included in these fundamental biopolymers and phytochemicals aid in stabilizing the produced nanoparticle. Plant-based nanoparticle synthesis is the preferred method

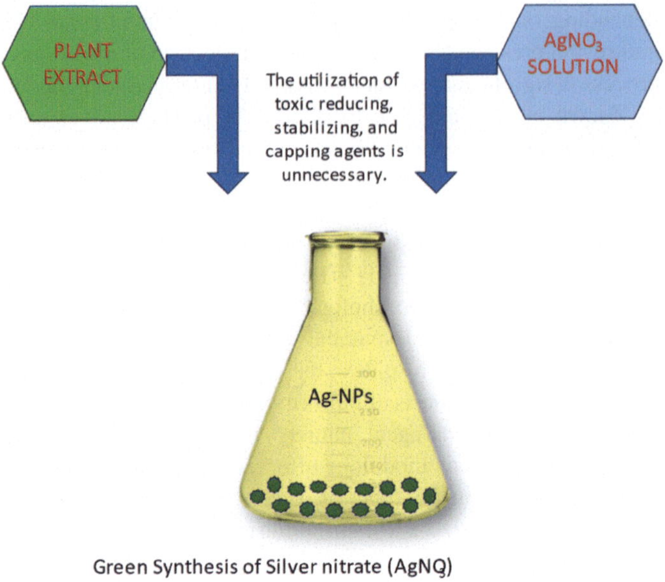

Fig. 1 This figure illustrates the process of green synthesis of silver nanoparticles (AgNPs) using plant extracts. It provides a step-by-step visual representation of how plant extracts act as reducing and stabilizing agents to transform silver ions into AgNPs. The figure details the selection of plant materials, extraction methods, reaction conditions, and the formation of nanoparticles, highlighting key factors such as concentration, temperature, and pH. Additionally, it visually depicts the role of phytochemicals present in the extracts in the reduction and stabilization of the nanoparticles. This diagram serves as an educational tool to understand the eco-friendly and sustainable approach to synthesizing AgNPs through biological means

among the three biological pathways due to several advantageous factors. This is because there are more plants, they are safe and non-toxic, and there is a wide range of phytoconstituents that can be used as effective reducing agents [98]. The synthesis of nanoparticles occurs through a three-step procedure that includes the reduction of metal ions, as evidenced by observable colour changes, followed by the clustering of particles, and ultimately the stability of the resulting nanoparticles [99]. It depicts the streamlined methodology employed in the manufacture of nanoparticles. Various physical factors, such as temperature, duration, and the ratio of precursor chemical to biomaterial, as well as chemical factors like pH, molarity of precursor chemical, and concentration of extracts, have an impact on the production of green nanoparticles. As an illustration, the promotion of smaller-sized nanoparticles is facilitated by a slightly alkaline pH as a result of accelerated reactions, but at a higher pH, the aggregation of nanoparticles takes place [100]. Hence, it is imperative to adjust the processing conditions in order to achieve favourable morphological characteristics and composition, which are crucial for enhancing the productivity and efficiency of nanoparticle synthesis on a large scale.

3.4 Synthesis of AgNPs via a Photochemical Process

In the realm of photochemical synthesis, researchers commonly employ two distinct techniques: the photo-physical (top to bottom) approach and the photochemical (bottom to up) approach. The primary methodology utilizes metals in their elemental form, whereas the subsequent approach utilizes their corresponding ionic precursors. The direct photo-reduction approach is a widely utilized method in the synthesis of nanoparticles (NPs), encompassing various metallic materials. Photosensitization is a well-established methodology utilized in the fabrication process of silver nanoparticles (AgNPs). This technique involves the reduction of metallic ions through the activation of excited species generated via photochemical mechanisms [101]. The direct photoreduction of silver nitrate in the presence of sodium citrate at ambient temperature has been observed to exhibit facilitation by various light sources. These light sources encompass a spectrum of wavelengths, including blue, ultraviolet (UV), white, orange, green, and cyan [102, 103]. The optical properties associated with the size and form of nanoparticles (NPs) are influenced by different light sources. Stable silver nanoparticles (AgNPs) can be generated using a straightforward and consistent UV photoactivation method in an aqueous Triton X-100 medium [104]. The molecules of Triton X-100 exhibit a dual functionality reminiscent of stabilizers commonly employed in nanoparticle systems. Specifically, they serve as both capping agents, which prevent the aggregation and growth of nanoparticles, and reducing agents, which facilitate the reduction of metal ions to form nanoparticles. The dispersion-controlled technique is a method that involves the use of a surfactant solution to aid in the growth and restoration of nanoparticles, thereby ensuring the maintenance of their size distribution. A distinct inquiry was conducted to examine the successful production of silver nanoparticles (AgNPs) by employing carboxymethylated chitosan and a water-based alkaline solution of silver nitrate, with the assistance of UV light exposure. The carboxymethylated chitosan (CMCTS) derivative, which is both biocompatible and water-soluble, has demonstrated a range of properties that render it highly suitable for numerous applications. One potential application of this substance lies in its capacity to function as a reducing agent for Ag^+ ions, thereby facilitating the reduction of silver ions to silver nanoparticles (NPs). Additionally, it exhibits stabilizing properties, rendering it suitable for the stabilization of silver nanoparticles with a thickness ranging from 2 to 8 nm [104].

The pertinent factors in photochemical synthesis techniques encompass several key aspects. Firstly, these techniques provide a means of conducting photo-induced processing in a sterilized manner, thereby ensuring enhanced accessibility and improved resolution of use. Secondly, they facilitate the effective generation of reducing agents and the production of nanoparticles through the utilization of photoirradiation. Lastly, these techniques offer notable flexibility in their application. The photochemical synthesis technique enables the formulation of nanoparticles in many distinct environments, including emulsions, surfactant micelles, polymer films, cells, and glasses [105].

4 Characterisation of Nanoparticles AgNPs

The process of AgNPs synthesis, a crucial stage involves the characterization of nanoparticles. Characterization refers to the process of quantifying the physical and chemical attributes of nanoparticles. Ensuring the suitability of nanoparticles for application in the biomedical industry is an imperative and pivotal stage. The characterization of silver nanoparticles plays a vital role in ensuring their safety and usefulness. Minor alterations in the desirable characteristics of nanoparticles can potentially yield significant ramifications for their therapeutic efficacy. Various techniques can be employed to characterize nanoparticles, with transmission electron microscopy (TEM) and UV-visible spectroscopy being the most prevalent methods. Additional commonly used techniques in this field encompass X-ray diffraction, Fourier-transform infrared spectroscopy (FTIR), nuclear magnetic resonance (NMR) techniques, energy dispersive X-ray spectroscopy, Brunauer–Emmett–Teller (BET) analysis, small-angle X-ray scattering, X-ray photon spectroscopy, and selected area diffraction. UV-visible spectroscopy is a rapid, straightforward, and effective method for characterizing nanoparticles. The phenomenon of surface plasmon resonance (SPR) is shown by silver nanoparticles. The phenomenon observed in silver atoms, wherein the absorption of light energy leads to the resonance of electrons between the valence band and conduction band, is responsible for the emergence of a surface plasmon resonance (SPR) band. The absorption band's peak, also known as the absorbance maximum, is influenced by several elements, such as the shape and size of the nanoparticles. The effective characterization of the synthesized nanoparticles can be achieved by comparing the acquired absorbance maximum value with the corresponding value reported in the literature. Typically, transmission electron microscopy (TEM) is commonly used in the characterization of nanoparticles due to its ability to provide insights into both physical features, such as shape and size, and chemical properties [106]. Transmission electron microscopy (TEM) utilizes an electron beam as a means of characterization. The electron beam is allowed to traverse the silver nanoparticle sample, resulting in electron-nanoparticle interactions. The translation of electron interactions into visual representations facilitates the subsequent analysis and extraction of pertinent properties.

5 Conditions to Ensure Efficient Antimicrobial Action

Different ways of making nanoparticles result in nanoparticles with slightly different shape, size, and distribution, it is important to understand the properties that affect the antimicrobial effectiveness of silver nanoparticles before looking at the different ways of making them. The size, shape, particle composition, cell type, and type of reducing agents used during their synthesis are a few factors that affect the cytotoxicity of silver nanoparticles. Hence, it is crucial to ensure that the synthesized silver nanoparticles exhibit uniformity in four key characteristics, namely size, shape, morphology, and functionality. In terms of size, research has indicated that smaller nanoparticles exhibit greater stability and demonstrate enhanced antibacterial efficacy.

Li et al. conducted a study to investigate the correlation between size and antibacterial activity. The researchers examined the activity of silver nanoparticles with several diameters (5, 15, and 55 nm). Their findings indicated that the nanoparticle with the smallest diameter (5 nm) exhibited the highest level of biological activity [107]. Numerous studies found in the existing literature have supported this association. Additionally, it has been determined that although all silver nanoparticles exhibit exceptional antibacterial properties, the cytotoxicity is further heightened when a large concentration of small silver nanoparticles is employed [108]. The correlation between size and biological activity has been attributed to the phenomenon that smaller nanoparticles possess a greater surface area, facilitating a more substantial interaction between the nanoparticle and the microorganism. Smaller nanoparticles exhibit an enhanced ability to traverse the cellular membranes of microorganisms, resulting in membrane disruption and increased permeability. Consequently, this process ultimately culminates in cellular lysis. The utilization of smaller nanoparticles also results in enhanced intracellular presence, hence augmenting cytotoxicity.

The form of a nanoparticle affects its bioavailability and antibacterial properties. Nanoparticles are present in diverse morphologies, encompassing spherical, triangular, and hexagonal shapes. The maximum toxicity of hexagonal nanoparticles has been discovered [109]. The relationship between toxicity and shape has been attributed to variations in surface area among different nanoparticle shapes. Therefore, it is clear that a nanoparticle's size and shape have a significant impact on its cytotoxicity and antimicrobial efficacy. Therefore, it is imperative that the selected approach for nanoparticle synthesis yield a substantial quantity of silver nanoparticles with precise dimensions and morphology in order to guarantee their efficacy as an antibacterial agent.

6 Comparison of Phytotoxicity Between Chemically Synthesized and Green Synthesized of AgNPs

The investigation of phytotoxicity in higher plants is crucial for the development of a comprehensive toxicity profile for nanoparticles. The seed germination and root elongation tests are common acute phytotoxicity assays. It has many benefits, such as being sensitive, easy, cheap, and able to be used with volatile substances or materials [110]. The results of our study show that the concentration of both chemically made and green synthesized silver nanoparticles (AgNPs) had a big effect on the germination of seeds and the length of roots and shoots (data not shown). This effect was observed across all six species examined, namely Lolium, wheat, bean, common vetch, canola, and lettuce. The solutions containing green and chemical AgNPs exhibited varying levels of phytotoxicity across varied concentrations. The

IC50 values, representing the concentrations at which 50% inhibition occurs, were shown to be higher for seed germination, root length, and shoot length in seeds or seedlings exposed to green-synthesized AgNPs compared to those exposed to chemically synthesized AgNPs. The inhibitory effect of chemically synthesized AgNPs on seed germination and seedling growth was found to be more pronounced compared to green-produced AgNPs. Hence, it can be inferred that AgNPs synthesized by green methods exhibit a lower level of phytotoxicity compared to AgNPs generated through chemical means. This disparity in phytotoxicity can be attributed to variations in the preparation techniques employed rather than differences in particle size or diameter.

In the investigation conducted by Krishnaraj et al. [111], the focus was directed towards examining the interactions between silver nanoparticles and *Bacopa monnieri* (Linn.) Wettst. plants cultivated in an aqueous medium. The silver nanoparticles utilized in this study were synthesized through the amalgamation of $AgNO_3$ with aqueous extracts derived from *Acalypha indica* Linn. leaves. The observed impact of green silver nanoparticles (AgNPs) on the development and metabolism of *Bacopa monnieri* (Linn.) was noted by the researchers. It was observed that the presence of a minimal level of stress had an effect on the aforementioned plant species. The investigation revealed that the application of 100 ppm of green AgNPs did not elicit any discernible impact on the germination efficacy, length of the roots, or shoot size. A comprehensive suppression of growth was observed upon exposure to a concentration of 100 ppm of $AgNO_3$. The examination of root and stem anatomy under a light microscope revealed the detection of structure-related abnormalities, as reported by the researchers. Furthermore, the verification of the existence of silver in the root and stem tissues of *Bacopa monnieri* was conducted through the utilization of an atomic absorption spectrophotometer. The experimental findings indicate that the utilization of green silver nanoparticles (AgNPs) resulted in enhanced plant growth and reduced toxicity compared to silver nitrate ($AgNO_3$). This observation can be attributed to the higher concentrations of protein and carbohydrates present in the green AgNPs, as well as the lower levels of total phenol content, catalase, and peroxidase activities. In a study conducted by Panda et al. [112], the researchers observed that silver nanoparticles, synthesized using the fragrant spath of the male inflorescence of screw pine (*Pandanus odorifer* Forssk.), exhibited reduced cytotoxicity levels when compared to Ag^+ ions alone in the *Allium cepa* assay system.

In contrast, the study conducted by Amooaghaie et al. [113] aimed to investigate the harmful effects of chemical silver nanoparticles (AgNPs) and silver nitrate ($AgNO_3$) on the processes of seed germination and seedling growth in *Brassica nigra*. The researchers discovered that both chemical AgNPs and $AgNO_3$ had a negative impact on seed germination, lipase activity, soluble and reducing sugar concentrations in germinating seeds, and seedling growth. However, the detrimental effects were more significant for chemical AgNPs compared to $AgNO_3$. In the study conducted by Qian et al. [114], it was observed that chemical AgNPs had a more pronounced inhibitory effect on the root elongation of Arabidopsis thaliana compared to $AgNO_3$. The researchers provided evidence that chemical AgNPs can accumulate in leaves and disrupt chloroplast function, leading to changes in the transcription of antioxidant enzymes and aquaporin genes. These effects were observed to be more pronounced with AgNPs compared to $AgNO_3$, indicating that AgNPs exhibit greater toxicity than $AgNO_3$. Based on the findings of the literature research, this study presents a novel comparison between the phytotoxic effects of chemical silver nanoparticles (AgNPs) and green silver nanoparticles, which has not been previously explored. Previous studies have frequently reported on the toxicity of chemical or green silver nanoparticles (AgNPs) in comparison to silver ions (Ag^+) Fig. 2).

In conclusion, the data we have collected, along with the existing literature, substantiates the assertion that green-synthesized AgNPs are a safer alternative to chemically synthesized AgNPs when considering their application in the field of agriculture. There is a potential for AgNPs synthesized through the utilization of plant extracts to possess a thin layer of capping organic material derived from the plant leaf broth. Consequently, this organic coating may result in a reduction in their toxicity in comparison to AgNPs synthesized using chemical procedures. Previous studies have demonstrated that the phytotoxicity caused by nanoparticles (NPs) can be influenced by their surface modification [115]. However, our research has not yet examined the effects of these surface modifications. Therefore, future investigations should focus on addressing this subject. El-Temsah and Joner [116] and Mazumdar

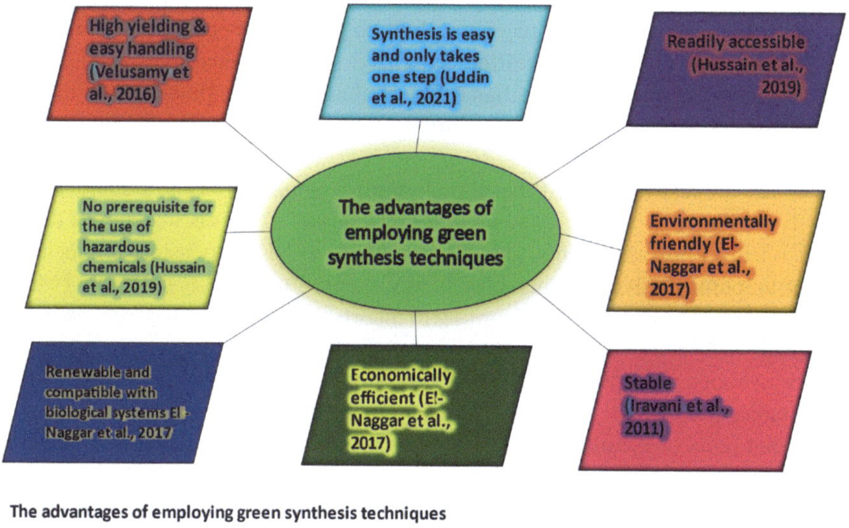

Fig. 2 This diagram offers a detailed insight into the advantages of employing green synthesis techniques for nanoparticle production, particularly silver nanoparticles (AgNPs). It visually compares green synthesis methods with conventional chemical synthesis, highlighting key benefits such as reduced environmental impact, lower toxicity, and the use of sustainable resources. The figure illustrates how plant extracts, algae, and other biological materials serve as reducing and capping agents, replacing harsh chemicals. Additionally, it showcases the energy efficiency, cost-effectiveness, and potential for large-scale production inherent in green synthesis methods. This diagram serves as an educational tool to emphasize the eco-friendly and sustainable aspects of green nanoparticle synthesis

et al. [117] both highlighted the phytotoxic properties of chemical AgNPs, namely their ability to hinder seed germination and impede the growth of seedlings. There is a chance that aqueous media will occupy the roughly 10 mm-long intracellular spaces within the seed coat parenchyma. This presence of aqueous media may aid in the facilitation of the transportation of soluble nutrients and tiny particles to the embryo. The observed suppression of seed germination by the tiny AgNPs may be attributed to their strong inhibitory effects. The study by Yin et al. [118] looked at how silver nanoparticles (AgNPs) were taken up by the grass species *Lolium multiflorum* and what bad effects they had. It was observed that the concentration of silver (Ag) in both the root and shoot of the plants rose as the exposure to silver nanoparticles (AgNP) increased. The growth of seedlings was found to be hindered by the presence of AgNPs.

Seedlings exposed to 40 mg L^{-1} GA-coated silver nanoparticles (AgNPs) exhibited inhibited root hair development as well as notable characteristics such as heavily vacuolated and collapsed cortical cells and damaged epidermis and root cap. On the other hand, seedlings that were subjected to equivalent amounts of $AgNO_3$ or supernatants derived from ultra-centrifuged AgNPs solutions did not exhibit any such abnormalities. The authors together proposed that the observed growth inhibition and cell damage can be attributable either to the nanoparticles themselves or to the capacity of AgNPs to release dissolved Ag that interacts with important biological receptors. In their study, Lee et al. [119] conducted an evaluation of the effects of citrate-stabilized silver nanoparticles (AgNPs) on the plant species Arabidopsis thaliana. The evaluation encompassed three distinct aspects: physiological phytotoxicity, cellular accumulation, and subcellular transport of AgNPs.

The researchers discovered that the phytotoxic impact of AgNPs could not be entirely attributed to the presence of silver ions alone. This conclusion was drawn based on the observation that plants exposed to AgNP suspensions collected a greater amount of silver compared to plants exposed to $AgNO_3$ solutions (where Ag is symbolic of silver ions). According to reports, nanoparticles have the potential to engage with plant roots through many methods, including adsorption onto the surface of roots, absorption into the cell walls of roots, and uptake into root cells. Both studies propose the potential movement of substances to the shoots and leaves from other parts of the plant [120, 121]. This research may provide an explanation for the observed retardation in root and shoot length caused by AgNPs. There is a prevailing belief that nanoparticles have the potential to disperse across the apoplast, which refers to the intercellular space. Once in this place, they have the ability to be absorbed by apoplastic membranes or gain entry into the xylem at points of injury, as suggested by Nowack and Bucheli in 2007. The toxicity of nanoparticles and their uptake into roots have been documented by researchers. This phenomenon can be attributed to various factors, such as the inherent characteristics of the nanoparticles, including their chemical composition, concentration, size, aggregation, and functionalization. Additionally, experimental conditions such as temperature, exposure duration, and method of exposure (e.g., seeds, seedlings, or cell suspensions), as well as the specific plant species involved, have been found to influence the uptake and toxicity of nanoparticles [120, 121]. According to our research, it has been observed that the toxicity of nanoparticles may also be attributed to the methods employed in their manufacture.

Based on the evaluation of seed germination and seedling growth across the six species, it was observed that lettuce and canola exhibited the highest susceptibility and tolerance, respectively, to the stress induced by AgNPs. According to the findings of [122], our data substantiated that the impact of AgNPs on seed germination was not definitively associated with taxonomy, nor could it be reliably predicted based on seed size. The inhibitory doses (IC50) of both green and chemical silver nanoparticles (AgNPs) were higher for seed germination compared to root elongation in all species, as shown in Table 1. This suggests that root growth is more susceptible to nanoparticle toxicity than seed germination. The seed coat assumes a crucial function in safeguarding the embryo against detrimental external influences. According to Lin and Xing [123], seed coats possess the ability to exhibit selective permeability. Hence, it is plausible that contaminants, which clearly impede root growth, may not have an impact on germination if they are unable to permeate the seed coverings. According to Lin and Xing [123], roots are typically the primary site of exposure to elevated levels of contaminants, making them more susceptible to the manifestation of toxic symptoms compared to shoots. In the conducted investigation, it was seen that shoots exhibited a greater IC50 value compared to roots. Additionally, it was noted that shoots were able to grow to a certain extent despite the inhibition of root elongation caused by the presence of green and chemical AgNPs.

In conclusion, this chapter shows that the utilization of N. sativa leaves can serve as a cost-effective alternative to conventional approaches. In previous studies, researchers have utilized seed extract from N. sativa to facilitate the environmentally friendly synthesis of gold nanoparticles (Au) [124] and silver nanoparticles (Ag) [125]. The utilization of cumin seeds in both fresh and dried forms for medicinal purposes, as well as their application as a natural food additive and spice, is evident. However, the leaves of cumin are not commonly employed in agriculture or animal feeding due to their high concentration of secondary metabolites, particularly phenolics (Bourgou et al. 2008). The current approach is characterized by its simplicity, safety, cost-effectiveness, lack of toxicity, and environmentally benign nature in comparison to the use of harmful chemicals. Moreover, this study represents a novel attempt to compare the cytotoxic and phytotoxic effects of chemically manufactured silver nanoparticles (AgNPs) with those synthesized using green methods. As anticipated, the results indicate that green synthesized AgNPs exhibit a higher level of safety compared to their chemically synthesized counterparts. This study presents a novel prospect for the synthesis of Ag nanoparticles utilizing natural ingredients, hence offering potential benefits for many applications. The results of our study have the potential to be applied in several fields, such as drug formulation, biomedical research, and agriculture, in the future.

7 Conclusion

Nanoparticles are a type of particles that have the ability to improve the qualities of plants, as well as boost seed growth and seed germination. Nanoparticles that are synthesized using environmentally friendly methods exhibit reduced cost. In contrast, chemical procedures such as the sol–gel technique and polymer precursor approach are associated with higher costs. This is the primary factor contributing to the preference for green-synthesized nanoparticles over chemically synthesized nanoparticles. The utilization of chemically synthesized nanoparticles has been explored for their potential efficacy and utility in promoting seed germination. However, our analysis of several scholarly literature indicates that these nanoparticles may have adverse effects on seed growth, perhaps exhibiting hazardous properties. Green-synthesized nanoparticles offer a superior method for enhancing seed germination while minimizing or eliminating harmful consequences. These nanoparticles are environmentally friendly, economically viable, and exhibit remarkable antibacterial capabilities. As previously stated, the impacts of chemically synthesized nanoparticles are contingent upon both the soil composition and the specific crop species. Various research has provided evidence supporting the notion that nanoparticles synthesized by green methods may exhibit reduced efficacy, yet possess the advantageous qualities of sustainability and eco-friendliness. Based on the aforementioned results, it can be observed that the utilization of chemically synthesized nanoparticles entails greater hazards compared to green synthesized nanoparticles, as evidenced by their heightened phytotoxicity effects on seed germination and seed growth. Ultimately, the analysis of gathered data reveals that in the future, the advantages of green-synthesized nanoparticles will surpass those of chemically synthesized nanoparticles, hence leading to a shift in their usage.

8 The Future Prospective Areas of Investigation

To date, numerous studies have been conducted to investigate the impact of various nanoparticles on plant growth parameters, including but not limited to seed germination, root length, and shoot length. Numerous studies have demonstrated the beneficial impact of nanoparticles on plant growth and their potential use in the agricultural domain. Various studies have also documented the detrimental impacts of nanoparticles on the process of seed germination, with the extent of these effects being contingent upon whether the nanoparticles were produced by environmentally friendly (green) methods or chemical methods. Green-synthesized nanoparticles have been shown to predominantly exhibit a beneficial impact on seed germination and agricultural practices. Conversely, chemically synthesized nanoparticles have been found to exert a detrimental influence on plant growth and agricultural endeavors. The impact of various nanoparticles on plant growth exhibits variability and is contingent upon factors such as the method of synthesis (green or chemical), as well as the size and concentration of the solution employed. Furthermore, further research is warranted to investigate the diverse impacts of metal and metal oxide nanoparticles on augmenting crop yield and improving the overall quality of agriculturally significant crops. Furthermore, it is imperative to conduct an investigation into the various concentrations of growth media and the interplay between nanoparticles and minerals in order to have a comprehensive understanding of their respective impacts. These studies have the potential to deepen our understanding of the impact of nanoparticles at a more profound level and contribute to the improvement of plant quality. However, research investigating the full extent of the advantageous impacts of nanoparticles on plants is still lacking.

References

1. Bachheti RK, Bachheti A, Husen A (2024) Metal and metal-oxide based nanomaterials. In: Synthesis, agricultural, biomedical and environmental interventions. Springer Nature Singapore Pte Ltd., Singapore
2. Husen A (2022) Engineered nanomaterials for sustainable agricultural production, soil improvement and stress management. Elsevier Inc., Cambridge, USA. https://doi.org/10.1016/C2021-0-00054-7
3. Husen A (2023) Nanomaterials and nanocomposites exposures to plants. In: Response, interaction, phytotoxicity and defense mechanisms. Springer Nature Singapore Pte Ltd., Singapore. https://doi.org/10.1007/978-981-99-2419-6
4. Husen A, Bachheti RK, Bachheti A (2023) Current trends in green nano-emulsions. In: Food, agriculture and biomedical sectors. Springer Nature Singapore Pte Ltd., Singapore
5. Husen A, Iqbal M (2019) Nanomaterials and plant potential. Springer International Publishing AG, Cham, Switzerland. https://doi.org/10.1007/978-3-030-05569-1
6. Husen A, Jawaid M (2020) Nanomaterials for agriculture and forestry applications. Elsevier Inc., Cambridge, USA. https://doi.org/10.1016/C2018-0-02349-X
7. Husen A, Siddiqi KS (2023) Advances in smart nanomaterials and their applications. Elsevier Inc., Cambridge, USA. https://doi.org/10.1016/C2021-0-02202-1
8. Servin AD, White JC (2016) Nanotechnology in agriculture: next steps for understanding engineered nanoparticle exposure and risk. NanoImpact 1:9–12. https://doi.org/10.1016/j.impact.2015.12.002
9. Parandhaman T, Dey MD, Das SK (2019) Biofabrication of supported metal nanoparticles: exploring the bioinspiration strategy to mitigate the environmental challenges. Green Chem 21(20):5469–5500. https://doi.org/10.1039/C9GC02291K
10. Singh J, Dutta T, Kim KH, Rawat M, Samddar P, Kumar P (2018) 'Green' synthesis of metals and their oxide nanoparticles: applications for environmental remediation. J Nanobiotechnology 16:84
11. Tarannum N, Divya D, Gautam YK (2019) Facile green synthesis and applications of silver nanoparticles: a state-of-the-art review. RSC Adv 9:34926–34948
12. Saratale RG, Saratale GD, Shin HS, Jacob JM, Pugazhendhi A, Bhaisare M et al (2018) New insights on the green synthesis of metallic nanoparticles using plant and waste biomaterials: current knowledge, their agricultural and environmental applications. Environ Sci Pollut Res Int 25:10164–10183
13. Roy A, Bulut O, Some S, Mandal AK, Yilmaz MD (2019) Green synthesis of silver nanoparticles: biomolecule-nanoparticle organizations targeting antimicrobial activity. RSC Adv 9:2673–2702

14. Alavi M, Karimi N, Valadbeigi T (2019) Antibacterial, antibiofilm, antiquorum sensing, antimotility, and antioxidant activities of green fabricated Ag, Cu, TiO_2, ZnO, and Fe_3O_4 NPs via Protoparmeliopsis muralis lichen aqueous extract against multidrug-resistant bacteria. ACS Biomater Sci Eng 5:4228–4243
15. Sarkar S, Kotteeswaran V (2018) Green synthesis of silver nanoparticles from aqueous leaf extract of pomegranate (Punica granatum) and their anticancer activity on human cervical cancer cells. Adv Nat Sci Nanosci Nanotechnol 9:025014
16. Cyril N, George JB, Joseph L, Raghavamenon AC, PS V (2019) Assessment of antioxidant, antibacterial and anti-proliferative (lung cancer cell line A549) activities of green synthesized silver nanoparticles from Derris trifoliata. Toxicol Res (Camb) 8:297–308
17. Hamelian M, Zangeneh MM, Amisama A, Varmira K, Veisi H (2018) Green synthesis of silver nanoparticles using Thymus kotschyanus extract and evaluation of their antioxidant, antibacterial and cytotoxic effects. Appl Organomet Chem 32:e4458
18. Hajebi S, Tabrizi MH, Moghaddam MN, Shahraki F, Yadamani S (2019) Rapeseed flower pollen bio-green synthesized silver nanoparticles: a promising antioxidant, anticancer and antiangiogenic compound. J Biol Inorg Chem 24:395–404
19. Mortazavi-Derazkola S, Ebrahimzadeh MA, Amiri O, Goli HR, Rafiei A, Kardan M et al (2019) Facile green synthesis and characterization of Crataegus microphylla extract-capped silver nanoparticles (CME@Ag-NPs) and its potential antibacterial and anticancer activities against AGS and MCF-7 human cancer cells. J Alloys Compd 153186:820
20. Samari F, Salehipoor H, Eftekhar E, Yousefinejad S (2018) Low-temperature biosynthesis of silver nanoparticles using mango leaf extract: catalytic effect, antioxidant properties, anticancer activity and application for colorimetric sensing. New J Chem 42:15905–15916
21. Ulaeto SB, Mathew GM, Pancrecious JK, Nair JB, Rajan TPD, Maiti KK et al (2019) Biogenic Ag nanoparticles from neem extract: their structural evaluation and antimicrobial effects against Pseudomonas nitroreducens and Aspergillus unguis (NII 08123). ACS Biomater Sci Eng 6:235–245
22. Ebrahimzadeh MA, Naghizadeh A, Amiri O, Shirzadi-Ahodashti M, MortazaviDerazkola S (2019) Green and facile synthesis of Ag nanoparticles using Crataegus pentagyna fruit extract (CP-AgNPs) for organic pollution dyes degradation and antibacterial application. Bioorg Chem 94:103425
23. Rao KJ, Paria S (2015) Aegle marmelos leaf extract and plant surfactants mediated green synthesis of Au and Ag nanoparticles by optimizing process parameters using Taguchi method. ACS Sustain Chem Eng 3:483–491
24. Su D-L, Li P-J, Ning M, Li G-Y, Shan Y (2019) Microwave assisted green synthesis of pectin based silver nanoparticles and their antibacterial and antifungal activities. Mater Lett 244:35–38
25. Asghar MA, Zahir E, Shahid SM, Khan MN, Asghar MA, Iqbal J et al (2018) Iron, copper and silver nanoparticles: green synthesis using green and black tea leaves extracts and evaluation of antibacterial, antifungal and aflatoxin B1 adsorption activity. Lebensmittel-Wissenschaft and Technologie 90:98–107
26. Singh PK, Bhardwaj K, Dubey P, Prabhune A (2015) UV-assisted size sampling and antibacterial screening of Lantana camara leaf extract synthesized silver nanoparticles. RSC Adv 5:24513–24520
27. Siddiqui MN, Redhwi HH, Achilias DS, Kosmidou E, Vakalopoulou E, Ioannidou MD (2017) Green synthesis of silver nanoparticles and study of their antimicrobial properties. J Polym Environ 26:423–433
28. Kathiravan V (2018) Green synthesis of silver nanoparticles using different volumes of Trichodesma indicum leaf extract and their antibacterial and photocatalytic activities. Res Chem Intermed 44:4999–5012
29. Francis S, Joseph S, Koshy EP, Mathew B (2017) Synthesis and characterization of multifunctional gold and silver nanoparticles using leaf extract of Naregamia alata and their applications in the catalysis and control of mastitis. New J Chem 41:14288–14298

30. Rui M, Ma C, Tang X, Yang J, Jiang F, Pan Y et al (2017) Phytotoxicity of silver nanoparticles to peanut (Arachis hypogaea L.): physiological responses and food safety. ACS Sustain Chem Eng 5:6557–6567
31. Geisler-Lee J, Brooks M, Gerfen J, Wang Q, Fotis C, Sparer A et al (2014) Reproductive toxicity and life history study of silver nanoparticle effect, uptake and transport in Arabidopsis thaliana. Nanomaterials 4:301–318
32. Yan A, Chen Z (2019) Impacts of silver nanoparticles on plants: a focus on the phytotoxicity and underlying mechanism. Int J Mol Sci 20:1003
33. Yang J, Jiang F, Ma C, Rui Y, Rui M, Adeel M et al (2018) Alteration of crop yield and quality of wheat upon exposure to silver nanoparticles in a life cycle study. J Agric Food Chem 66:2589–2597
34. Acharya P, Jayaprakasha GK, Crosby KM, Jifon JL, Patil BS (2019) Green-synthesized nanoparticles enhanced seedling growth, yield, and quality of onion (Allium cepa L.). ACS Sustain Chem Eng 7:14580–14590
35. Marslin G, Siram K, Maqbool Q, Selvakesavan RK, Kruszka D, Kachlicki P et al (2018) Secondary metabolites in the green synthesis of metallic nanoparticles. Materials (Basel) 11:940
36. Hulla J, Sahu S, Hayes A (2015) Nanotechnology: history and future. Hum Exp Toxicol 34(12):1318–1321. https://doi.org/10.1177/0960327115603588
37. Zhang XF, Liu ZG, Shen W, Gurunathan S (2016) Silver nanoparticles: synthesis, characterization, properties, applications, and therapeutic approaches. Int J Mol Sci 17:1534
38. Marandi M, Torabi N (2020) The constructive role of ZnSe passivating layer on the photovoltaic performance of the fast-fabricated CdS/CdSe quantum dot sensitized solar cells. Opt Mater 105:109918
39. Zhou R, Fu S, Li H, Yuan D, Tang B, Zhou G (2019) Experimental study on thermal performance of copper nanofluids in a miniature heat pipe fabricated by wire electrical discharge machining. Appl Therm Eng 160:113989
40. Rahimi M, Noruzi EB, Sheykhsaran E, Ebadi B, Kariminezhad Z, Molaparast M, Mehrabani MG, Mehramouz B, Yousefi M, Ahmadi R et al (2020) Carbohydrate polymer-based silver nanocomposites: recent progress in the antimicrobial wound dressings. Carbohydr Polym 231:115696
41. Diaz-Alvarez A, Higuchi R, Li Q, Shingaya Y, Nakayama T (2020) Associative routing through neuromorphic nanowire net-works. AIP Adv 10:025134
42. Liu Y, Wang S, Wang Z, Yao Q, Fang S, Zhou X, Yuan X, Xie J (2020) The in situ synthesis of silver nanoclusters inside a bacterial cellulose hydrogel for antibacterial applications. J Mater Chem B 8:4846–4850
43. Yu H, Zhu Y, Mu B, Hui A, Wang A (2021) Removal of a cationic dye from aqueous solution by a porous adsorbent templated from eco-friendly Pickering MIPEs using chitosan-modified semi-coke particles. New J Chem 45:3848–3856
44. Murgunde BK, Mulla R, Rabinal MK (2020) A rapid synthesis of silver nanoparticle foam by ultrasonication. J Porous Mater 27:1727–1733
45. Proessl C, Kuebler M, Nowroozi MA, Paul S, Clemens O, Kramm UI (2021) Investigation of the thermal removal steps of capping agents in the synthesis of bimetallic iridium-based catalysts for the ethanol oxidation reaction. Phys Chem Chem Phys 23:563–573
46. Kuntyi OI, Kytsya AR, Mertsalo IP, Mazur AS, Zozula GI, Bazylyak LI, Topchak RV (2019) Electrochemical synthesis of silver nanoparticles by reversible current in solutions of sodium polyacrylate. Colloid Polym Sci 297:689–695
47. Massironi A, Morelli A, Grassi L, Puppi D, Braccini S, Maisetta G, Esin S, Batoni G, Della Pina C, Chiellini F (2019) Ulvan as novel reducing and stabilizing agent from renewable algal biomass: application to green synthesis of silver nanoparticles. Carbohydr Polym 203:310–321
48. Mohaghegh S, Osouli-Bostanabad K, Nazemiyeh H, Javadzadeh Y, Parvizpur A, Barzegar-Jalali M, Adibkia K (2020) A comparative study of eco-friendly silver nanoparticles synthesis using Prunus domestica plum extract and sodium citrate as reducing agents. Adv Powder Technol 31:1169–1180

49. Al-Awadi AS, El-Toni AM, Alhoshan M, Khan A, Labis JP, Al-Fatesh A, Abasaeed AE, Al-Zahrani SM (2019) Impact of precursor sequence of addition for one-pot synthesis of Cr-MCM-41 catalyst nanoparticles to enhance ethane oxidative dehy-drogenation with carbon dioxide. Ceram Int 45:1125–1134
50. Bachhav PA, Shroff RM, Shirkhedkar AA (2020) Silver nanoparticles: a comprehensive review on mechanism, synthesis and biomedical applications. Asian J Pharm Res 10
51. Zarlaida F, Adlim M (2017) Gold and silver nanoparticles and indicator dyes as active agents in colorimetric spot and strip tests for mercury (II) ions: a review. Microchim Acta 184:45–58
52. Xue Y, Zhan X, Sun S, Karuppagounder SS, Xia S, Dawson VL, Dawson TM, Laterra J, Zhang J, Ying M (2019) Synthetic mRNAs drive highly efficient iPS cell differentiation to dopaminergic neurons. Stem Cells Transl Med 8:112–123
53. Hoang V-T, Dinh NX, Pham TN, Hoang TV, Tuan PA, Huy TQ, Le A-T (2021) Scalable electrochemical synthesis of novel biogenic silver nanoparticles and its application to high-sensitive detection of 4-nitrophenol in aqueous system. Adv Polym Technol 6646219
54. Mallakpour S, Hussain CM, Gulati S, Kumar S, Diwan A, Singh P, Mongia A (2021) "Nanosil-ver": a versatile and newgeneration nanoproduct in biomedical applications. https://link.springer.com/referenceworkentry/10.1007%2F978-981-15-6453-6_48-1. Accessed 18 June 2021
55. Gao H, Yang H, Wang C (2017) Controllable preparation and mechanism of nano-silver mediated by the microemulsion system of the clove oil. Results Phys 7:3130–3136
56. Lin F, Zheng J, Guo W, Zhu Z, Wang Z, Dong B, Lin C, Huang B, Lu B (2019) Smart cellulose-derived magnetic hydrogel with rapid swelling and deswelling properties for remotely controlled drug release. Cellulose 26:6861–6877
57. Isogai A (2020) Cellulose nanofibers: recent progress and future prospects. J Fiber Sci Technol 76:310–326
58. Iftikhar M, Zahoor M, Naz S, Nazir N, Batiha GES, Ullah R, Bari A, Hanif M, Mahmood HM (2020) Green synthesis of silver nanoparticles using Grewia optiva leaf Aqueous extract and isolated compounds as reducing agent and their bio-logical activities. J Nanomater 8949674
59. Zarei Z, Razmjoue D, Karimi J (2020) Green synthesis of silver nanoparticles from Caralluma tuberculata extract and its antibac-terial activity. J Inorg Organomet Polym Mater 30:4606–4614
60. Keijok WJ, Pereira RHA, Alvarez LAC, Prado AR, da Silva AR, Ribeiro J, De Oliveira JP, Guimarães MCC (2019) Controlled biosynthesis of gold nanoparticles with Coffea arabica using factorial design. Sci Rep 9:16019
61. Sundarrajan SK, Pottail L (2021) Green synthesis of bimetallic Ag@Au nanoparticles with aqueous fruit latex extract of Artocarpus heterophyllus and their synergistic medicinal efficacies. Appl Nanosci 11:971–981
62. Farah MA, Ali M, Chen S-M, Li Y, Al-Hemaid FM, Abou-Tarboush FM, Al-Anazi KM, Lee J (2016) Silver nanoparticles synthesized from Adenium obesum leaf extract induced DNA damage, apoptosis and autophagy via generation of reactive oxygen species. Colloids Surf B Biointerfaces 141:158–169
63. Aref MS, Salem SS (2020) Bio-callus synthesis of silver nanoparticles, characterization, and antibacterial activities via Cinnamomum camphora callus culture. Biocatal Agric Biotechnol 27:101689
64. Gopinath V, MubarakAli D, Priyadarshini S, Priyadharsshini NM, Thajuddin N, Velusamy P (2012) Biosynthesis of silver nanoparticles from Tribulus terrestris and its antimicrobial activity: a novel biological approach. Colloids Surf B Biointerfaces 96:69–74
65. Jahagirdar AS, Shende S, Gade A, Rai M (2020) Bioinspired synthesis of copper nanoparticles and its efficacy on seed viability and seedling growth in mungbean (Vigna radiata L.). Curr Nanosci 16:246–252
66. Robles-Martínez M, Patiño-Herrera R, Pérez-Vázquez FJ, Montejano-Carrizales JM, González JFC, Pérez E (2020) Mentha piperita as a natural support for silver nanoparticles: a new anti-Candida albicans treatment. Colloid Interface Sci Commun 35:100253

67. Mahdavi-Ourtakand M, Jafari P, Safaeijavan R (2017) Antibacterial activity of biosynthesized silver nanoparticles from fruit ex-tracts of Bunium persicum Boiss. Int J Bio-Inorg Hybr Nanomater 6:245–251
68. Manikandan V, Velmurugan P, Jayanthi P, Park JH, Chang WS, Park YJ, Cho M, Oh BT (2018) Biogenic synthesis from Prunus × yedoensis leaf extract, characterization, and photocatalytic and antibacterial activity of TiO_2 nanoparticles. Res Chem Intermed 44:2489–2502
69. Dawodu FA, Onuh CU, Akpomie KG, Unuabonah EI (2019) Synthesis of silver nanoparticle from Vigna unguiculata stem as adsorbent for malachite green in a batch system. SN Appl Sci 1:1–346
70. Goswami T, Bheemaraju A, Sharma AK, Bhandari S (2021) Perylenetetracarboxylic acid-incorporated silver nanocluster for cost-effective visible-light-driven photocatalysis and catalytic reduction. Colloid Polym Sci 299:925–936
71. Li BJ, Li S, Ma BR, Jing Y, Zhang ZM, Li ZJ, Zhao SY (2020) Progress in the preparation of silver nanometer antibacterial material. Mater Sci Forum 996:76–81
72. Verma D, Mudgal B, Chaudhary P, Mahakur B, Mitra D, Pant K, Mohapatra PKD, Thapliyal A, Janmeda P (2020) Medicinal plant of Uttarakhand (India) and their benefits in the treatment of tuberculosis: current perspectives. J BioSci Biotechnol 9:75–85
73. Abdullah MR, Haque ME, Sarwar AKMG, Ashrafuzzaman M, Rahman MM (2020) Diversity of underutilized fruits and their uses in Karnaphuli range, Rangamati, Bangladesh. Int J For Ecol Environ 1:10–20
74. Baydoun SA, Kanj D, Raafat K, Aboul Ela M, Chalak L, Arnold-Apostolides N (2017) Ethnobotanical and economic importance of wild plant species of Jabal Moussa Bioreserve, Lebanon. J Ecosys Ecograph S7:1
75. Grkovic T, Akee RK, Thornburg CC, Trinh SK, Britt JR, Harris MJ, Evans JR, Kang U, Ensel S, Henrich CJ et al (2020) National Cancer Institute (NCI) program for natural products discovery: rapid isolation and identification of biologically active natural products from the NCI prefractionated library. ACS Chem Biol 15:1104–1114
76. Muhammad A, Kashere MA (2020) Neem, Azadirachta indica L. (A. Juss): an eco-friendly botanical insecticide for managing farmers' insects pest problems—a review. FUDMA J Sci 4:484–491
77. Ali S, Bansal S, Mishra RP (2020) Fumaria indica (L), a famous medicinal herb of tribal regions of Jabalpur, Madhya Pradesh: broad spectrum antibacterial and phytochemical profiling against some pathogenic microorganisms. Pharmacogn J 12:619–623
78. Duguma HT (2020) Wild edible plant nutritional contribution and consumer perception in Ethiopia. Int J Food Sci 1–16
79. Pollo LA, Martin EF, Machado VR, Cantillon D, Wildner LM, Bazzo ML, Waddell SJ, Biavatti MW, Sandjo LP (2021) Search for antimicrobial activity among fifty-two natural and synthetic compounds identifies anthraquinone and polyacetylene classes that inhibit Mycobacterium tuberculosis. Front Microbiol 11:3619
80. Sawicka B, Otekunrin OA, Skiba D, Bienia B, Cwintal M (2020) Plant-derived stimulants and psychoactive substances—social and economic aspects. Clin Med Res 5:313–335
81. Fahmy HM, Eldin RES, Abu Serea ES, Gomaa NM, Aboelmagd GM, Salem SA, Elsayed ZA, Edrees A, Shams-Eldin E, Shalan AE (2020) Advances in nanotechnology and antibacterial properties of biodegradable food packaging materials. RSC Adv 10:20467–20484
82. Ahmed MS, Soundhararajan R, Akther T, Kashif M, Khan J, Waseem M, Srinivasan H (2019) Biogenic AgNPs synthesized via endophytic bacteria and its biological applications. Environ Sci Pollut Res 26:26939–26946
83. Cobos M, De-La-Pinta I, Quindós G, Fernández MJ, Fernández MD (2020) Graphene oxide–silver nanoparticle nanohybrids: synthesis, characterization, and antimicrobial properties. Nanomaterials 10:376
84. Hafeez M, Shaheen R, Akram B, Abdin ZU, Haq S, Mahsud S, Ali S, Khan RT (2020) Green synthesis of cobalt oxide nanoparticles for potential biological applications. Mater Res Express 7:025019

85. Khan AU, Malik N, Khan M, Cho MH, Khan MM (2017) Fungi-assisted silver nanoparticle synthesis and their applications. Bioprocess Biosyst Eng 41:1–20
86. Elbaghdady K, EL-Shatoury EIN, Abdallah O, Khalil M (2018) Biogenic production of silver nanoparticles by Enterobacter clo-acae Ism26. Turk J Biol 42:319–328
87. Jadoun S, Arif R, Jangid NK, Meena RK (2020) Green synthesis of nanoparticles using plant extracts: a review. Environ Chem Lett 19:355–374
88. Wong XY, Sena-Torralba A, Álvarez-Diduk R, Muthoosamy K, Merkoçi A (2020) Nanomaterials for nanotheranostics: tuning their properties according to disease needs. ACS Nano 14:2585–2627
89. Shanab SMM, Partila AM, Ali HEA, Abdullah MA (2020) Impact of gamma-irradiated silver nanoparticles biosynthesized from Pseudomonas aeruginosa on growth, lipid, and carbohydrates of Chlorella vulgaris and Dictyochloropsis splendida. J Radiat Res Appl Sci 14:70–81
90. Zuorro A, Iannone A, Natali S, Lavecchia R (2019) Green synthesis of silver nanoparticles using bilberry and red currant waste extracts. Processes 7:193
91. Baiocco D, Lavecchia R, Natali S, Zuorro A (2016) Production of metal nanoparticles by agro-industrial wastes: a green opportunity for nanotechnology. Chem Eng Trans 47:67–72
92. Amgoth C, Singh A, Santhosh R, Yumnam S, Mangla P, Karthik R, Guping T, Banavoth M (2019) Solvent assisted size effect on AuNPs and significant inhibition on K562 cells. RSC Adv 9:33931–33940
93. Rónavári A, Igaz N, Adamecz D, Szerencsés B, Molnar C, Kónya Z, Pfeiffer I, Kiricsi M (2021) Green silver and gold nanoparticles: biological synthesis approaches and potentials for biomedical applications. Molecules 26:844
94. Sana SS, Kumbhakar DV, Pasha A, Pawar SC, Grace AN, Singh RP, Nguyen V, Van Le Q VNQ, Peng W (2020) Crotalaria verrucosa leaf extract mediated synthesis of zinc oxide nanoparticles: assessment of antimicrobial and anticancer activity. Molecules 25:4896. https://doi.org/10.3390/m01ecu1es25214896
95. Kausar H, Mehmood A, Khan RT, Ahmad KS, Hussain S, Nawaz F, Iqbal MS, Nasir M, Ullah TS (2022) Green synthesis and characterization of copper nanoparticles for investigating their effect on germination and growth of wheat. PLoS ONE 17(6 June). https://doi.org/10.1371/journal.pone.0269987
96. Altaf M, Zeyad MT, Hashm Md A, Manoharadas S, Hussain SA, Abuhaslle Mohammed MSA, Almuzainl AM (2021) Effective inhibition and eradication of pathogenic biofilms by titanium dioxide nanoparticles synthesized using Carum copticum extract. RSC Adv 19248–19257. https://doi.org/10.1039/d1ra02876frsc.li/rsc-advances
97. Ibrahim E, Zhang M, Zhang Y, Hossain A, Qiu W, Chen Y, Wang Y, Wu W, Sun G, Li B (2020) Green-synthesization of silver nanoparticles using endophytic bacteria isolated from garlic and its antifungal activity against wheat fusarium head blight pathogen Fusarium graminearum. Nanomaterials 10:219. https://doi.org/10.3390/nan010020219; Iravani S (2011) Green synthesis of metal nanoparticles using plants. Green Chem 13:2638–2650
98. Uddin S, Safdar L, Bin, Anwar S, Iqbal J, Laila S, Abbasi BA, Saif MS, Ali M, Rehman A, Basit A, Wang Y, Quraishi UM (2021) Green synthesis of nickel oxide nanoparticles from berberis balochistanica stem for investigating bioactivities. Molecules 26(6):1–14. https://doi.org/10.3390/molecules26061548
99. Some S, Bulut O, Biswas K, Kumar A, Roy A, Sen IK, Mandal A, Franco OL, İnce İA, Neog K, Das S, Pradhan S, Dutta S, Bhattacharjya D, Saha S, Das Mohapatra PK, Bhuimali A, Unni BG, Kati A, Ocsoy I (2019) Effect of feed supplementation with biosynthesized silver nanoparticles using leaf extract of Morus indica L. V1 on Bombyx mori L. (Lepidoptera: Bombycidae). Sci Rep 9(1):1–13. https://doi.org/10.1038/s41598-019-50906-6
100. Haroon M, Zaidi A, Ahmed B, Rizvi A, Khan MS, Musarrat J (2019) Effective inhibition of phytopathogenic microbes by eco-friendly leaf extract mediated silver nanoparticles (AgNPs). Indian J Microbiol 59(3):273–287. https://doi.org/10.1007/s12088-019-00801-5
101. Vu XH, Dien ND, Pham TTH, Trang TT, Ca NX, Tho PT, Vinh ND, Van Do P (2020) The sensitive detection of methylene blue using silver nanodecahedra prepared through a photochemical route. RSC Adv 10:38974–38988

102. Gui R, Jin H (2019) Recent advances in synthetic methods and applications of photo-luminescent molecularly imprinted polymers. J Photochem Photobiol C Photochem Rev 41:100315
103. Soliwoda K, Tomaszewska E, Socha E, Krzyczmonik P, Ignaczak A, Orlowski P, Krzyzowska M, Celichowski G, Grobelny J (2017) The role of tannic acid and sodium citrate in the synthesis of silver nanoparticles. J Nanopart Res 19:1–15
104. Al-Saedy MA, Ali AA, Sabah R, Hamed WM, Kafi M (2020) Preparation method of silver nano particles. J Adv Sci Technol 3:1–8
105. Singh S (2017) Synthesis of nanoparticles for biomedical applications. In: Dhawan A, Anderson D, Shanker R (eds) Nanotoxicology: experimental and computational perspectives. Royal Society of Chemistry, London, UK, pp 39–93
106. Smith DJ (2015) Characterization of nanomaterials using transmission electron microscopy. Nanocharacterization 2:1–29
107. Li J, Rong K, Zhao H, Li F, Lu Z, Chen R (2013) Highly selective antibacterial activities of silver nanoparticles against Bacillus subtilis. J Nanosci Nanotechnol 6806–6813
108. Liu HL, Dai SA, Fu KY, Hsu SH (2010) Antibacterial properties of silver nanoparticles in three different sizes and their nanocomposites with a new waterborne polyurethane. Int J Nanomed 5:1017–1028
109. El-Zahry MR, Mahmoud A, Refaat IH, Mohamed HA, Bohlmann H, Lendl B (2015) Antibacterial effect of various shapes of silver nanoparticles monitored by SERS. Talanta 138:183–189
110. Wang XD, Sun C, Gao SX et al (2001) Validation of germination rate and root elongation as indicator to assess phytotoxicity with Cucumis sativus. Chemosphere 44:1711–1721
111. Krishnaraj C, Jagan EG, Ramachandran R et al (2012) Effect of biologically synthesized silver nanoparticles on Bacopa monnieri (Linn.) Wettst. plant growth metabolism. Process Biochem 47:651–658
112. Panda KK, Acharya VMM, Krishnaveni R et al (2011) In vitro biosynthesis and genotoxicity bioassay of silver nanoparticles using plants. Toxicol In Vitro 25:1097–1105
113. Amooaghaie R, Saeri MR, Azizi M (2015) Synthesis, characterization and biocompatibility of silver nanoparticles synthesized from Nigella sativa leaf extract in comparison with chemical silver nanoparticles. Ecotoxicol Environ Saf 120:400–408
114. Qian H, Peng X, Han X, Ren J, Sun L, Fu Z (2013) Comparison of the toxicity of silver nanoparticles and silver ion on the growth of terrestrial plant model Arabidopsis thaliana. J Environ Sci 25:1947–1955. https://doi.org/10.1016/S1001-0742(12)60301-5
115. Kouvaris P, Delimitis A, Zaspalis V et al (2012) Green synthesis and characterization of silver nanoparticles produced using Arbutus unedo leaf extract. Mater Lett 76:18–20
116. El-Temsah YS, Joner EJ (2012) Impact of Fe and Ag nanoparticles on seed germination and differences in bioavailability during exposure in aqueous suspension and soil. Environ Toxicol 27(1):42–49
117. Mazumdar H, Ahmed GU (2011) Synthesis of silver nanoparticles and its adverse effect on seed germinations in Oryza sativa, Vigna radiata and Brassica campestris. Int J Adv Biotechnol Res 2(4):404–413
118. Yin L, Cheng Y, Espinasse B et al (2011) More than the ions: the effects of silver nanoparticles on Lolium multiflorum. Environ Sci Technol 45(6):2360–2367
119. Lee CW, Mahendera S, Zodorow K et al (2010) Developmental phytotoxicity of metal oxide nanoparticles to Arabidopsis thaliana. Environ Toxicol Chem 29:669–675
120. Ma X, Geiser-Lee J, Deng Y et al (2010) Interactions between engineered nanoparticles (ENPs) and plants: phytotoxicity, uptake and accumulation. Sci Total Environ 408:3053–3061
121. Nowack B, Bucheli TD (2007) Occurrence, behavior and effects of nanoparticles in the environment. Environ Pollut 150:5–22
122. Yin L, Colman BP, MC-Gill BM et al (2012) Effect of silver nanoparticle exposure on germination and early growth of eleven wetland plants. PLoS ONE 7(10):e47674. https://doi.org/10.1371/journal.pone.0047674

123. Lin D, Xing B (2007) Phytotoxicity of nanoparticles: inhibition of seed germination and root growth. Environ Pollut 150:243–250
124. Katti K, Chanda N, Shukla R, Zambre A, Suibramanian T, Kulkarni RR, Kannan R, Katti KV (2009) Green nanotechnology from cumin phytochemicals: generation of biocompatible gold nanoparticles. Int J Green Nanotechnol Biomed 1(1):B39–B52
125. Sangeetha J, Sandhya J, Philip J (2014) Biosynthesis and functionalization of silver nanoparticles using Nigella sativa, Dioscorea alata and Ferula asafetida. Sci Adv Mater 6:1–10

Silver Nanoparticles as the Sterilant in Large-Scale Micropropagation of Plants

Yasemin Kemeç Hürkan and Kaan Hürkan

Abstract In tissue cultures, surface sterilization is one of the most crucial concerns. The subsequent steps cannot be completed without success in the surface sterilization phase. Because of this, the most important factor affecting a culture's efficacy is the surface sterilization of plant parts (explants). When surface sterilizing explants, sterilization agents like sodium hypochlorite, ethanol, calcium hypochlorite, hydrogen peroxide, mercury chloride, different surfactants (such as Tween 20), and plant protection combinations are commonly utilized. These sterilant are often poisonous; they discolour objects and prevent growth and development in vitro. Nowadays, nanotechnology is applied successfully in numerous fields. In plant tissue culture, it is evident that silver nanoparticles (Ag NPs) are widely employed in antibacterial and antiseptic applications. One technique for producing Ag NPs other than physical and chemical synthesis is biosynthesis. When it comes to all known forms of microorganisms in the literature, silver ions are the most efficient antimicrobial and antibacterial compounds. They are not only better than other antimicrobial agents, but they are also safe for the environment and human health.

Keywords Ag NPs · Plant tissue culture · Microorganism · Sterilization in micropropagation · Nanoparticle · Nanotechnology

1 Introduction

Plant cell, organ cultures, and tissue are used extensively in plant biotechnology fundamental research as well as in the propagation of commercially useful species [15]. Rapidly multiplying explants have shown that micropropagation works well

Y. K. Hürkan (✉)
Iğdır, Türkiye
e-mail: kemecyasemin@gmail.com

K. Hürkan
Department of Agricultural Biotechnology, Faculty of Agriculture, Iğdır University, 76000 Iğdır, Türkiye
e-mail: kaanhurkan@hotmail.com

A. Husen (ed.), *Plant Response to Silver Nanoparticles*, Smart Nanomaterials Technology, https://doi.org/10.1007/978-981-97-7352-7_17

for producing disease-free plants and secondary chemicals (Fig. 1). Furthermore, there are still drawbacks to micropropagation, including the potential for microbial contaminations (bacteria and fungus) that prevent plant development in the culture media [9].

Due to its lengthy and dependable history, autoclaves are typically employed in micropropagation labs to sterilize the culture material. The media and culture vessels (glass containers, nylon bags, plastic boxes, etc.) are usually autoclaved for 20–30 min at 121 °C and 15 pressure to sterilize them. Although autoclaving is an efficient method, it can result in the development of hazardous breakdown products such as 5-(hydroxymethyl)-2-furaldehyde and phenolics, and it is a significant financial burden for numerous micropropagation facilities [10, 43]. Chemical sterilization using substances like sodium hypochlorite (NaClO), hydrogen peroxide, and chlorine dioxide can take the role of physical sterilization via autoclaving. Although chemical sterilization is often hazardous to explants and reduces the effectiveness of multiplication, it may be a very efficient disinfectant [7, 8, 28, 42].

Nanoparticles (NPs) are atomic or molecular aggregates that are not bigger than 100 nm in at least one dimension and have special physicochemical features (a high surface-to-volume ratio, high reactivity, and simple cell absorption) [38]. According to Cvjetko et al. [12], copper, carbon nanotubes, gold, aluminium, iron, silver, zinc oxide, zinc, silica, and titanium dioxide are the most manufactured NPs worldwide. Numerous disciplines use them, including biology, electronics, optics, medicine, energetics, and the material sciences [18].

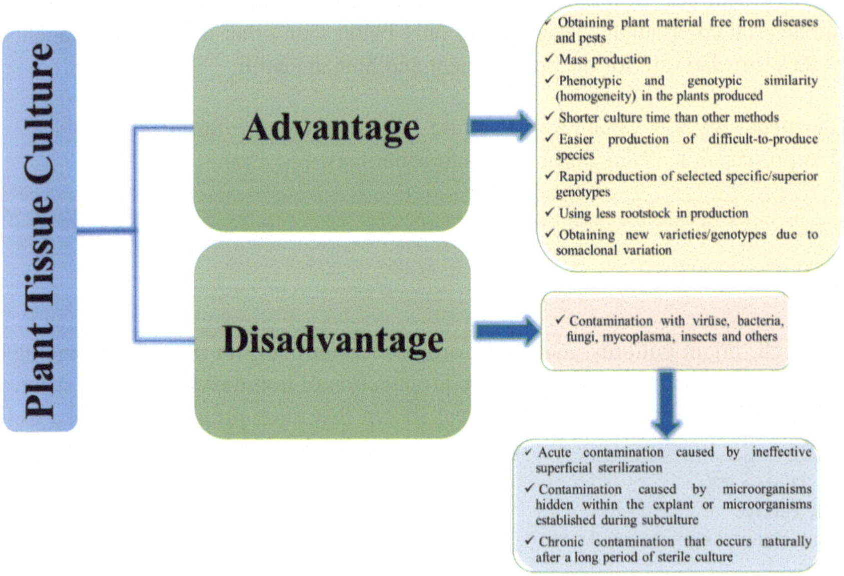

Fig. 1 Advantages and disadvantages of plant tissue culture

In recent years, the interaction between NPs and plant growth has piqued the curiosity of scientists worldwide. Numerous studies have been conducted to investigate the effects of metal nanoparticles on plants [21, 33, 35]. Research has shown that nanomaterials can affect plant growth and development in both positive and negative ways, depending on their composition, structure, concentrations, and plant species [20, 37]. AgNP has been shown to have antimicrobial qualities in the past, and explants in vitro culture tolerate their inclusion in medium well [35].

In addition to being demonstrated to enhance organogenesis, somatic embryogenesis, genetic transformation, callus induction, metabolite synthesis and soma clonal variation, silver nanoparticles (AgNPs) also demonstrate antibacterial efficacy against a number of plant infections [2, 5, 20, 23]. A study by Kim et al. [20] found that adding silver nanoparticles to the culture media significantly reduces microbial contamination on the surface of the plants, enhances the morphology of explants, and completely removes bacterial contamination. Both approaches yield excellent outcomes, particularly when tissue culture is included.

2 Sterilization Methods Commonly Used in Plant Tissue Culture

In plant tissue culture, microbial infection of plants has a negative impact. In addition to increasing culture mortality and competing for resources, these microorganisms can cause inconsistent growth, tissue necrosis, decreased shoot proliferation, and diminished roots [27]. Certain internal pollutants are not usually visible in the early phases of culture because they are difficult to remove and become evident in later subcultures [29]. Rinsing under running water for 30–90 min and disinfecting with alcohols, fungicides, $HgCl_2$, and NaOCl, among other surface sterilizing agents, effectively remove these external contaminants.

While some of these sterilizers are useful in disinfecting explants from field-grown plants, others are not. To save time and effort when using tissue culture procedures, it might be essential to follow an accurate and thorough sterilizing protocol. After sterilizing, the explants must, nonetheless, still be viable. Various disinfectants are applied to plant tissues to remove contaminants. To minimize damage to the explants and increase their survival, it is essential to standardize the concentration, duration of exposure of the explants to the various sterilities, and the order in which these sterilities are used because some of these agents are toxic to plant tissues [13].

Aseptic conditions are mandatory for the successful performance and continuity of plant cell and tissue culture methods. In vitro environments where plant explants are cultured provide optimum conditions for the development and reproduction of fungi and bacteria. Therefore, the most important practice before starting culture is the sterilization of the explant. Unsuccessful surface sterilization prevents the continuity of the culture. Although ethyl alcohol, sodium or hydrogen peroxide, calcium hypochlorite, mercuric chloride, and silver nitrate are used for surface sterilization, ethyl alcohol and sodium hypochlorite are mostly used for this purpose (Fig. 2).

The sterilizing process may vary according to the species and collecting season. Mercuric chloride, or $HgCl_2$, must be disposed of carefully since it is very harmful to people and plants alike. Since mercury is a phytotoxic mineral, it is crucial to thoroughly wash the plant material several times to get rid of any remnants of the mineral. NaOCl, or sodium hypochlorite, is a frequently used solution for surface sterilization. It is diluted to the right quantities and is sold as commercial bleach products like Clorox and Fast. Plants submerged in this mixture for 5–20 min. Because of their phytotoxicity, it is necessary to confirm that each type of explant has a balance between duration and concentration. Although ethanol is a potent sterilizer, it is also quite harmful to plants. Explants are therefore often only exposed to it for a few minutes or seconds. Alcohol will harm tissue more severely the more sensitive it is. Because the tissue that will be developed or explanted is confined inside the surface sterilized structure, alcohol may be used to treat tissues like seeds, unopened flower buds or dormant buds for lengthy periods of time. Usually, a 70% ethanol pre-treatment is used before the application of a second sterilizing agent. Generally speaking, ethanol sterilizes surfaces, although it can dehydrate tissue. Despite being

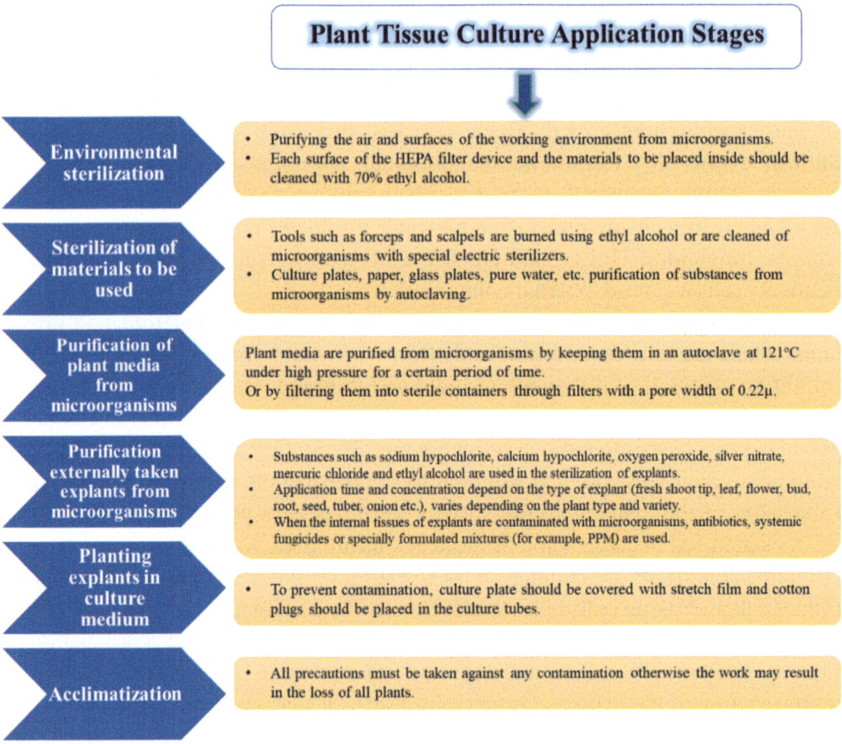

Fig. 2 Stages of propagation of plants with plant tissue culture method

exceedingly strong, toxic, and deadly for both people and plants, mercuric chloride is an effective way to get rid of bacterial infections.

Products that are non-toxic to both humans and plants are also available. PPM™ (Plant preservative mixture) can be given as an example. As an alternative to traditional fungicides, antibiotics, and sterilizing agents, PPM™ is a broad-spectrum biocide that is used to reduce contamination in plant tissue culture. PPM inhibits many vital enzymes in the electron transport chain and citric acid cycle by entering the microbial cell wall. This is how PPM regulates microorganisms. PPM is a useful product, but it needs to choose the right concentration based on the kind of plant [25].

3 The Importance of Silver Nanoparticles in Plant Tissue Culture

Plant tissue culture is a fundamental and crucial component of plant biotechnology, and developments in this field have a big influence on a lot of other biotechnological fields. In the realms of medicine and the environment today, nanomaterials particularly Nano silver [NS] are often employed as antibacterial agents. Since ancient times, silver has been known to have antibacterial properties against a wide variety of microbes. Today, silver nanoparticles, which may be as tiny as 1 nm, are used in many pharmacological and medical applications [11]. Silver has a startling array of peculiar physical, chemical, and biological characteristics at the nanoscale. According to Lansdown [22], the main characteristics of NS include its extensive antimicrobial spectrum, availability, economical production technique, heat resistance, simplicity of application, wide variety of applications, and absence of resistant microorganisms (Fig. 3). It is also environmentally friendly and non-toxic. Only a small number of papers have mentioned the use of NS in plant biotechnology, in contrast to the focus on novel uses of NS in medicine.

One of the most well-established and researched uses of NPs is their antibacterial ability; the most well-known examples of this include Ag, ZnO, and TiO_2 NPs. The dimensions, size, distribution, and kind of NPs determine how well they remove microbial contaminants from plant tissue cultures. Studies have reported that NPs negatively affect the survival and subsequent regeneration of explants during application periods. Hence, investigating the effects of types and doses of NPs is quite important in order to determine the optimal dose with minimal phytotoxicity. The synergistic effects of NPs can further boost their efficiency when used in combination with antibiotics or sterilizing treatments [3, 17].

A broad variety of nanoparticles, such as silicon (Si), SiO_2, silver (Ag), gold (Au), titanium dioxide (TiO_2), nickel (Ni), magnesium oxide (MgO), CuO, aluminium oxide (Al_2O_3), zinc oxide (ZnO), and iron oxide (Fe_3O_4), have been reported to exhibit antimicrobial activities against various microorganisms [6]. Abdi et al. [1] looked at the application of silver nanoparticles (Ag NPs) in Valeriana officinalis to

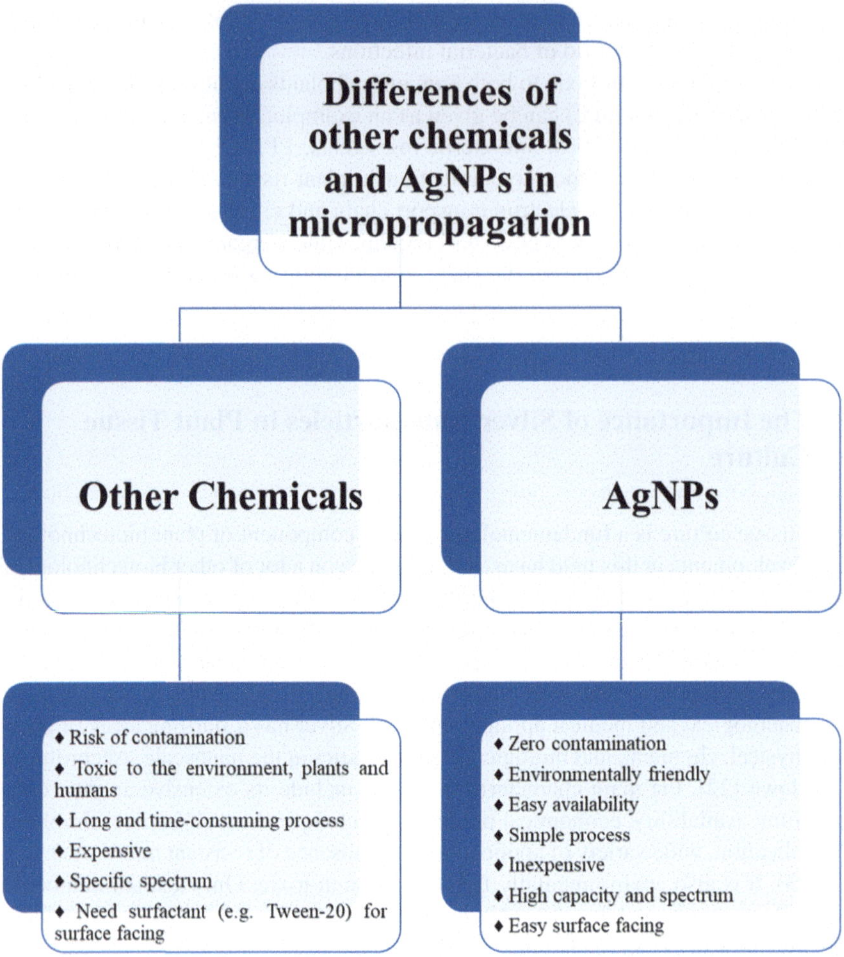

Fig. 3 Advantages and disadvantages of using other chemicals and AgNPs in plant tissue culture

control microbial contamination for the first time. Furthermore, the Ag NPs treatment in the tissue culture medium had no effect on shoot multiplication or subsequent roots. According to Fakhrfeshani et al. [14], the juvenile capitulum of *Gerbera jamesonii* treated with 200 mg L^{-1} Ag NPs and 1.5% NaOCl had eradicated all microbiological pollutants favourably and had no adverse effects on micropropagation. In Gouran et al. [16] study, they effectively decontaminated V*itis vinifera* leaf explants using 100 mg L^{-1} Ag NPs and 95% EtOH. After 20 min of treatment with a 200 mg L^{-1} Ag NPs solution, the number of bacteria present in the Rosa hybrida explants decreased [36]. However, Kalsaitkar et al. [19] found that adding 160 mg L^{-1} Ag NPs to the MS medium dramatically decreased internal bacterial contamination in callus cultures of *Bacopa monnieri*. Ag NPs and thymol together suppressed the development of microorganisms in *Cynodon dactylon*. Nodal explants treated with 100 mg L^{-1} thymol and 200 mg L^{-1} Ag NPs solution for 60 min were used to start contamination-free cultures [40]. In a similar vein, Arab et al. [4] examined the Ag NP immersion and addition (100 and 150 mg L^{-1}) to the MS medium; they found that in G × N15 rootstocks (a hybrid of almond and peach), internal and exterior contamination was decreased. After adding 50–200 mg L^{-1} of Ag NPs to the MS medium, it was recently discovered that there was no contamination in the shoot cultures of *Vanilla planifolia* [39].

Mahna et al. [26] investigated how Ag NPs treatment affected tomato cotyledons, potato leaves, and Arabidopsis seeds surface decontamination. It was discovered that explants treated with 100 mg L^{-1} Ag NPs for 1–5 min were the most effective for completely disinfecting the seeds, cotyledons, or leaves and did not negatively impact the viability of the explants. But when they were treated with the maximum concentration of Ag NPs, there was a decrease in seed germination as well as in leaf and cotyledon survival. It is quite challenging to remove microbial contamination from in vitro cells when it comes to woody plants. Only a small percentage of the explants that were exposed to EtOH and Clorox for 60 min before being treated with 100 mg L^{-1} Ag NPs did not experience any microbial contamination. Nonetheless, internal pollutants in the olive explants were well reduced by adding 4 mg L^{-1} Ag NPs to the media [30]. The contamination rate was lowered from 61.5 to 11.3% by immersing different *Araucaria excelsa* explants in Ag NPs (200 mg L^{-1}) for 180 min [34]. In order to investigate contamination-free cultures, tobacco, and potato shoot buds were grown on MS media containing TiO_2 NPs or Ag NPs [31, 32].

4 Conclusions

Plant tissue culture is currently directly relevant to the commercial world, and its usage in fundamental sciences including genetics, cell biology, biotechnology, and biochemistry is proof of its value. In addition to offering a means of mass replication, tissue culture enables the creation of genetically engineered and disease-free plants. It also offers a pathway for the synthesis of secondary metabolites. Despite all this approach's benefits, a few methodological issues, primarily related to explant contamination, prevent it from being fully utilized as an effective methodology for biotechnological research. As an aside, external and internal contamination of micropropagation is a typical worry since bacteria and fungi may grow considerably faster than plant cells and take all the nutrients, preventing plant development. While repetitive subcultures and meristem culture—cultivating the pathogen-free portion of the plant—are two strategies to lessen the risk of bacterial and fungal contaminations during in vitro propagation, a more efficient way to sterilize plant tissues still seems necessary to avoid labour-intensive trial and error and lengthy decontamination processes. Although it has been reported that antibiotics are generally phytotoxic and have an inhibitory effect on explant survival, callus induction, regeneration, and multiplication, antifungal and antibiotic agents may be used to treat persistent fungal and bacterial contaminations and lessen their effects [24, 41]. On the other hand, inadequate sterilization of ex vitro (taken from field orchard greenhouse) explants might result in a scarcity of specimens for further investigation. Nowadays, silver nanoparticles have shown useful in reducing microbial contaminants in explants, without having any phytotoxic effects on plants.

References

1. Abdi G, Salehi H, Khosh-Khui M (2008) Nano silver: a novel nanomaterial for removal of bacterial contaminants in valerian (Valeriana officinalis L.) tissue culture. Acta Physiol Plant 30:709–714
2. Adebomojo A, AbdulRahaman A (2020) Surface sterilization of Ocimum seeds and tissues with biosynthesized nanosilver and its effects on callus induction. In: IOP conference series: materials science and engineering, vol 1. IOP Publishing, pp 012024
3. Applerot G, Lellouche J, Perkas N, Nitzan Y, Gedanken A, Banin E (2012) ZnO nanoparticle-coated surfaces inhibit bacterial biofilm formation and increase antibiotic susceptibility. RSC Adv 2:2314–2321
4. Arab MM, Yadollahi A, Hosseini-Mazinani M, Bagheri S (2014) Effects of antimicrobial activity of silver nanoparticles on in vitro establishment of G × N15 (hybrid of almond × peach) rootstock. J Genet Eng Biotechnol 12:103–110
5. Azeez L, Lateef A, Wahab A, Rufai M, Salau A, Ajayi E, Ajayi M, Adegbite M, Adebisi B (2019) Phytomodulatory effects of silver nanoparticles on Corchorus olitorius: its antiphytopathogenic and hepatoprotective potentials. Plant Physiol Biochem 136:109–117
6. Beyth N, Houri-Haddad Y, Domb A, Khan W, Hazan R (2015) Alternative antimicrobial approach: nano-antimicrobial materials. Evid Based Complement Alternat Med 2015:1–16
7. Brondani GE, deOliveira LS, Bergonci T, BrondaniAE FFAM, da Silva ALL, Goncalves AN (2013) Chemical sterilization of culture medium: a low-cost alternative to in vitro establishment of plants. Sci For 41:257–264
8. Cardoso JC, Gerald LTS, Silva JAT (2018) Micropropagation in the twenty-first century. In: Loyola-Vargas V, Ochoa-Alejo N (eds) Plant cell culture protocols. Springer, Switzerland, pp 17–46. https://doi.org/10.1007/978-1-61779-818-4
9. Cassells AC (1991) Problems in tissue culture: culture contamination. In: Debergh PC, Zimmerman RH (eds) Micropropagation: technology and application. Springer, Dordrecht, pp 31–44. https://doi.org/10.1007/978-94-009-2075-0
10. Chen C (2016) Cost analysis of plant micropropagation of Phalaenopsis. Plant Cell Tiss Org Cult 126:167–175. https://doi.org/10.1007/s11240-016-0987-4
11. Chen X, Schluesener HJ (2008) Nanosilver: a nanoproduct in medical application. Toxicol Lett 176:1–12
12. Cvjetko P, Zovko M, Peharec Štefanić P, Biba R, Tkalec M, Domijan AM, Vinković Vrček I, Letofsky-Papst I, Šikić S, Balen B (2018) Phytotoxic effects of silver nanoparticles in tobacco plants. Environ Sci Pollut Res 25:5590–5602. https://doi.org/10.1007/s11356-017-0928-8
13. Dhingra MK, Naik PS, Chandra R, Randhawa GJ (1992) Tissue culture technique for potato health, conservation, micropropagation and improvement. Tech Bull 39
14. Fakhrfeshani M, Bagheri A, Sharifi A (2012) Disinfecting effects of nano silver fluids in Gerbera (Gerbera jamesonii) capitulum tissue culture. J Biol Environ Sci 6:121–127
15. Gamborg OL (2002) Plant tissue culture: biotechnology milestones in vitro. Cell Dev Biol - Plant 38:84–92
16. Gouran A, Jirani M, Mozafari AA, Saba MK, Ghaderi N, Zaheri S (2014) Effect of silver nanoparticles on grapevine leaf explants sterilization at in vitro conditions. In: 2nd National conference on nanotechnology from theory to application, Isfahan, Iran, pp 1–6

17. Guzman M, Dille J, Godet S (2012) Synthesis and antibacterial activity of silver nanoparticles against gram-positive and gram-negative bacteria. Nanomed Nanotechnol Biol Med 8:37–45
18. Hou J, Wu Y, Li X, Wei B, Li S, Wang X (2018) Toxic effects of different types of zinc oxide nanoparticles on algae, plants, invertebrates, vertebrates and microorganism. Chemosphere 193:852–860. https://doi.org/10.1016/j.chemosphere.2017.11.077
19. Kalsaitkar P, Tanna J, Kumbhare A, Akre S, Warade C, Gandhare N (2014) Silver nanoparticles induced effect on in-vitro callus production in Bacopa monnieri. Asian J Biol Life Sci 3:167–172
20. Kim DH, Gopal J, Sivanesan I (2017) Nanomaterials in plant tissue culture: the disclosed and undisclosed. RSC Adv 7(58):36492–36505
21. Kole C, Kumar DS, Khodakovskaya MV (eds) (2016) Plant nanotechnology: principles and practices. Springer, Switzerland
22. Lansdown AB (2002) Silver. I: its antibacterial properties and mechanism of action. J Wound Care 11:125–130
23. Lateef A, Folarin B, Oladejo S, Akinola P, Beukes L, Gueguim-Kana E (2018) Characterization, antimicrobial, antioxidant, and anticoagulant activities of silver nanoparticles synthesized from Petiveria alliacea L. leaf extract. Prep Biochem Biotechol 48(7):646–652
24. Leifert C, Camotta H, Waites WM (1992) Effect of combinations of antibiotics on micropropagated Clematis, Delphinium, Hosta, Iris and Photinia. Plant Cell Tissue Organ Cult 29:153–160
25. Mahmoud SN, Al-Ani NK (2016) Effect of different sterilization methods on contamination and viability of nodal segments of Cestrum nocturnum L. Int J Res Stud Biosci 4(1):4–9
26. Mahna N, Vahed SZ, Khani S (2013) Plant in vitro culture goes nano: nanosilver-mediated decontamination of ex vitro explants. J Nanomed Nanotechnol 4:1
27. Oyebanji OB, Nweke O, Odebunmi O, Galadima NB, Idris MS, Nnodi UN, Afolabi AS, Ogbadu GH (2009) Simple, effective and economical explant-surface sterilization protocol for cowpea, rice and sorghum seeds. AJB 8(20)
28. Pais AK, Silva AP, Souza JC, Teixeira SL, Ribeiro JM, Peixoto AR, Paz CD (2016) Sodium hypochlorite sterilization of culture medium in micropropagation of Gerbera hybrid cv. Essandre. Afr J Biotech 15:1995–1998. https://doi.org/10.5897/AJB2016.15405
29. Reed BM, Mentzer J, Tanprasert P, Yu X (1997) Internal bacterial contamination of micropropagated hazelnut: identification and antibiotic treatment. In: Pathogen and microbial contamination management in micropropagation, pp 169–174
30. Rostami AA, Shahsavar A (2009) Olive "Mission" explants. Asian J Plant Sci 8:505–509
31. Safavi K (2012) Evaluation of using nanomaterial in tissue culture media and biological activity. In: 2nd International conference on ecological, environmental and biological sciences (EEBS'2012), Bali, Indonesia, pp 5–8
32. Safavi K, Mortazaeinezahad F, Esfahanizadeh M, Asgari MJ (2011) In vitro antibacterial activity of nanomaterial for using in tobacco plants tissue culture. World Acad Sci Eng Technol 79:372–373
33. Sanzari I, Leone A, Ambrosone A (2019) Nanotechnology in plant science: to make a long story short. Front Bioeng Biotechnol 7:1–12. https://doi.org/10.3389/fbioe.2019.00120
34. Sarmast M, Salehi H, Khosh-Khui M (2011) Nano silver treatment is effective in reducing bacterial contaminations of Araucaria excelsa R. Br. var. glauca explants. Acta Biol Hung 62:477–484
35. Sarmast MK, Salehi H (2016) Silver nanoparticles: an influential element in plant nanobiotechnology. Mol Biotechnol 58:441–449. https://doi.org/10.1007/s12033-016-9943-0
36. Shokri S, Babaei A, Ahmadian M, Hessami S, Arab MM (2014) The effects of different concentrations of nano silver on elimination of bacterial contaminations and phenolic exudation of Rose (Rosa hybrida L.) in vitro culture. Int J Farm Allied Sci 3:50–54
37. Siddiqui MH, Al-Whaibi MH, Firoz M, Al-Khaishany MY (2015) Role of nanoparticles in plants. In: Siddiqui MH (ed) Nanotechnology and plant sciences. Springer, Basel, pp 19–35
38. Singh A, Singh NB, Afzal S, Singh T, Hussain I (2018) Zinc oxide nanoparticles: a review of their biological synthesis, antimicrobial activity, uptake, translocation and biotransformation in plants. J Mater Sci 53:185–201. https://doi.org/10.1007/s10853-017-1544-1

39. Spinoso-Castillo JL, Chavez-Santoscoy RA, Bogdanchikova N, Pérez-Sato JA, Morales-Ramos V, Bello-Bello JJ (2017) Antimicrobial and hormetic effects of silver nanoparticles on in vitro regeneration of vanilla (Vanilla planifolia Jacks. ex Andrews) using a temporary immersion system. Plant Cell Tissue Organ Cult 129:195–207
40. Taghizadeh M, Solgi M (2014) The application of essential oils and silver nanoparticles for sterilization of Bermuda grass explants in in vitro culture. Int J Horticult Sci Technol 1:131–140
41. Teixeirada Silva JA, Nhut DT, Tanaka M, Fukai S (2003) The effect of antibiotics on the in vitro growth response of chrysanthemum and tobacco stem transverse thin cell layers (tTCLS). SciHortic 97:397–410
42. Vargas DP, Formoso RS, Dutra LF, Newton AM, Santos J, Ueno B (2016) Chemical sterilization of in vitro culture for peach rootstock. Colloq Agr 12:1–6
43. Wang XJ, Hsiao KC (1995) Sugar degradation during autoclaving: effects of duration and solution volume on breakdown of glucose. Physiol Plant 94:415–418. https://doi.org/10.1111/j.1399-3054.1995.tb00947.x

Environmental Safety and Ethical Issues Related to Use of Silver Nanoparticles in Plants and Soil System

Divya Kapoor, Mayur Mukut Murlidhar Sharma, Sheetal Yadav, Pankaj Sharma, and Azamal Husen

Abstract Nanoparticles have transitioned from being merely a subject of scientific curiosity to becoming an indispensable component of our daily lives and the ecosystem. Nanotechnology was well established in the past few decades and has now become an inevitable part of industries and agriculture. Nanomaterials are widely used in each sector, especially in agricultural practices. Among various types of nanoparticles, silver nanoparticles (AgNPs) are extensively utilized in agriculture for enhancing crop production, disease resistance, and longevity. The adverse effects of nanomaterials are not concealed, and there is a pressing need for extensive research into their long-term implications. This chapter summarizes the characterization and uptake of silver nanoparticles, their fate in the environment, toxicity, ethical issues, and environment safety strategies related to plant and soil system.

Keywords AgNPs · Nano-toxicity · Nano-safety · Sustainable agriculture · Nanoparticle accumulation

D. Kapoor · S. Yadav · P. Sharma (✉)
Department of Microbiology, CCS Haryana Agricultural University, Hisar, Haryana 125004, India
e-mail: psk14495@gmail.com; sharmap@uic.edu

M. M. M. Sharma
Department of Agriculture and Industry, Kangwon National University, Chuncheon, Korea

P. Sharma
Department of Ophthalmology and Visual Sciences, University of Illinois Chicago, Chicago 60612, USA

A. Husen
Wolaita Sodo University, PO Box 138, Wolaita, Ethiopia

Department of Biotechnology, Graphic Era (Deemed to be University), Dehradun, Uttarakhand 248002, India

A. Husen (ed.), *Plant Response to Silver Nanoparticles*, Smart Nanomaterials Technology, https://doi.org/10.1007/978-981-97-7352-7_18

1 Introduction

Nanotechnology is a burgeoning field focused on the development and utilization of nanomaterials, often referred to as "twenty-first century materials," which have found widespread application across various sectors, particularly agriculture, owing to their ability to overcome size-related challenges [23, 70, 71]. Ever since Nobel laureate Richard P. Feynman introduced the concept of "nanotechnology" in his renowned 1959 lecture "There's Plenty of Room at the Bottom" [13], the field has seen numerous revolutionary advancements. Nanoparticles (NPs) comprise a broad category of materials consisting of particulate substances with at least one dimension less than 100 nm [28]. With their unique chemical and physical properties, engineered nanomaterials (ENMs), due to their incredibly small size, have permeated every sector of our daily lives. Out of the diverse array of engineered nanomaterials (ENMs), silver nanoparticles (AgNPs) stand out as the most widely used [78]. In order to track the proliferation of nanotechnology in consumer markets, the Woodrow Wilson International Center for Scholars and the Project on Emerging Nanotechnology established the Nanotechnology Consumer Product Inventory (CPI) in 2005, cataloging 54 products. AgNPs are the most popular advertised nanomaterial in the CPI, present in 438 products (24%) [78].

The incorporation of nanotechnology into agriculture heralds a significant departure from traditional farming methods, offering innovative remedies for enduring obstacles. Through the utilization of nanomaterials, sensors, and devices, farmers are equipped with unparalleled understanding of the complex molecular interactions within soil, crops, and pests, facilitating meticulous and effective management of resources [3, 5, 17–20, 51, 52, 59–61, 65, 67–69, 87–91]. Nanoscale delivery mechanisms enable precise deployment of agrochemicals, reducing ecological footprint while optimizing effectiveness. Additionally, nanomaterials empower the cultivation of robust crops by enhancing nutrient absorption, bolstering stress resilience, and pioneering pest control approaches. Furthermore, the emergence of nanogenomics expedites breeding initiatives, fostering the creation of high-yield, nutrient-dense crop cultivars customized to distinct environmental parameters. Simultaneously, nanosensors provide instantaneous monitoring of crop vitality, pest invasions, and environmental factors, enabling preemptive actions and maximizing crop yield. Nevertheless, as with any nascent technology, a thorough assessment of ethical, environmental, and regulatory ramifications is crucial to guarantee prudent implementation and enduring benefits. By means of cooperative research, inventive solutions, and conscientious management, nanotechnology has the potential to cultivate a robust, productive, and eco-conscious agricultural landscape for forthcoming generations [32, 33, 57, 58, 62–64, 66]. Incorporating AgNPs further exemplifies the innovative potential of nanotechnology in agriculture. These nanoparticles, known for their antimicrobial properties, offer a sustainable alternative for pest management and disease control, reducing reliance on conventional chemical pesticides. Additionally, AgNPs have been explored for enhancing nutrient uptake in plants

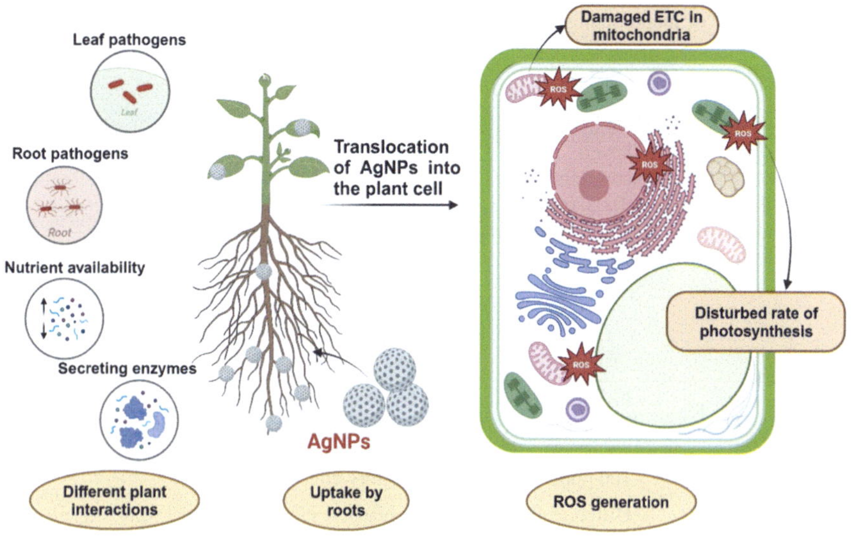

Fig. 1 A portrayal depicting the different positive and negative effects of silver nanoparticles on plant systems

and promoting growth, contributing to improved crop yields and quality (Fig. 1 and Table 1).

AgNPs have a broad antimicrobial activity thus they can be used in households, healthcare, and agriculture [44, 50]. AgNPs were initially introduced in the agricultural domain based on their documented efficacy in plant growth stimulation [39], fungicidal activity against fungal pathogens [1], and facilitation of fruit ripening processes [80]. However, the escalating use of AgNPs raises concerns about their potential release into the environment during synthesis, incorporation into products, handling, and the subsequent recycling or disposal of these products. The increasing utilization of AgNPs unavoidably heightens the likelihood of their discharge into the environment throughout the processes of AgNP synthesis, integration into products, and the subsequent handling, recycling, or disposal of said products [31]. Due to their high usage in agriculture, AgNPs tend to stream as biosolids along with shallow waters such as lakes, and streams [72]. Many studies have shown the accumulation of AgNPs in water bodies and soil environments. Examination of wastewater from a sewage treatment facility revealed the presence of AgNPs measuring 9.3 nm in size, with a concentration of 1900 ng/L [16]. Unfortunately, these concentrations are steadily increasing without interruption. In addition to the particulate form of silver, AgNPs can undergo alterations into silver oxide or silver sulfide through oxidation or sulfidation processes, respectively, each of which can affect soil and living organisms in diverse manners [74]. Consequently, it is imperative to investigate the environmental behavior of nanoparticles and explore potential methods for their remediation.

Table 1 Review of literature highlighting positive and negative effects of silver nanoparticles on plants

Plant	Concentration of AgNP	Effect	Refs.
Corn (*Zea mays* L.), watermelon (*Citrullus lanatus*, *Matsum.* and *Nakai*) and zucchini (*Cucurbita pepo* L.)	Seven concentrations (0.05, 0.1, 0.5, 1, 1.5, 2, and 2.5 mg/mL)	Significant enhancement of the germination percentage values was observed in comparison with untreated seeds. AgNPs showed a toxic effect on corn root elongation, whereas watermelon and zucchini seedling growth were positively affected by certain concentrations of AgNPs	[2]
Common bean (*Phaseolus vulgaris* L.) and corn (*Zea mays* L.)	Five levels of silver nanoparticles (20, 40, 60, 80, and 100 ppm) were used	It had a stimulating effect on the growth of the plantlets, while the enhanced concentrations induced an inhibitory effect	[55]
Jasmine rice, *Oryza sativa* L. cv. *KDML 105*	Different sizes of AgNPs (20, 30–60, 70–120, and 150 nm diameter), at different concentrations (0.1, 1, 10, 100, and 1000 mg/L)	The negative effects of AgNPs were supported by leaf cell deformation when rice seeds were treated with 150 nm-diameter AgNP at the concentration of 10 or 100 mg/L during seed germination	[73]
Potato dextrose agar (PDA), malt extract agar, and corn meal agar plates	WA-CV-WA13B, WA-AT-WB13R, and WA-PR-WB13R silver nanoparticles (AgNPs) at concentrations of 10, 25, 50, and 100 ppm	Treatment with WA-CV-WB13R AgNPs resulted in maximum inhibition of most fungi	[27]
Oryza sativa	1000 μg/mL	The particles penetrated inside the root cell and damaged the cell wall as well as vacuoles	[35, 36]
Phaseolus radiatus and *Sorghum bicolor*	Citrate-coated AgNPs at an average size of 10 nm	*P. radiatus* and *S. bicolor* showed a concentration-dependent growth inhibition effect	[30]
Common wetland plants	1, 10 or 40 mg Ag/L	PVP-AgNP had no effect on germination while 40 mg Ag L^{-1} GA-AgNP exposure significantly reduced the germination rate of three species and enhanced the germination rate of one species	[84]

(continued)

Table 1 (continued)

Wheat	25, 50, 75 and 100 mg/L	AgNPs protect wheat plants against heat stress and improve plant root length (5 and 5.4%), shoot length (22.2 and 26.1%), root number (6.6 and 7.5%), plant fresh weight (1.3 and 2%) and plant dry weight (0.36 and 0.60%) in 50 and 75 mg/L AgNPs, respectively	[21]
Tomato plants	0, 25, 50, 75 and 100 mg/L	The greater increases in amino acid content were observed at 75 and 100 mg L^{-1} concentrations	[24]
Brassica juncea var. Varuna	100, 500, and 1000 mg/L	AgNPs induced a slight increase in root/shoot length and this effect was correlated with an increase in chlorophyll and protein content at all concentration as compared to control (without AgNPs)	[54]

2 AgNPs: Properties and Applications

The determination of cytotoxicity relies heavily on various physical and chemical characteristics of AgNPs, such as surface chemistry, size, distribution, shape, morphology, composition, coating, agglomeration, dissolution rate, reactivity in solution, efficiency of ion release, cell type, and the type of reducing agents employed during synthesis [38, 45, 47] As an illustration, the utilization of different *Bacillus* species to derive biological reducing agents enables the synthesis of AgNPs in diverse shapes, including spherical, rod-shaped, octagonal, hexagonal, triangular, flower-like, among others. There are multiple reports supporting the claim that the size of nanoparticles has an inverse relation with the toxicity it causes in plants that is smaller the size of the nanomaterial, the more the toxicity which has been attributed to their larger surface area [86]. Additionally, the shape of the nanomaterial also plays a crucial role in determining the toxicology. For instance, there are multiple shapes of AgNPs that are employed in multiple sectors such as nanocubes, nanoplates, nanorods, spherical nanoparticles, flower-like, and so on. Moreover, the coatings available on the nanomaterial (biological or chemical), and surface charge are essential determinants of the safety levels of AgNPs. For example, release of positive charge Ag^{+} particles renders them a better fungicidal activity than a negative surface charged nanoparticle [48].

AgNPs are employed in the fields for numerous positive reasons. One of them is enhancement in seed germination post its applications. Seed germination is a crucial stage in establishing agricultural plants and is paramount for crop quality [34]. AgNPs have the capability to integrate into seed coats, facilitating water uptake within seeds, which in turn promotes starch metabolism and seed germination. Additionally, they are utilized as nano-priming agents to improve the viability of aged rice seeds and enhance starch metabolism. In one of the studies, AgNPs were produced by employing leaf extract from *Moringa oleifera* and evaluated for their impact on seed germination and growth. Various concentrations of AgNPs ranging from 25 to 100 ppm were tested alongside a control group to augment wheat seed germination and seedling growth. Remarkably, the concentration of AgNPs at 100 ppm exhibited substantial growth compared to the control. Furthermore, a notable enhancement in dry root weight, fresh root weight, root elongation, and root length was observed at the concentration of 100 ppm of AgNPs [53]. AgNPs have demonstrated several beneficial effects on plant growth, however, certain plants have exhibited toxicity in response to their application, suggesting potential adverse effects. For instance, the application of AgNPs at varying concentrations, such as 1000, 1600, and 1200 μg/mL, was observed on Brassica campestris, Oryza sativa, and Vigna radiata, respectively. Root inhibition occurred at concentrations of 4500 and 6000 μg/mL of ions. Notably, root length was significantly reduced to 1 and 0.5% in Oryza sativa and Brassica campestris; Vigna radiata, respectively [35, 36].

Other than seed germination, AgNPs also show a positive impact on plant growth. AgNPs have a broad range of antimicrobial properties such as antiviral, antibacterial, and antifungal agents in agricultural contexts. They have reported on extending the shelf life of fruits, foliage, flowers, and vegetables while also promoting plant growth and metabolism. Although multiple factors play a crucial role in deciding its activity such as exposure time, plant species, composition, particle size, concentration, functionalization, and numerous other factors. In a research investigation, varying concentrations of AgNPs—0, 5, 10, and 20 mg/L—were employed. The application of these nanoparticles at a concentration of 5 mg/L notably augmented root development as novel biostimulants. Furthermore, they enhanced the concentration of nitrogen (N), phosphorus (P), and potassium (K) in leaves, thus conferring benefits in improving seedling performance during the initial developmental stages [15]. A study investigated the effects of AgNPs treatments (4 mg/L) on lotus crops through four different methods: control, soil treatment with 4 mg/L five days before planting (T1), treatment with 4 mg/L before planting (T2), and combined periodic foliar application, plant, and soil treatments at 4 mg/L (T3). The various methods of AgNPs application influenced parameters such as leaf diameter, plant height, dry and fresh leaf weight, and certain biochemical aspects [81]. AgNPs were synthesized using an extract derived from pistachio seed coat waste. The resulting AgNPs solution, with a concentration of 75 mg/L, was sprayed onto eggplants, resulting in enhanced carotenoid/chlorophyll contents and plant growth. Additionally, the incorporation of fly ash further stimulated plant growth. Specifically, AgNPs were sprayed at a concentration of 75 ppm onto soil amended with 20% fly ash, leading to a significant improvement in plant growth [26].

The accumulation of AgNPs in leaves and roots triggers cellular and tissue-level defense mechanisms, altering proteomic expression, antioxidant activities, and metabolism. This leads to botanical changes including the generation of reactive oxygen species, changes in chlorophyll levels, superoxide dismutase activities, increased levels of H_2O_2, glutathione, carotenoids, ascorbate, proline contents, and more. These processes result in the suppression of photosynthesis, irregular morphological modifications, and other symptoms [8]. A study showed a decrease in plant weight, shorter height, and lower biomass of wheat (T. aestivum) when treated with AgNPs at a concentration of 20, 200, and 2000 mg/kg [83]. AgNPs were synthesized via the citrate reduction method and stabilized using chitosan. Lactuca sativa L. (Asteraceae) seeds were exposed to varying AgNPs concentrations of 12.5, 25, 50, and 100 ppm to evaluate their potential toxicity. The percentage of germinated seeds and root morphological alterations were utilized as toxicity parameters. Notably, at a concentration of 100 ppm, there was a reduction in root growth compared to the positive control which was distilled water [10]. An observed decrease in root hair development was noted in Arabidopsis thaliana upon the application of AgNPs at a concentration of 0.2 μg/L [53].

AgNPs are the most crucial in agriculture that encompasses enhancing seed germination and plant growth. Other than that, AgNPs have found their way into several other important products including Nano-pesticides, Nano-fertilizers, food packaging, pest management, crops, and food protection (nano-antibacterial, nano-antifungal, and nano-antiviral agents). Pests pose a growing menace to the agricultural industry, diminishing crop yields and compromising crop quality. The use of synthetic pesticides on soil and plants further exacerbates environmental degradation. Nano pesticides represent an innovative approach for effectively controlling and managing pests of agricultural importance. AgNPs can be used as nanotechnology-based pesticide and show minimal side effects on non-targeted insects. AgNPs ranging from 50 to 100 nm were synthesized utilizing *Solanum lycopersicum*. These AgNPs were applied to *M. rosae* at concentrations of 200, 300, 400, and 500 ppm, with distilled water serving as the control. The mortality rate of *M. rosae* exhibited optimal efficacy at a concentration of 500 ppm. Additionally, the mortality rate was observed to be directly associated with higher concentrations [7]. AgNPs-based nano-fertilizers are also very effective in regulating the release of fertilizers to the plants in the right amount. The utilization of nano fertilizers derived from onion extracts has proven effective for tomato and brinjal plants. AgNPs function as nano fertilizers, gradually and efficiently releasing nutrients to mitigate nutrient loss. This aids in enhancing nutrient absorption from the soil by the plants [85].

Similarly, nano-based pesticides are also effective in providing high and proper doses to the plants to fight against multiple pest infestations. Mortality tests, repellent activity assessments, antifeedant tests, and ovipositional deterrence evaluations have confirmed that AgNPs formulated with malathion exhibit maximum pesticidal activity against *Tribolium castaneum*. This beetle species is a significant pest in stored grains and has shown resistance to conventional synthetic insecticides. AgNPs and malathion were applied in varying concentrations (50, 75, and 100 ppm) to test disks. The mortality rates for AgNPs and malathion at 75 ppm were 75% and 95%, respectively. Thus, AgNPs have demonstrated effectiveness in controlling pests and beetles [4]. The agricultural sector grapples with issues of low yield and crop diseases, which are currently receiving considerable attention. Hence, researchers should delve into this area to investigate the effectiveness of AgNPs in enhancing plant growth and agricultural practices.

3 Environmental Safety Issues

The widespread utilization of AgNPs in contemporary society has not been thoroughly evaluated in relation to their environmental risks and the equilibrium between nanoparticles and biodiversity [29]. Lately, there has been a notable increase in the utilization of AgNPs as nano-pesticides and nano-fertilizers which raises more concerns toward their judicious use in the environment. AgNPs are regarded as highly toxic based on their L(E) C50 values, which represent the concentration causing 50% mortality in tested organisms when administered at once, with levels below 0.1 mg L^{-1} for environmentally relevant organisms [9]. Conversely, the predicted environmental concentrations (PECs) of AgNPs in the environment typically fall within the range of 0.03–0.08 mg L^{-1} [12]. However, the World Health Organization (WHO) and Environmental Protection Agency (EPA) have reported the maximum contamination limit (MCL) for the toxic range of silver is 0.1 mg L^{-1} [75, 82]. Unfortunately, naturally occurring silver concentrations in the environment, especially in surface waters, are usually minimal but are progressively increasing at elevated levels due to runoff and wastewater from urban and industrial areas.

It is alarming to comprehend that AgNPs can undergo oxidation, leading to the generation of Ag ions that may be released into the soil, exhibiting greater reactivity than the particulate form. This ionic form is widely recognized as highly toxic to living organisms. Failure to recognize and address this issue promptly may result in unforeseen devastation. Ag ion contamination could inhibit plant growth, pollute freshwater sources, and consequently pose deleterious effects on human health. Therefore, it is of utmost importance to timely recognize the harmful effects of using AgNPs. In the early 1980s, the U.S. Environmental Protection Agency (EPA) took many measures to limit the use of AgNPs in the environment. Regulations have been implemented to restrict the total concentration of Ag in aquatic systems to a range of 1.2–1.3 ppb [79]. Numerous studies are underway to investigate the long-term side effects of AgNPs usage. Within aquatic systems, there is a heightened risk of bioaccumulation across all trophic levels, resulting in severe toxic effects [77]. On contamination with AgNPs, these can be transformed into other forms such as chloride and nitrate in aquatic biosystems [40]. Out of which nitrate is the most toxic form that also interacts with other particles leading to high toxicity. After mercury, silver is considered the most toxic element in the environment. Although they are toxic by different mechanisms but are both fatal to numerous living organisms. The silver particles can bind to the cell organelles and these binding interactions can lead to their entry inside the cell such as through calcium channels and further cause cell death [42, 43].

3.1 Effect on Plants

Nanomaterials are used to improve the overall plant health and economy of the food sector, but the unsustainable use of the NPs can lead to unprecedented harm to the quality of crops and can impact human health subsequently. AgNPs when find their way into the wastewater or any other surface waters can be taken up by the plants and can lead to accumulation inside the plants. Due to the extremely small size of the AgNPs, they can travel both short (cell to cell) and long distances (roots to shoots through xylem cells) in plants. This finally leads to AgNPs on the food table of humans via the food chain. There are numerous studies reporting the toxic effects of AgNPs on plant health. In a study, researchers found Ag ions as well as particulate forms in the leaves of tomato (*L. esculentum*) when treated with 10, 20, and 30 mg L^{-1} of 20 nm AgNPs for a week via roots [42, 43]. This study provides evidence that AgNPs can be taken up by plants and stored. This can cause long lasting effects at molecular and morphological levels in plants. Qian et al. [49] showed that *A. thaliana* when exposed to AgNPs lead to disruption of the thylakoid membrane and chloroplast structure that disrupted the photosynthesis rate and metabolism of the plant. Further one of the most important crops, *Oryza sativa* was affected by the use of AgNPs at 0.5–1 mg. There was a significant lowering of mitochondrial membrane potential [41]. There are many studies indicating that exposure to AgNPs also leads to an affect number of gene expressions. At the physiological level, plant responses are influenced by changes in gene and protein expression. Kaveh et al. [25] investigated the impact of AgNPs on gene expression, revealing upregulation of 286 genes and downregulation of 81 genes in Arabidopsis thaliana seedlings exposed to up to 20 mg/L of 20 nm AgNPs for 10 days. Additionally, proteomics studies indicate that AgNPs alter the expression of proteins involved in signal transduction, defense mechanisms, and responses to oxidative stress [37].

This suggests that the impact of AgNPs on plants is linked to the generation of reactive oxygen species (ROS) and elevated production of antioxidant enzymes, leading to a reduction in shoot and root growth, as well as other pigments. Consequently, this diminishes photosynthesis, plant biomass, and ultimately, crop productivity.

In conclusion, the impact of AgNPs on plants can lead to either the stimulation or suppression of growth, development, or defense mechanisms, influenced by factors such as the type, size, and concentration of AgNPs, along with the plant species and duration of exposure. It is crucial to consider plant–microbe interactions and the effects of AgNPs on microorganisms when implementing phytotechnology strategies.

3.2 Effect on Soil Health

AgNPs have shown antimicrobial properties and thus employed in various crops but when leached in soil can also lead to higher toxicity in soil. Additionally, it can affect the microbial diversity and activity of the soil microorganisms [14, 46, 56, 76]. Depending on multiple factors like microne, soil, and concentration of AgNPs, it can affect both enzymatic and non-enzymatic activity of the soil microbiota. Jain et al. [22] reported that exposure to 2.5 μM AgNPs encouraged the enzymatic activity of numerous fungi and bacteria, whereas 25 and 50 μM contact had a contradictory effect. The presence of AgNPs disrupts the functioning of membrane transporters responsible for the movement of vital elements like potassium and chloride within cells. Furthermore, AgNPs and Ag^{+} that penetrate bacterial cells disrupt DNA replication [11] and impede bacterial growth. Besides soil bacteria and fungi, AgNPs can also influence decomposers. Research on the plant-decomposing invertebrate *Limnephilus* sp. indicated that the leaf-shredding behavior of this organism is impacted by both the size and concentration of AgNPs [6]. Invertebrates fulfill a crucial function in soil structure and play a significant role in triggering plant defense mechanisms. It's crucial to recognize that alterations in the microbial population and activity within the soil influence soil characteristics, subsequently impacting plant growth, physiological functions, and molecular responses. These factors influence a plant's ability to eliminate nanoparticles from its surroundings. The implications of AgNPs being released into the environment and their toxic effects on nearby organisms underscore the necessity of adopting sustainable approaches to mitigate their adverse impacts on living organisms. Phytotechnology, involving the use of plants to remove or trap pollutants, offers a promising and environmentally sound approach that can be implemented in nanoparticle-contaminated environments.

4 Ethical Dimensions of AgNPs

The use of AgNPs is on the rise and subsequently raises numerous ethical concerns, chiefly regarding their potential environmental and health impacts. The ethical issues associated with the use of AgNPs arise due to their impact on the environment, plant health, soil health, and finally humans. AgNPs released into the environment disrupt the ecosystems by affecting the balance of various organisms, including microorganisms, plants, and animals. Many studies have shown the changes in microbial communities due to exposure to AgNPs. They further interfere with nutrient cycling, alter microbial composition, and disrupt food chains, leading to long-term ecological consequences. Due to their nanosized, the chances of Bioaccumulation and Biomagnification are huge. AgNPs accumulate in organisms through various pathways, including direct ingestion or absorption through the skin. Once absorbed, they can biomagnify in food chains, potentially reaching toxic levels in higher trophic levels, such as predatory fish or mammals. This subsequently leads to long-term effects of AgNPs when not fully studied for their toxicity levels depending on their size, shape, plant species, and environmental factors. Their persistence in the environment and potential for transformation into more toxic forms raise concerns about their lasting impact on ecosystems and biodiversity. As discussed earlier as well that

Ag ions are more toxic than their particulate forms and other forms such as nitrate are far more toxic and lethal.

These small ions can surpass the biological barriers leading to cellular damage and adverse health effects. Many studies have shown that AgNPs can accumulate in various organs, including the liver, kidneys, and brain, raising concerns about potential systemic toxicity and chronic health effects. Even people in the industries that are involved in the synthesis of these nanoparticles get exposed to them. Workers involved in the production or handling of AgNPs may face occupational health risks due to inhalation or dermal exposure. Ethical considerations include ensuring workplace safety and providing adequate protective measures for these workers. The widespread use of AgNPs as antimicrobial agents can exert selective pressure on microbial populations, favoring the emergence of resistant strains. This can compromise the effectiveness of AgNPs in combating infections and contribute to the broader issue of antibiotic resistance. Silver-resistant genes acquired by bacteria in response to exposure to AgNPs can persist in the environment, potentially transferring to other bacteria and contributing to the spread of antibiotic resistance genes. Ethical considerations regarding the use of AgNPs also include the need for robust risk assessments to evaluate the safety of products containing AgNPs for both human health and the environment. Regulatory agencies must ensure that appropriate testing methodologies are used to assess potential hazards and risks associated with these nanoparticles. Furthermore, regulatory frameworks should promote transparency and accountability in the development, manufacturing, and marketing of products containing AgNPs. Stakeholders should have access to information about the potential risks and benefits of these products to make informed decisions. Most importantly, ethical considerations should include the provision of clear and accurate information to consumers about products containing AgNPs. This includes labeling requirements that indicate the presence of AgNPs and provide information about their intended use, potential risks, and proper disposal. In certain contexts, such as medical devices or consumer products with direct skin contact, ensuring informed consent involves educating consumers about the risks associated with AgNPs and allowing them to make autonomous decisions about their use.

Ethical considerations in healthcare include ensuring equitable access to healthcare products containing AgNPs, such as wound dressings or antimicrobial coatings on medical devices. Addressing healthcare disparities requires strategies to ensure that vulnerable populations have access to these products, regardless of socioeconomic status or geographic location. Cost considerations also play a role in equity and access, particularly in healthcare settings. Ethical frameworks should address affordability issues to ensure that essential healthcare products containing AgNPs are accessible to all individuals, including those in low-resource settings. Furthermore, communication strategies to inform the public about the potential risks and benefits of AgNPs. Transparent communication can help build trust and facilitate informed public discourse on the responsible development and use of these nanoparticles.

Addressing these ethical issues requires collaboration among stakeholders, including researchers, policymakers, industry representatives, healthcare professionals, and civil society organizations. Implementing comprehensive risk management strategies and ethical frameworks can help ensure the responsible development and use of AgNPs while minimizing their potential negative impacts on the environment, human health, and society.

5 Conclusion

In conclusion, the use of AgNPs in plants and soil systems presents both opportunities and challenges in terms of environmental safety and ethical considerations. While AgNPs offer potential benefits such as enhanced plant growth and pest control, their release into the environment raises concerns about ecological impact and human health. The potential for AgNPs to accumulate in soil, affect microbial communities, and disrupt ecosystem dynamics underscores the importance of thorough risk assessment and regulatory oversight. Moreover, ethical considerations regarding the responsible use of AgNPs, including their long-term effects on biodiversity and ecosystem stability, must be carefully addressed. Moving forward, a multidisciplinary approach that integrates scientific research, environmental monitoring, and ethical decision-making is essential to ensure the safe and sustainable application of AgNPs in agriculture and environmental management.

References

1. Alavi SV, Dehpour AA (2009, April) Evaluation of the nanosilver colloidal solution in comparison with the registered fungicide to control greenhouse cucumber downy mildew disease in the north of Iran. In: VI International postharvest symposium, vol 877, pp 1643–1646
2. Almutairi ZM, Alharbi A (2015) Effect of silver nanoparticles on seed germination of crop plants. Int J Nucl Quantum Eng 9(6):689–693
3. Anamika, Patra A, Shehzad S, Rani A, Sharma P, Mohammad KF, Mehta S (2023) Microbial elicitors for priming plant defense mechanisms. In: Sustainable agriculture reviews 60: microbial processes in agriculture, 2023 Feb 22. Springer Nature Switzerland, Cham, pp 175–196
4. AS AA (2019) Comparative bioassay of silver nanoparticles and malathion on infestation of red flour beetle, Tribolium castaneum. J Basic Appl Zool 80(1):1–10
5. Bachheti RK, Bachheti A, Husen A (2024) Metal and metal-oxide based nanomaterials. In: Synthesis, agricultural, biomedical and environmental interventions. Springer Nature Singapore Pte Ltd., Singapore
6. Batista D, Pascoal C, Cássio F (2017) How do physicochemical properties influence the toxicity of silver nanoparticles on freshwater decomposers of plant litter in streams? Ecotoxicol Environ Saf 140:148–155
7. Bhattacharyya A, Prasad R, Buhroo AA, Duraisamy P, Yousuf I, Umadevi M et al (2016) One-pot fabrication and characterization of silver nanoparticles using Solanum lycopersicum: an eco-friendly and potent control tool against rose aphid, Macrosiphum rosae. J Nanosci 2016

8. Budhani S, Egboluche NP, Arslan Z, Yu H, Deng H (2019) Phytotoxic effect of silver nanoparticles on seed germination and growth of terrestrial plants. J Environ Sci Health C 37(4):330–355
9. Burić P, Jakšić Ž, Štajner L, Sikirić MD, Jurašin D, Cascio C, Calzolai L, Lyons DM (2015) Effect of silver nanoparticles on Mediterranean sea urchin embryonal development is species specific and depends on moment of first exposure. Mar Environ Res 111:50–59
10. de Paiva Pinheiro SK, de Medeiros Chaves M, Miguel TBR, de Freitas Barros FC, Farias CP, Ferreira OP, de Castro Miguel E (2020) Toxic effects of silver nanoparticles on the germination and root development of lettuce (Lactuca sativa). Aust J Bot 68(2):127–136
11. Dong Y, Zhu H, Shen Y, Zhang W, Zhang L (2019) Antibacterial activity of silver nanoparticles of different particle size against Vibrio natriegens. PLoS ONE 14:e0222322
12. Fernandes JP, Mucha AP, Francisco T, Gomes CR, Almeida CMR (2017) Silver nanoparticles uptake by salt marsh plants—implications for phytoremediation processes and effects in microbial community dynamics. Mar Pollut Bull 119:176–183
13. Feynman R (2018) There's plenty of room at the bottom. In: Feynman and computation. CRC Press, pp 63–76
14. Grün A-L, Straskraba S, Schulz S, Schloter M, Emmerling C (2018) Long-term effects of environmentally relevant concentrations of silver nanoparticles on microbial biomass, enzyme activity, and functional genes involved in the nitrogen cycle of loamy soil. J Environ Sci 69:12–22
15. Guzmán-Báez GA, Trejo-Téllez LI, Ramírez-Olvera SM, Salinas-Ruíz J, Bello-Bello JJ, Alcántar-González G et al (2021) Silver nanoparticles increase nitrogen, phosphorus, and potassium concentrations in leaves and stimulate root length and number of roots in tomato seedlings in a hormetic manner. Dose-Response 19(4):15593258211044576
16. Hoque ME, Khosravi K, Newman K, Metcalfe CD (2012) Detection and characterization of silver nanoparticles in aqueous matrices using asymmetric-flow field flow fractionation with inductively coupled plasma mass spectrometry. J Chromatogr A 1233:109–115
17. Husen A (2022) Engineered nanomaterials for sustainable agricultural production, soil improvement and stress management. Elsevier Inc., Cambridge, USA. https://doi.org/10.1016/C2021-0-00054-7
18. Husen A (2023) Nanomaterials and nanocomposites exposures to plants. In: Response, interaction, phytotoxicity and defense mechanisms. Springer Nature Singapore Pte Ltd., Singapore. https://doi.org/10.1007/978-981-99-2419-6
19. Husen A, Bachheti RK, Bachheti A (2023) Current trends in green nano-emulsions. In: Food, agriculture and biomedical sectors. Springer Nature Singapore Pte Ltd., Singapore
20. Husen A, Iqbal M (2019) Nanomaterials and plant potential. Springer International Publishing AG, Cham, Switzerland. https://doi.org/10.1007/978-3-030-05569-1
21. Iqbal M, Raja NI, Mashwani ZUR, Hussain M, Ejaz M, Yasmeen F (2019) Effect of silver nanoparticles on growth of wheat under heat stress. Iran J Sci Technol, Trans A: Sci 43:387–395
22. Jain A, Kumar S, Seena S (2019) Can low concentrations of metal oxide and Ag loaded metal oxide nanoparticles pose a risk to stream plant litter microbial decomposers? Sci Total Environ 653:930–937
23. Kapoor D, Yadav S, Sharma MM, Sharma P (2023) Interaction between metal nanoparticles and PGPR on the plant growth and development. In: Nanomaterials and nanocomposites exposures to plants: response, interaction, phytotoxicity and defense mechanisms. Springer Nature Singapore, Singapore, pp 327–351
24. Karami Mehrian S, Heidari R, Rahmani F (2015) Effect of silver nanoparticles on free amino acids content and antioxidant defense system of tomato plants. Indian J Plant Physiol 20:257–263
25. Kaveh R, Li Y-S, Ranjbar S, Tehrani R, Brueck CL, Van Aken B (2013) Changes in Arabidopsis thaliana gene expression in response to silver nanoparticles and silver ions. Environ Sci Technol 47:10637–10644
26. Khan M, Khan AU, Moon IS, Felimban R, Alserihi R, Alsanie WF, Alam M (2021) Synthesis of biogenic silver nanoparticles from the seed coat waste of pistachio (Pistacia vera) and their effect on the growth of eggplant. Nanotechnol Rev 10(1):1789–1800

27. Kim SW, Jung JH, Lamsal K, Kim YS, Min JS, Lee YS (2012) Antifungal effects of silver nanoparticles (AgNPs) against various plant pathogenic fungi. Mycobiology 40(1):53–58
28. Laurent S, Forge D, Port M, Roch A, Robic C, Vander Elst L, Muller RN (2008) Magnetic iron oxide nanoparticles: synthesis, stabilization, vectorization, physicochemical characterizations, and biological applications. Chem Rev 108(6):2064–2110
29. Lazim ZM, Salmiati, Samaluddin AR, Salim MR, Arman NZ (2020) Toxicity of silver nanoparticles and their removal applying phytoremediation system to water environment: an overview. J Environ Treat Tech 8:978–984
30. Lee WM, Kwak JI, An YJ (2012) Effect of silver nanoparticles in crop plants Phaseolus radiatus and Sorghum bicolor: media effect on phytotoxicity. Chemosphere 86(5):491–499
31. Lombi E, Donner E, Scheckel KG, Sekine R, Lorenz C, Von Goetz N, Nowack B (2014) Silver speciation and release in commercial antimicrobial textiles as influenced by washing. Chemosphere 111:352–358
32. Loyal A, Pahuja SK, Rani N, Pooja, Srivastava RK, Sharma P (2023) Titanium oxide nanoparticles: plant response, interaction, phytotoxicity, and defence mechanisms. In: Nanomaterials and nanocomposites exposures to plants: response, interaction, phytotoxicity and defense mechanisms. Springer Nature Singapore, Singapore, pp 263–284
33. Loyal A, Pahuja SK, Sharma P, Malik A, Srivastava RK, Mehta S (2023) Potential environmental and human health implications of nanomaterials used in sustainable agriculture and soil improvement. In: Engineered nanomaterials for sustainable agricultural production, soil improvement and stress management. Academic Press, pp 387–412
34. Mahakham W, Sarmah AK, Maensiri S, Theerakulpisut P (2017) Nanopriming technology for enhancing germination and starch metabolism of aged rice seeds using phytosynthesized silver nanoparticles. Sci Rep 7(1):8263
35. Mazumdar H, Ahmed GU (2011) Phytotoxicity effect of silver nanoparticles on Oryza sativa. Int J ChemTech Res 3(3):1494–1500
36. Mazumdar H, Ahmed GU (2011) Synthesis of silver nanoparticles and its adverse effect on seed germinations in Oryza sativa, Vigna radiata and Brassica campestris. Int J Adv Biotechnol Res 2(4):404–413
37. Mirzajani F, Askari H, Hamzelou S, Schober Y, Römpp A, Ghassempour A, Spengler B (2014) Proteomics study of silver nanoparticles toxicity on Oryza sativa L. Ecotoxicol Environ Saf 108:335–339
38. Misra RD, Girase B, Depan D, Shah JS (2012) Hybrid nanoscale architecture for enhancement of antimicrobial activity: immobilization of silver nanoparticles on thiol-functionalized polymer crystallized on carbon nanotubes. Adv Eng Mater 14:B93–B100. https://doi.org/10.1002/adem.201180081
39. Monica RC, Cremonini R (2009) Nanoparticles and higher plants. Caryologia 62(2):161–165
40. Moreno-Garrido I, Pérez S, Blasco J (2015) Toxicity of silver and gold nanoparticles on marine microalgae. Mar Environ Res 111:60–73
41. Nair PMG, Chung IM (2014) Physiological and molecular level effects of silver nanoparticles exposure in rice (Oryza sativa L.) seedlings. Chemosphere 112:105–113
42. Noori A, Donnelly T, Colbert J, Cai W, Newman LA, White JC (2020) Exposure of tomato (Lycopersicon esculentum) to silver nanoparticles and silver nitrate: physiological and molecular response. Int J Phytoremediat 22:40–51
43. Noori A, Ngo A, Gutierrez P, Theberge S, White JC (2020) Silver nanoparticle detection and accumulation in tomato. J Nanopart Res 22:131
44. Pandian AMK, Karthikeyan C, Rajasimman M, Dinesh MG (2015) Synthesis of silver nanoparticle and its application. Ecotoxicol Environ Saf 121:211–217
45. Park MV, Neigh AM, Vermeulen JP, de la Fonteyne LJ, Verharen HW, Briedé JJ, van Loveren H, de Jong WH (2011) The effect of particle size on the cytotoxicity, inflammation, developmental toxicity and genotoxicity of silver nanoparticles. Biomaterials 32:9810–9817. https://doi.org/10.1016/j.biomaterials.2011.08.085
46. Park S, Park HH, Ko Y-S, Lee SJ, Le TS, Woo K, Ko G (2017) Disinfection of various bacterial pathogens using novel silver nanoparticle-decorated magnetic hybrid colloids. Sci Total Environ 609:289–296

47. Powers CM, Badireddy AR, Ryde IT, Seidler FJ, Slotkin TA (2011) Silver nanoparticles compromise neurodevelopment in PC12 cells: critical contributions of silver ion, particle size, coating, and composition. Environ Health Perspect 119:37–44. https://doi.org/10.1289/ehp.1002337
48. Przemieniecki SW, Oćwieja M, Ciesielski S, Halecki W, Matras E, Gorczyca A (2022) Chemical structure of stabilizing layers of negatively charged silver nanoparticles as an effector of shifts in soil bacterial microbiome under short-term exposure. Int J Environ Res Public Health 19(21):14438
49. Qian H, Peng X, Han X, Ren J, Sun L, Fu Z (2013) Comparison of the toxicity of silver nanoparticles and silver ions on the growth of terrestrial plant model Arabidopsis thaliana. J Environ Sci 25:1947–1956
50. Rai M, Yadav A, Gade A (2009) Silver nanoparticles as a new generation of antimicrobials. Biotechnol Adv 27(1):76–83
51. Rani K, Rani A, Sharma P, Dahiya A, Punia H, Kumar S, Sheoran S, Banerjee A (2022) Legumes for agroecosystem services and sustainability. In: Advances in legumes for sustainable intensification. Academic Press, pp 363–380
52. Rani K, Sharma P, Kumar S, Wati L, Kumar R, Gurjar DS, Kumar D, Kumar R (2019) Legumes for sustainable soil and crop management. In: Sustainable management of soil and environment, pp 193–215
53. Sabir S, Arshad M, Satti SH (2018) Effect of green synthesized silver nanoparticles on seed germination and seedling growth in wheat. Int J Agron Agric Res 12:1–7
54. Sadak MS (2019) Impact of silver nanoparticles on plant growth, some biochemical aspects, and yield of fenugreek plant (Trigonella foenum-graecum). Bull Natl Res Cent 43:38. https://doi.org/10.1186/s42269-019-0077-y
55. Salama HM (2012) Effects of silver nanoparticles in some crop plants, common bean (Phaseolus vulgaris L.) and corn (Zea mays L.). Int Res J Biotechnol 3(10):190–197
56. Samarajeewa A, Velicogna J, Princz J, Subasinghe R, Scroggins R, Beaudette L (2017) Effect of silver nano-particles on soil microbial growth, activity and community diversity in a sandy loam soil. Environ Pollut 220:504–513
57. Sangwan S, Sharma P, Wati L, Mehta S (2023) Effect of chitosan nanoparticles on growth and physiology of crop plants. In: Engineered nanomaterials for sustainable agricultural production, soil improvement and stress management. Academic Press, pp 99–123
58. Sharma MM, Kapoor D, Rohilla R, Sharma P (2023) Nanomaterials and their toxicity to beneficial soil microbiota and fungi associated plants rhizosphere. In: Nanomaterials and nanocomposites exposures to plants: response, interaction, phytotoxicity and defense mechanisms. Springer Nature Singapore, Singapore, pp 353–380
59. Sharma MM, Sharma P, Kapoor D, Beniwal P, Mehta S (2021) Phytomicrobiome community: an agrarian perspective towards resilient agriculture. In: Plant performance under environmental stress: hormones, biostimulants and sustainable plant growth management, pp 493–534
60. Sharma P, Meyyazhagan A, Easwaran M, Sharma MM, Mehta S, Pandey V, Liu WC, Kamyab H, Balasubramanian B, Baskaran R, Klemeš JJ (2022) Hydrogen sulfide: a new warrior in assisting seed germination during adverse environmental conditions. Plant Growth Regul 98(3):401–420
61. Sharma P, Pandey V, Sharma MM, Patra A, Singh B, Mehta S, Husen A (2021) A review on biosensors and nanosensors application in agroecosystems. Nanoscale Res Lett 16:1–24
62. Sharma P, Patra A, Singh B, Mehta S (2023) Microbial rejuvenation of soils for sustainable agriculture. In: Sustainable agriculture reviews 60: microbial processes in agriculture. Springer Nature Switzerland, Cham, pp 293–323
63. Sharma P, Sangwan S, Kaur H, Patra A, Anamika, Mehta S (2023) Diversity and evolution of nitrogen fixing bacteria. In: Sustainable agriculture reviews 60: microbial processes in agriculture, 95–120
64. Sharma P, Sangwan S, Kaur H, Patra A, Mehta S (2023) Microbial remediation of agricultural residues. In: Sustainable agriculture reviews 60: microbial processes in agriculture. Springer Nature Switzerland, Cham, pp 325–358

65. Sharma P, Sangwan S, Kumari A, Singh S, Kaur H (2022) Impact of climate change on soil microorganisms regulating nutrient transformation. In: Plant stress mitigators: action and application. Springer Nature Singapore, Singapore, pp 145–172
66. Sharma P, Sangwan S, Mehta S (2023) Emerging role of phosphate nanoparticles in agriculture practices. In: Engineered nanomaterials for sustainable agricultural production, soil improvement and stress management. Academic Press, pp 71–97
67. Sharma P, Sharma MM, Anamika, Kapoor D, Rani K, Singh D, Barkodia M (2020) Role of microbes for attaining enhanced food crop production. In: Microbial biotechnology: basic research and applications, pp 55–78
68. Sharma P, Sharma MM, Malik A, Vashisth M, Singh D, Kumar R, Singh B, Patra A, Mehta S, Pandey V (2021) Rhizosphere, rhizosphere biology, and rhizospheric engineering. In: Plant growth-promoting microbes for sustainable biotic and abiotic stress management. Springer International Publishing, Cham, pp 577–624
69. Sharma P, Sharma MM, Patra A, Vashisth M, Mehta S, Singh B, Tiwari M, Pandey V (2020) The role of key transcription factors for cold tolerance in plants. In: Transcription factors for abiotic stress tolerance in plants. Academic Press, pp 123–152
70. Singh A, Sharma P, Kumari A, Kumar R, Pathak DV (2019) Management of root-knot nematode in different crops using microorganisms. In: Plant biotic interactions: state of the art, pp 85–99
71. Singh S, Sangwan S, Sharma P, Devi P, Moond M (2020) Nanotechnology for sustainable agriculture: an emerging perspective. J Nanosci Nanotechnol 21(6):3453–65; Nayana AR, Joseph BJ, Jose A, Radhakrishnan EK (2020) Nanotechnological advances with PGPR applications. In: Sustainable agriculture reviews 41: nanotechnology for plant growth and development, pp 163–80
72. Sun TY, Gottschalk F, Hungerbühler K, Nowack B (2014) Comprehensive probabilistic modelling of environmental emissions of engineered nanomaterials. Environ Pollut 185:69–76
73. Thuesombat P, Hannongbua S, Akasit S, Chadchawan S (2014) Effect of silver nanoparticles on rice (Oryza sativa L. cv. KDML 105) seed germination and seedling growth. Ecotoxicol Environ Saf 104:302–309
74. Torrent L, Marguí E, Queralt I, Hidalgo M, Iglesias M (2019) Interaction of silver nanoparticles with mediterranean agricultural soils: Lab-controlled adsorption and desorption studies. J Environ Sci 83:205–216
75. U.S. Environmnetal Proection Agency (EPA) (2009) National primary drinking water regulation table. EPA 816-F-09-0004. https://www.epa.gov/ground-water-and-drinking-water/national-primary-drinking-water-regulation-table
76. Uddin I, Ahmad K, Khan AA, Kazmi MA (2017) Synthesis of silver nanoparticles using Matricaria recutita (Babunah) plant extract and its study as mercury ions sensor. Sens Bio-Sens Res 16:62–67
77. Valenti LE, Giacomelli CE (2017) Stability of silver nanoparticles: agglomeration and oxidation in biological relevant conditions. J Nanopart Res 19:156
78. Vance ME, Kuiken T, Vejerano EP, McGinnis SP, Hochella MF Jr, Rejeski D, Hull MS (2015) Nanotechnology in the real world: redeveloping the nanomaterial consumer products inventory. Beilstein J Nanotechnol 6(1):1769–1780
79. Varner K, El-Badawy A, Feldhake D, Venkatapathy R (2010) State-of-the-science review: everything nanosilver and more. EPA/600/R-10/084:2010. US Environmental Protection Agency, Washington, DC, USA
80. Vinković T, Novák O, Strnad M, Goessler W, Jurašin DD, Parađiković N, Vrček IV (2017) Cytokinin response in pepper plants (Capsicum annuum L.) exposed to silver nanoparticles. Environ Res 156:10–18
81. Vu NQH, Hong HTK, Quang HT (2021, December) Effects of different treatments of silver nanoparticles (AgNPs) on the growth & physiological characteristics of lotus (Nelumbo nucifera). In: IOP conference series: earth and environmental science, vol 947, No 1. IOP Publishing, p 012038

82. World Health Organization (WHO) (2011) Guidelines for drinking-water quality, 4th edn. World Health Organization (WHO), Geneva, Switzerland, p 415. http://apps.who.int/iris/bitstream/handle/10665/44584/9789241548151_eng.pdf;jsessionid=7602427D0558C27BC51742431A74F67E?sequence=1
83. Yang J, Jiang F, Ma C, Rui Y, Rui M, Adeel M et al (2018) Alteration of crop yield and quality of wheat upon exposure to silver nanoparticles in a life cycle study. J Agric Food Chem 66(11):2589–2597
84. Yin L, Colman BP, McGill BM, Wright JP, Bernhardt ES (2012) Effects of silver nanoparticle exposure on germination and early growth of eleven wetland plants
85. Yuvaraj M, Subramanian KS (2020) Novel slow release nanocomposite fertilizers. In: Nanotechnology and the environment. IntechOpen
86. Zhang XF, Liu ZG, Shen W, Gurunathan S (2016) Silver nanoparticles: synthesis, characterization, properties, applications, and therapeutic approaches. Int J Mol Sci 17(9):1534. https://doi.org/10.3390/ijms17091534.PMID:27649147;PMCID:PMC5037809
87. Kapoor D, Sharma P, Sharma MM, Yadav S, Husen A (2024 Sep 11) Exploring soil microbiota and their role in plant growth, stress tolerance, disease control and nutrient immobilizer. Biocatal Agric Biotechnol 103358
88. Sharma MM, Kapoor D, Loyal A, Kumar R, Sharma P, Husen A (2024 May 1) Environmental toxicity of engineered carbon nanoparticles. In: Carbon-Based nanomaterials: Synthesis, agricultural, biomedical, and environmental interventions. Springer Nature Singapore, Singapore, pp 337–353
89. Sharma MM, Kapoor D, Sharma P, Husen A (2024 May 1) Role of Carbon Nanomaterials in the Prevention of Plant Disease. In: Carbon-Based Nanomaterials: Synthesis, Agricultural, Biomedical, and Environmental Interventions. Springer Nature Singapore, Singapore, pp 105–121
90. Loyal A, Pahuja SK, Vart D, Chander S, Kumar V, Rani N, Sharma P, Husen A (2024 Jan 1) Exploring the role of beneficial elements in enhancing the resistance of plants to biotic stress. In: Essential minerals in plant-soil systems, Elsevier, pp 327–346
91. Kapoor D, Sharma MM, Yadav S, Sharma P (2024 Jan 1) Applications of Trichoderma virens and biopolymer-based biostimulants in plant growth and productions. In: Biostimulants in plant protection and performance, Elsevier, pp 349–367

The manufacturer's authorised representative in the EU is Springer Nature Customer Service Centre GmbH, Europaplatz 3, 69115 Heidelberg, Germany. If you have any concerns regarding our products, please contact ProductSafety@springernature.com

Printed and bound by CPI Group (UK) Ltd, Croydon, CR0 4YY
01/10/2026
02225934-0011